Levels 1 & 2

BENCHMARK SERIES

Microsoft® Access®

365

2019 Edition

ita Rutkosky

Jan Davidson
Lambton College
Sarnia, Ontario

Audrey Roggenkamp
Pierce College Puyallup
Puyallup, Washington

Ian Rutkosky
Pierce College Puyallup
Puyallup, Washington

D1157516

PARADIGM
EDUCATION SOLUTIONS

St. Paul

Vice President, Content and Digital Solutions: Christine Hurney
Director of Content Development: Carley Fruzzetti
Developmental Editor: Jennifer Joline Anderson
Director of Production: Timothy W. Larson
Production Editor/Project Manager: Jen Weaverling
Senior Design and Production Specialist: Jack Ross
Cover and Interior Design: Valerie King
Copy Editor: Communicáto, Ltd.
Testers: Janet Blum, Lisa Hart
Indexer: Terry Casey
Vice President, Director of Digital Products: Chuck Bratton
Digital Projects Manager: Tom Modl
Digital Solutions Manager: Gerry Yumul
Senior Director of Digital Products and Onboarding: Christopher Johnson
Supervisor of Digital Products and Onboarding: Ryan Isdahl
Vice President, Marketing: Lara Weber McLellan
Marketing and Communications Manager: Selena Hicks

Cover Photo Credit: © lowball-jack/GettyImages
Interior Photo Credits: Follow the Index.

ISBN 978-0-76388-728-5 (print)
ISBN 978-0-76388-710-0 (digital)

© 2020 by Paradigm Publishing, LLC
875 Montreal Way
St. Paul, MN 55102
Email: CustomerService@ParadigmEducation.com
Website: ParadigmEducation.com

Printed in the United States of America

28 27 26 25 24 23 22 21 20 19   1 2 3 4 5 6 7 8 9 10 11 12

Brief Contents

Contents

Microsoft Access Level 2

Achieving Proficiency in Access

The Benchmark Series, *Microsoft® Access® 365, 2019 Edition*, is designed for students who want to learn how to use Microsoft's feature-rich data management tool to track, report, and share information. No prior knowledge of databases is required. After successfully completing a course in Microsoft Access using this courseware, students will be able to do the following:

- Create database tables to organize business or personal records.
- Modify and manage tables to ensure that data is accurate and up to date.
- Perform queries to assist with decision making.
- Plan, research, create, revise, and publish database information to meet specific communication needs.
- Examine a workplace scenario requiring the reporting and analysis of data, assess the information requirements, and then prepare the materials that achieve the goal efficiently and effectively.

Well-designed pedagogy is important, but students learn technology skills through practice and problem solving. Technology provides opportunities for interactive learning as well as excellent ways to quickly and accurately assess student performance. To this end, this course is supported with Cirrus, Paradigm's cloud-based training and assessment learning management system. Details about Cirrus as well as its integrated student courseware and instructor resources can be found on page xii.

Proven Instructional Design

The Benchmark Series has long served as a standard of excellence in software instruction. Elements of the series function individually and collectively to create an inviting, comprehensive learning environment that leads to full proficiency in computer applications. The following visual tour highlights the structure and features that comprise the highly popular Benchmark model.

Microsoft

Access Level 1

Microsoft

Access Level 2

g Access Tables

: between Tables

Tables

Unit 1

Advanced Tables, Relationships, Queries, and Forms

Chapter 1	Designing the Structure of Tables
Chapter 2	Building Relationships and Lookup Fields
Chapter 3	Advanced Query Techniques
Chapter 4	Creating and Using Custom Forms

Unit Openers display the unit's four chapter titles. Each level of the course contains two units with four chapters each.

Chapter Openers Present Learning Objectives

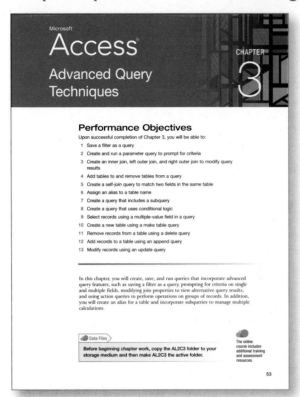

Chapter Openers present the performance objectives and an overview of the skills taught.

Data Files are provided for each chapter.

Activities Build Skill Mastery within Realistic Context

Multipart Activities provide a framework for instruction and practice on software features. An activity overview identifies tasks to accomplish and key features to use in completing the work.

Typically, a file remains open throughout all parts of the activity. Students save their work incrementally. At the end of the activity, students save, print, and then close the file.

Tutorials provide interactive, guided training and measured practice.

Quick Steps in the margins allow fast reference and review.

Hints offer useful tips on how to use features efficiently and effectively.

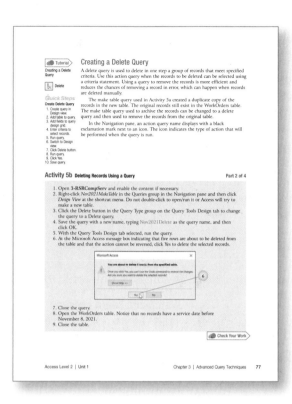

Between activity parts, the text presents instruction on the features and skills necessary to accomplish the next portion of the activity.

Step-by-Step Instructions guide students to the desired outcome for each project part. Screen captures illustrate what the screen should look like at key points.

Magenta Text identifies material to type.

Check Your Work model answer images are available in the online course, and students can use those images to confirm they have completed the activity correctly.

Chapter Review Tools Reinforce Learning

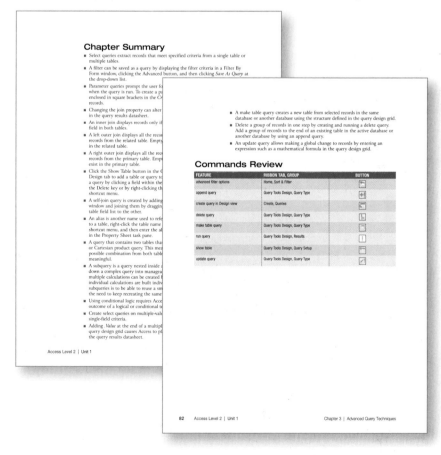

A **Chapter Summary** reviews the purpose and execution of key features.

A **Commands Review** summarizes visually the major software features and alternative methods of access.

The Cirrus Solution
Elevating student success and instructor efficiency

Powered by Paradigm, Cirrus is the next-generation learning solution for developing skills in Microsoft Office. Cirrus seamlessly delivers complete course content in a cloud-based learning environment that puts students on the fast track to success. Students can access their content from any device anywhere, through a live internet connection; plus, Cirrus is platform independent, ensuring that students get the same learning experience whether they are using PCs, Macs, or Chromebook computers.

Cirrus provides Benchmark Series content in a series of scheduled assignments that report to a grade book to track student progress and achievement. Assignments are grouped in modules, providing many options for customizing instruction.

Dynamic Training

The online Benchmark Series courses include interactive resources to support learning.

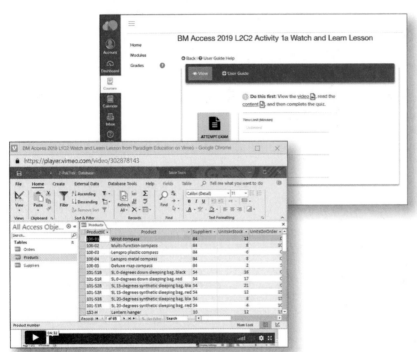

Watch and Learn Lessons include a video demonstrating how to perform the chapter activity, a reading to provide background and context, and a short quiz to check understanding of concepts and skills.

Guide and Practice Tutorials provide interactive, guided training and measured practice.

Hands On Activities enable students to complete chapter activities, compare their solutions against a Check Your Work model answer image, and submit their work for instructor review.

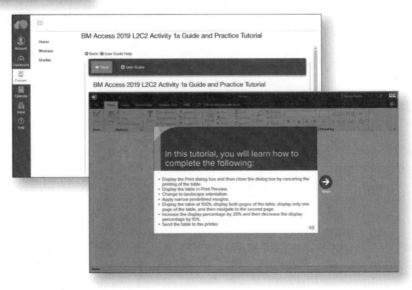

Chapter Review and Assessment

Review and assessment activities for each chapter are available for completion in Cirrus.

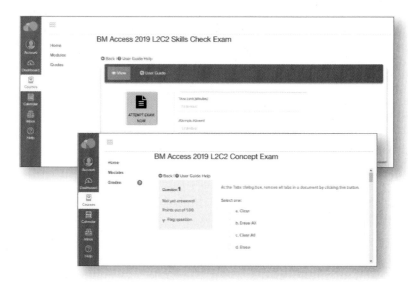

Concepts Check completion exercises assess their comprehension and recall of application features and functions as well as key terminology.

Skills Assessment Hands On Activity exercises evaluate the ability to apply chapter skills and concepts in solving realistic problems. Each is completed live in Access and is uploaded through Cirrus for instructor evaluation.

Visual Benchmark assessments test problem-solving skills and mastery of application features.

A **Case Study** requires analyzing a workplace scenario and then planning and executing a multipart project. Students search the web and/or use the program's Help feature to locate additional information required to complete the Case Study.

Exercises and **Projects** provide opportunities to develop and demonstrate skills learned in each chapter. Each is completed live in the Office application and is automatically scored by Cirrus. Detailed feedback and how-to videos help students evaluate and improve their performance.

Skills Check Exams evaluate students' ability to complete specific tasks. Skills Check Exams are completed live in the Office application and are scored automatically. Detailed feedback and instructor-controlled how-to videos help student evaluate and improve their performance.

Multiple-choice **Concepts Exams** assess understanding of key commands and concepts presented in each chapter.

Unit Review and Assessment

Review and assessment activities for each unit of each Benchmark course are also available for completion in Cirrus.

Assessing Proficiency exercises check mastery of software application functions and features.

Writing Activities challenge students to use written communication skills while demonstrating their understanding of important software features and functions.

Internet Research assignments reinforce the importance of research and information processing skills along with proficiency in the Office environment.

A **Job Study** activity at the end of Unit 2 presents a capstone assessment requiring critical thinking and problem solving.

Unit-Level Projects allow students to practice skills learned in the unit. Each is completed live in the Office application and automatically scored by Cirrus. Detailed feedback and how-to videos help students evaluate and improve their performance.

Student eBook

The Student eBook, accessed through the Cirrus online course, can be downloaded to any device (desktop, laptop, tablet, or smartphone) to make Benchmark Series content available anywhere students wish to study.

Instructor eResources

Cirrus tracks students' step-by-step interactions as they move through each activity, giving instructors visibility into their progress and missteps. With Exam Watch, instructors can observe students in a virtual, live, skills-based exam and join remotely as needed—a helpful option for struggling students who need one-to-one coaching, or for distance learners. In addition to these Cirrus-specific tools, the Instructor eResources for the Benchmark Series include the following support:

- Planning resources, such as lesson plans, teaching hints, and sample course syllabi
- Delivery resources, such as discussion questions and online images and templates
- Assessment resources, including live and annotated PDF model answers for chapter work and review and assessment activities, rubrics for evaluating student work, and chapter-based exam banks in RTF format

About the Authors

Nita Rutkosky began her career teaching business education at Pierce College in Puyallup, Washington, in 1978 and holds a master's degree in occupational education. In her years as an instructor, she taught many courses in software applications to students in postsecondary information technology certificate and degree programs. Since 1987, Nita has been a leading author of courseware for computer applications training and instruction. Her current titles include Paradigm's popular Benchmark Series, Marquee Series, and Signature Series. She is a contributor to the Cirrus online content for Office application courses and has also written textbooks for keyboarding, desktop publishing, computing in the medical office, and essential skills for digital literacy.

Jan Davidson started her teaching career in 1997 as a corporate trainer and postsecondary instructor and holds a Social Science degree, a writing certificate, and an In-Service Teacher Training certificate. Since 2001, she has been a faculty member of the School of Business and International Education at Lambton College in Sarnia, Ontario. In this role, she has developed curriculum and taught a variety of office technology, software applications, and office administration courses to domestic and international students in a variety of postsecondary programs. As a consultant and content provider for Paradigm Education solutions since 2006, Jan has contributed to textbook and online content for various titles. She has been author and co-author of Paradigm's Benchmark Series *Microsoft® Excel®*, Level 2, and *Microsoft® Access®*, Level 2 since 2013 and has contributed to the Cirrus online courseware for the series. Jan is also co-author of *Advanced Excel® 2016*.

Audrey Roggenkamp holds a master's degree in adult education and curriculum and has been an adjunct instructor in the Business Information Technology department at Pierce College in Puyallup, Washington, since 2005. Audrey has also been a content provider for Paradigm Education Solutions since 2005. In addition to contributing to the Cirrus online content for Office application courses, Audrey co-authors Paradigm's Benchmark Series, Marquee Series, and Signature Series. Her other available titles include *Keyboarding & Applications I and II* and *Using Computers in the Medical Office: Word, PowerPoint®, and Excel®*.

Ian Rutkosky has a master's degree in business administration and has been an adjunct instructor in the Business Information Technology department at Pierce College in Puyallup, Washington, since 2010. In addition to joining the author team for the Benchmark Series and Marquee Series, he has co-authored titles on medical office computing and digital literacy and has served as a co-author and consultant for Paradigm's Cirrus training and assessment software.

Microsoft®
Office

Getting Started in Office 365

Microsoft Office is a suite of applications for personal computers and other devices. These programs, known as *software*, include Word, a word processor; Excel, a spreadsheet editor; Access, a database management system; and PowerPoint, a presentation program used to design and present slideshows. Microsoft Office 365 is a subscription service that delivers continually updated versions of those applications. Specific features and functionality of Microsoft Office vary depending on the user's account, computer setup, and other factors. The Benchmark courseware was developed using features available in Office 365. You may find that with your computer and version of Office, the appearance of the software and the steps needed to complete an activity vary slightly from what is presented in the courseware.

Identifying Computer Hardware

The Microsoft Office suite can run on several types of computer equipment, referred to as *hardware*. You will need access to a laptop or a desktop computer system that includes a PC/tower, monitor, keyboard, printer, drives, and mouse. If you are not sure what equipment you will be operating, check with your instructor. The computer system shown in Figure G.1 consists of six components. Each component is discussed separately in the material that follows.

Figure G.1 Computer System

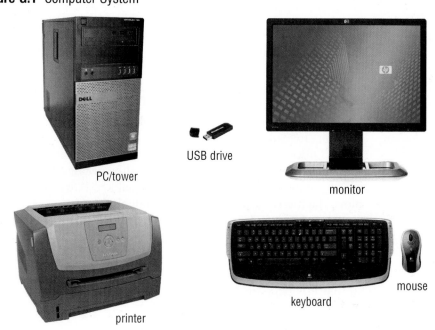

USB drive

PC/tower

monitor

printer

keyboard

mouse

Figure G.2 System Unit Ports

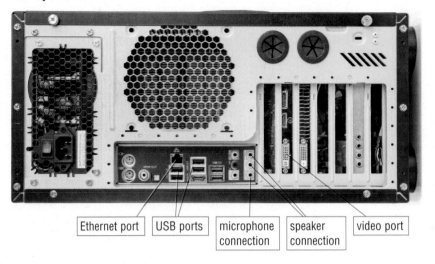

| Ethernet port | USB ports | microphone connection | speaker connection | video port |

System Unit (PC/Tower)

Traditional desktop computing systems include a system unit known as the *PC (personal computer)* or *tower*. This is the brain of the computer, where all processing occurs. It contains a Central Processing Unit (CPU), hard drives, and video cards plugged into a motherboard. Input and output ports are used for attaching peripheral equipment such as a keyboard, monitor, printer, and so on, as shown in Figure G.2. When a user provides input, the PC computes it and outputs the results.

Monitor

Hint Monitor size is measured diagonally and is generally the distance from the bottom left corner to the top right corner of the monitor.

A computer monitor looks like a television screen. It displays the visual information output by the computer. Monitor size can vary, and the quality of display for monitors varies depending on the type of monitor and the level of resolution.

Keyboard

The keyboard is used to input information into the computer. The number and location of the keys on a keyboard can vary. In addition to letters, numbers, and symbols, most computer keyboards contain function keys, arrow keys, and a numeric keypad. Figure G.3 shows a typical keyboard.

The 12 keys at the top of the keyboard, labeled with the letter *F* followed by a number, are called *function keys*. Use these keys to perform functions within each of the Office applications. To the right of the regular keys is a group of special or dedicated keys. These keys are labeled with specific functions that will be performed when you press the key. Below the special keys are arrow keys. Use these keys to move the insertion point in the document screen.

Some keyboards include mode indicator lights to indicate that a particular mode, such as Caps Lock or Num Lock, has been turned on. Pressing the Caps Lock key disables the lowercase alphabet so that text is typed in all caps, while pressing the Num Lock key disables the special functions on the numeric keypad so that numbers can be typed using the keypad. When you select these modes, a light appears on the keyboard.

Figure G.3 Keyboard

Drives and Ports
==

An internal hard drive is a disk drive that is located inside the PC and that stores data. External hard drives may be connected via USB ports for additional storage. Ports are the "plugs" on the PC, and are used to connect devices to the computer, such as the keyboard and mouse, the monitor, speakers, USB flash drives and so on. Most PCs will have a few USB ports, at least one display port, audio ports, and possibly an ethernet port (used to physically connect to the internet or a network).

Printer
==

An electronic version of a file is known as a *soft copy*. If you want to create a hard copy of a file, you need to print it. To print documents, you will need to access a printer, which will probably be either a laser printer or an ink-jet printer. A laser printer uses a laser beam combined with heat and pressure to print documents, while an ink-jet printer prints a document by spraying a fine mist of ink on the page.

Mouse
==

Most functions and commands in the Microsoft Office suite are designed to be performed using a mouse or a similar pointing device. A mouse is an input device that sits on a flat surface next to the computer. You can operate a mouse with your left or right hand. Moving the mouse on the flat surface causes a corresponding pointer to move on the screen, and clicking the left or right mouse buttons allows you to select various objects and commands.

Using the Mouse The applications in the Microsoft Office suite can be operated with the keyboard and a mouse. The mouse generally has two buttons on top, which you press to execute specific functions and commands. A mouse may also contain a wheel, which can be used to scroll in a window or as a third button. To use the mouse, rest it on a flat surface or a mouse pad. Put your hand over it with your palm resting on top of the mouse and your index finger resting on the left mouse button. As you move your hand, and thus the mouse, a corresponding pointer moves on the screen.

When using the mouse, you should understand four terms — *point*, *click*, *double-click*, and *drag*. To *point* means to position the mouse pointer on a desired item, such as an option, button, or icon. With the mouse pointer positioned on the item, *click* the left mouse button once to select the item. (In some cases you may *right-click*, which means to click the right mouse button, but generally, *click* refers to the left button.) To complete two steps at one time, such as choosing and then executing a function, *double-click* the left mouse button by tapping it twice in quick succession. The term *drag* means to click and hold down the left mouse button, move the mouse pointer to a specific location, and then release the button. Clicking and dragging is used, for instance, when moving a file from one location to another.

Hint Instructions in this course use the verb *click* to refer to tapping the left mouse button and the verb *press* to refer to pressing a key on the keyboard.

Using the Mouse Pointer The mouse pointer will look different depending on where you have positioned it and what function you are performing. The following are some of the ways the mouse pointer can appear when you are working in the Office suite:

- The mouse pointer appears as an I-beam (called the *I-beam pointer*) when you are inserting text in a file. The I-beam pointer can be used to move the insertion point or to select text.

- The mouse pointer appears as an arrow pointing up and to the left (called the *arrow pointer*) when it is moved to the Title bar, Quick Access Toolbar, ribbon, or an option in a dialog box, among other locations.

- The mouse pointer becomes a double-headed arrow (either pointing left and right, pointing up and down, or pointing diagonally) when you perform certain functions such as changing the size of an object.
- In certain situations, such as when you move an object or image, the mouse pointer displays with a four-headed arrow attached. The four-headed arrow means that you can move the object left, right, up, or down.

- When a request is being processed or when an application is being loaded, the mouse pointer may appear as a moving circle. The moving circle means "please wait." When the process is completed, the circle is replaced with a normal mouse pointer.
- When the mouse pointer displays as a hand with a pointing index finger, it indicates that more information is available about an item. The mouse pointer also displays as a hand with a pointing index finger when you hover over a hyperlink.

Touchpad

If you are working on a laptop computer, you may be using a touchpad instead of a mouse. A *touchpad* allows you to move the mouse pointer by moving your finger across a surface at the base of the keyboard (as shown in Figure G.4). You click and right-click by using your thumb to press the buttons located at the bottom of the touchpad. Some touchpads have special features such as scrolling or clicking something by tapping the surface of the touchpad instead of pressing a button with a thumb.

Figure G.4 Touchpad

Touchscreen

Smartphones, tablets, and touch monitors all use touchscreen technology (as shown in Figure G.5), which allows users to directly interact with the objects on the screen by touching them with fingers, thumbs, or a stylus. Multiple fingers or both thumbs can be used on most touchscreens, giving users the ability to zoom, rotate, and manipulate items on the screen. While many activities in this textbook can be completed using a device with a touchscreen, a mouse or touchpad might be required to complete a few activities.

Figure G.5 Touchscreen

Choosing Commands

A *command* is an instruction that tells an application to complete a certain task. When an application such as Word or PowerPoint is open, the *ribbon* at the top of the window displays buttons and options for commands. To select a command with the mouse, point to it and then click the left mouse button.

Notice that the ribbon is organized into tabs, including File, Home, Insert, and so on. When the File tab is clicked, a *backstage area* opens with options such as opening or saving a file. Clicking any of the other tabs will display a variety of commands and options on the ribbon. Above the ribbon, buttons on the Quick Access Toolbar provide fast access to frequently used commands such as saving a file and undoing or redoing an action.

Using Keyboard Shortcuts and Accelerator Keys

As an alternative to using the mouse, keyboard shortcuts can be used for many commands. Shortcuts generally require two or more keys. For instance, in Word, press and hold down the Ctrl key while pressing P to display the Print backstage area, or press Ctrl + O to display the Open backstage area. A complete list of keyboard shortcuts can be found by searching the Help files in any Office application.

Office also provides shortcuts known as *accelerator keys* for every command or action on the ribbon. These accelerator keys are especially helpful for users with motor or visual disabilities or for power users who find it faster to use the keyboard than click with the mouse. To identify accelerator keys, press the Alt key on the keyboard. KeyTips display on the ribbon, as shown in Figure G.6. Press the keys indicated to execute the desired command. For example, to begin checking

Figure G.6 Word Home Tab KeyTips

the spelling and grammar in a document, press the Alt key, press the R key on the keyboard to display the Review tab, and then press the letter C and the number 1 on the keyboard to open the Editor task pane.

Choosing Commands from a Drop-Down List

Some buttons include arrows that can be clicked to display a drop-down list of options. Point and click with the mouse to choose an option from the list. Some options in a drop-down list may have a letter that is underlined. This indicates that typing the letter will select the option. For instance, to select the option *Insert Table*, type the letter I on the keyboard.

If an option in a drop-down list is not available to be selected, it will appear gray or dimmed. If an option is preceded by a check mark, it is currently active. If it is followed by an ellipsis (...), clicking the option will open a dialog box.

Choosing Options from a Dialog Box or Task Pane

Some buttons and options open a *dialog box* or a task pane containing options for applying formatting or otherwise modifying the data in a file. For example, the Font dialog box shown in Figure G.7 contains options for modifying the font and adding effects. The dialog box contains two tabs—the Font tab and the Advanced tab. The tab that displays in the front is the active tab. Click a tab to make it active or press Ctrl + Tab on the keyboard. Alternately, press the Alt key and then type the letter that is underlined in the tab name.

Figure G.7 Word Font Dialog Box

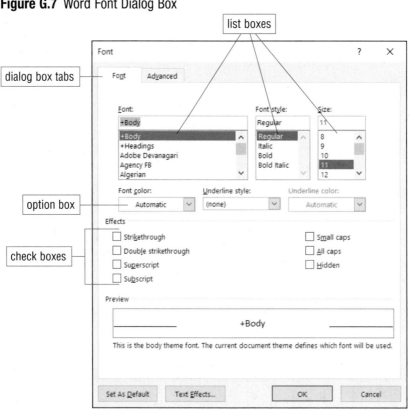

To choose an option from a dialog box using the mouse, position the arrow pointer on the option and then click the left mouse button. To move forward from option to option using the keyboard, you can press the Tab key. Press Shift + Tab to move back to a previous option. If the option displays with an underlined letter, you can choose it by pressing the Alt key and the underlined letter. When an option is selected, it is highlighted in blue or surrounded by a dotted or dashed box called a *marquee*. A dialog box contains one or more of the following elements: list boxes, option boxes, check boxes, text boxes, command buttons, radio buttons, and measurement boxes.

List Boxes and Option Boxes The fonts available in the Font dialog box, shown in Figure G.7 (on the previous page), are contained in a *list box*. Click an option in the list to select it. If the list is long, click the up or down arrows in the *scroll bar* at the right side of the box to scroll through all the options. Alternately, press the up or down arrow keys on the keyboard to move through the list, and press the Enter key when the desired option is selected.

Option boxes contain a drop-down list or gallery of options that opens when the arrow in the box is clicked. An example is the *Font color* option box in Figure G.8. To display the different color options, click the arrow at the right side of the box. If you are using the keyboard, press Alt + C.

Check Boxes Some options can be selected using a check box, such as the effect options in the dialog box in Figure G.7. If a check mark appears in the box, the option is active (turned on). If the check box does not contain a check mark, the option is inactive (turned off). Click a check box to make the option active or inactive. If you are using the keyboard, press Alt + the underlined letter of the option.

Text Boxes Some options in a dialog box require you to enter text. For example, see the Find and Replace dialog box shown in Figure G.8. In a text box, type or edit text with the keyboard, using the left and right arrow keys to move the insertion point without deleting text and use the Delete key or Backspace key to delete text.

Command Buttons The buttons at the bottom of the dialog box shown in Figure G.8 are called *command buttons*. Use a command button to execute or cancel a command. Some command buttons display with an ellipsis (...), which means another dialog box will open if you click that button. To choose a command button, click with the mouse or press the Tab key until the command button is surrounded by a marquee and then press the Enter key.

Figure G.8 Excel Find and Replace Dialog Box

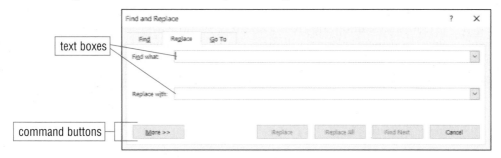

Figure G.9 Word Insert Table Dialog Box

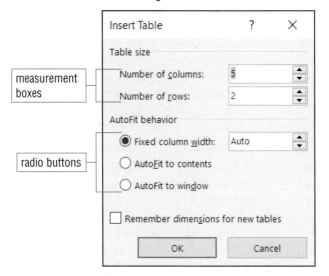

measurement boxes

radio buttons

Radio Buttons The Insert Table dialog box shown in Figure G.9 contains an example of *radio buttons*. Only one radio button can be selected at any time. When the button is selected, it is filled with a dark circle. Click a button to select it, or press and hold down the Alt key, press the underlined letter of the option, and then release the Alt key.

Measurement Boxes A *measurement box* contains an amount that can be increased or decreased. An example is shown in Figure G.9. To increase or decrease the number in a measurement box, click the up or down arrow at the right side of the box. Using the keyboard, press and hold down the Alt key and then press the underlined letter for the option, press the Up Arrow key to increase the number or the Down Arrow key to decrease the number, and then release the Alt key.

Choosing Commands with Shortcut Menus

The Office applications include shortcut menus that contain commands related to different items. To display a shortcut menu, point to the item for which you want to view more options with the mouse pointer and then click the right mouse button, or press Shift + F10. The shortcut menu will appear wherever the insertion point is positioned. In some cases, the Mini toolbar will also appear with the shortcut menu. For example, if the insertion point is positioned in a paragraph of text in a Word document, clicking the right mouse button or pressing Shift + F10 will display the shortcut menu and Mini toolbar, as shown in Figure G.10.

To select an option from a shortcut menu with the mouse, click the option. If you are using the keyboard, press the Up or Down Arrow key until the option is selected and then press the Enter key. To close a shortcut menu without choosing an option, click outside the menu or press the Esc key.

Figure G.10 Shortcut Menu and Mini Toolbar

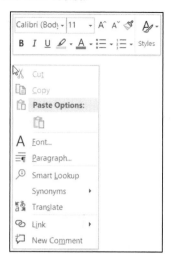

Working with Multiple Applications

As you learn the various applications in the Microsoft Office suite, you will notice many similarities between them. For example, the steps to save, close, and print are virtually the same whether you are working in Word, Excel, or PowerPoint. This consistency greatly enhances your ability to transfer knowledge learned in one application to another within the suite. Another benefit to using Microsoft Office is the ability to have more than one application open at the same time and to integrate content from one program with another. For example, you can open Word and create a document, open Excel and create a worksheet, and then copy a worksheet from the workbook into Word.

The Windows taskbar at the bottom of the screen displays buttons representing all the programs that are currently open. For example, Figure G.11 shows the taskbar with Word, Excel, Access, and PowerPoint open. To move from one program to another, click the taskbar button representing the desired application.

Maintaining Files and Folders

Windows includes a program named File Explorer that can be used to maintain files and folders. To open File Explorer, click the folder icon on the Windows taskbar. Use File Explorer to complete tasks such as copying, moving, renaming, and deleting files and folders and creating new folders. Some file management tasks can also be completed within Word, Excel, PowerPoint, or Access by clicking File and then *Open* or *Save As* and then clicking the *Browse* option to browse folders and files in a dialog box.

Directions and activities in this course assume that you are managing files and folders stored on a USB flash drive or on your computer's hard drive. If you are using your OneDrive account or another cloud-based storage service, some of the file and folder management tasks may vary.

Figure G.11 Windows Taskbar with Word, Excel, Access, and PowerPoint Open

Creating and Naming a Folder

Files (such as Word documents, Excel workbooks, PowerPoint presentations, and Access databases) are easier to find again when they are grouped logically in folders. In File Explorer and in the Open or Save As dialog box, the names of files and folders are displayed in the Content pane. Each file has an icon showing what type of file it is, while folders are identified with the icon of a folder. See Figure G.12 for an example of the File Explorer window.

Create a new folder by clicking the New folder button at the top of the File Explorer window or in the dialog box. A new folder displays with the name *New folder* highlighted. Type a name for the folder to replace the highlighted text, and then press the Enter key. Folder names can include numbers, spaces, and some symbols.

Selecting and Opening Files and Folders

Select files or folders in the window to be managed. To select one file or folder, simply click on it. To select several adjacent files or folders, click the first file or folder, hold down the Shift key, and then click the last file or folder. To select files or folders that are not adjacent, click the first file or folder, hold down the Ctrl key, click any other files or folders, and then release the Ctrl key. To deselect, click anywhere in the window or dialog box.

When a file or folder is selected, the path to the folder displays in the Address bar. If the folder is located on an external storage device, the drive letter and name may display in the path. A right-pointing arrow displays to the right of each folder name in the Address bar. Click the arrow to view a list of subfolders within a folder.

Double-click a file or folder in the Content pane to open it. You can also select one or more files or folders, right-click, and then click the *Open* option in the shortcut menu.

Figure G.12 File Explorer Window

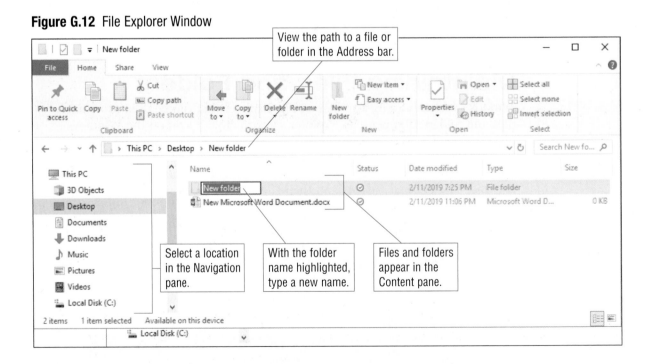

Deleting Files and Folders

Deleting files and folders is part of file maintenance. To delete a file or folder, select it and then press the Delete key. Alternatively, use the Delete button on the Home tab of the File Explorer window, or click the Organize button and then *Delete* in the dialog box. You can also right-click a file or folder and then choose the *Delete* option in the shortcut menu.

Files and folders deleted from the hard drive of the computer are automatically sent to the Recycle Bin, where they can easily be restored if necessary. If a file or folder is stored in another location, such as an external drive or online location, it may be permanently deleted. In this case, a message may appear asking for confirmation. To confirm that the file or folder should be deleted, click Yes.

To view the contents of the Recycle Bin, display the Windows desktop and then double-click the *Recycle Bin* icon. Deleted items in the Recycle Bin can be restored to their original locations, or the Recycle Bin can be emptied to free up space on the hard drive.

Moving and Copying Files and Folders

A file or folder may need to be moved or copied to another location. In File Explorer, select the file or folder and then click the Copy button at the top of the window, use the keyboard shortcut Ctrl + C, or right-click the file and select *Copy* in the shortcut menu. Navigate to the destination folder and then click the Paste button, use the keyboard shortcut Ctrl + P, or right-click and select *Paste*. If a copy is pasted to the same folder as the original, it will appear with the word *Copy* added to its name. To copy files in the Open or Save As dialog box, use the Organize button drop-down list or right-click to access the shortcut menu.

To move a file or folder, follow the same steps, but select *Cut* instead of *Copy* or press Ctrl + X instead of Ctrl + C. Files can also be dragged from one location to another. To do this, open two File Explorer windows. Click a file or folder and drag it to the other window while holding down the left mouse button.

Renaming Files and Folders

To rename a file or folder in File Explorer, click its name to highlight it and then type a new name, or right-click the file or folder and then select *Rename* at the shortcut menu. You can also select the file or folder and then click the Rename button on the Home tab of the File Explorer window or click *Rename* from the Organize button drop-down list at the Open or Save As dialog box. Type in a new name and then press the Enter key.

Viewing Files and Folders

Change how files and folders display in the Content pane in File Explorer by clicking the View tab and then clicking one of the view options in the Layout group. View files and folders as large, medium, or small icons; as tiles; in a list; or with details or information about the file or folder content. At the Open or Save As dialog box, click the Change your view button arrow and a list displays with similar options for viewing folders and files. Click to select an option in the list or click the Change your view button to see different views.

Displaying File Extensions Each file has a file extension that identifies the program and what type of file it is. Excel files have the extension *.xlsx*; Word files end with *.docx,* and so on. By default, file extensions are turned off. To view file extensions, open File Explorer, click the View tab, and then click the *File name extensions* check box to insert a check mark. Click the check box again to remove the check mark and stop viewing file extensions.

Displaying All Files The Open or Save As dialog box in an Office application may display only files specific to that application. For example, the Open or Save As dialog box in Word may only display Word documents. Viewing all files at the Open dialog box can be helpful in determining what files are available. Turn on the display of all files at the Open dialog box by clicking the file type button arrow at the right side of the *File Name* text box and then clicking *All Files* at the drop-down list.

Managing Files at the Info Backstage Area

The Info backstage area in Word, Excel, and PowerPoint provide buttons for managing files such as uploading and sharing a file, copying a path, and opening File Explorer with the current folder active. To use the buttons at the Info backstage area, open Word, Excel, or PowerPoint and then open a file. Click the File tab and then click the *Info* option. If a file is opened from the computer's hard drive or an external drive, four buttons display near the top of the Info backstage area as shown in Figure G.13.

Click the Upload button to upload the open file to a shared location such as a OneDrive account. Click the Share button and a window displays indicating that the file must be saved to OneDrive before it can be shared and provides an option that, when clicked, will save the file to OneDrive. Click the Copy Path button and a copy of the path for the current file is saved in a temporary location. This path can be pasted into another file, an email, or any other location where you want to keep track of the file's path. Click the Open file location button and File Explorer opens with the current folder active.

Figure G.13 Info Backstage Buttons

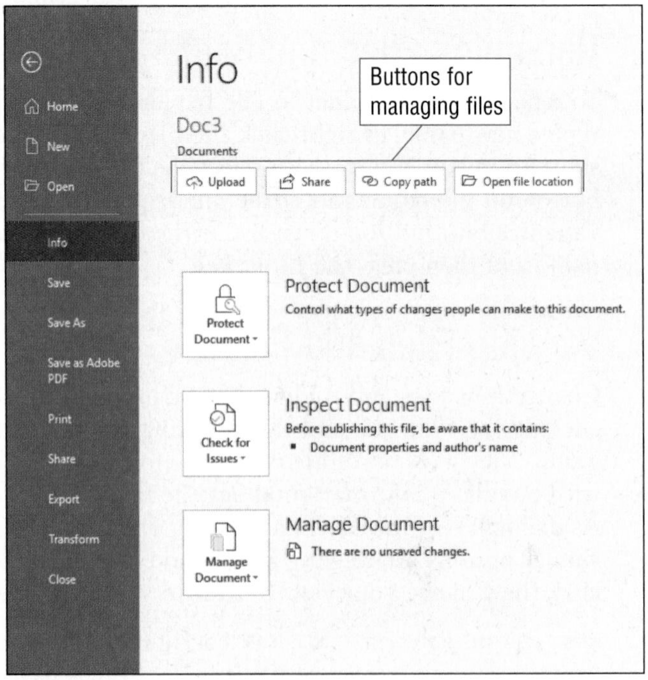

If you open Word, Excel, or PowerPoint and then open a file from OneDrive, only two buttons display—Share and Open file location. Click the Share button to display a window with options for sharing the file with others and specifying whether the file can be viewed and edited, or only viewed. Click the Open file location button to open File Explorer with the current folder active.

Customizing Settings

Before beginning computer activities in this textbook, you may need to customize your monitor's settings and change the DPI display setting. Activities in the course assume that the monitor display is set at 1920 × 1080 pixels and the DPI set at 125%. If you are unable to make changes to the monitor's resolution or the DPI settings, the activities can still be completed successfully. Some references in the text might not perfectly match what you see on your screen, so you may not be able to perform certain steps exactly as written. For example, an item in a drop-down gallery might appear in a different column or row than what is indicated in the step instructions.

Before you begin learning the applications in the Microsoft Office suite, take a moment to check the display settings on the computer you are using. Your monitor's display settings are important because the ribbon in the Microsoft Office suite adjusts to the screen resolution setting of your computer monitor. A computer monitor set at a high resolution will have the ability to show more buttons in the ribbon than will a monitor set to a low resolution. The illustrations in this textbook were created with a screen resolution display set at 1920 × 1080 pixels, as shown in Figure G.14.

Figure G.14 Word Ribbon Set at 1920 x 1080 Screen Resolution

Activity 1 **Adjusting Monitor Display**

Note: The resolution settings may be locked on lab computers. Also, some laptop screens and small monitors may not be able to display in a 1920 × 1080 resolution or change the DPI setting.

1. At the Windows desktop, right-click in a blank area of the screen.
2. In the shortcut menu, click the *Display settings* option.

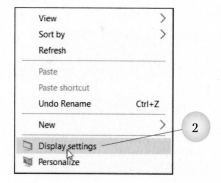

3. At the Settings window with the *Display* option selected, scroll down and look at the current setting displayed in the *Resolution* option box. If your screen is already set to 1920 × 1080, skip ahead to Step 6.

4. Click the Resolution option box and then click the *1920 × 1080* option. **Note: Depending on the privileges you are given on a school machine, you may not be able to complete Steps 4–5. If necessary, check with your instructor for alternative instructions.**

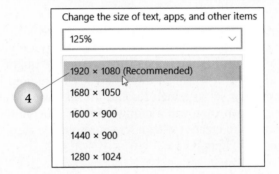

5. Click the Keep Changes button.
6. At the Settings window, take note of the current DPI percentage next to the text *Change the size of text, apps, and other items*. If the percentage is already set to 125%, skip to Step 8.
7. Click the option box below the text *Change the size of text, apps, and other items,* and then click the *125%* option in the drop-down list

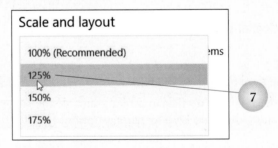

8. Click the Close button to close the Settings window.

Retrieving and Copying Data Files

While working through the activities in this course, you will often be using data files as starting points. These files are provided through your Cirrus online course, and your instructor may post them in another location such as your school's network drive. You can download all the files at once (described in the activity below), or download only the files needed for a specific chapter.

Activity 2 **Downloading Files to a USB Flash Drive**

Note: In this activity, you will download data files from your Cirrus online course. Make sure you have an active internet connection before starting this activity. Check with your instructor if you do not have access to your Cirrus online course.

1. Insert your USB flash drive into an available USB port.
2. Navigate to the Course Resources section of your Cirrus online course. *Note: The steps in this activity assume you are using the Chrome browser. If you are using a different browser, the following steps may vary.*
3. Click the Student Data Files link in the Course Resources section. A zip file containing the student data files will automatically begin downloading from the Cirrus website.
4. Click the button in the lower left corner of the screen once the files have finished downloading.

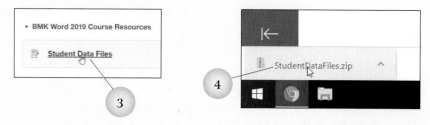

5. Right-click the *StudentDataFiles* folder in the Content pane.
6. Click the *Copy* option in the shortcut menu.
7. Click the USB flash drive that displays in the Navigation pane at the left side of the File Explorer window.
8. Click the Home tab in the File Explorer window.
9. Click the Paste button in the Clipboard group.

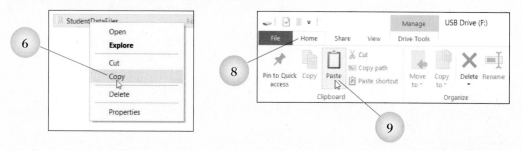

10. Close the File Explorer window by clicking the Close button in the upper right corner of the window.

Microsoft® Access® Level 1

Unit 1

Creating Tables and Queries

Microsoft®

Access®

Managing and Creating Access Tables

CHAPTER

1

Performance Objectives

Upon successful completion of Chapter 1, you will be able to:

1 Open and close objects in a database

2 Insert and delete records in a table

3 Insert, delete, and move fields in a table

4 Hide, unhide, freeze, and unfreeze fields

5 Adjust table column width

6 Preview and print a table

7 Change page layout

8 Design and create a table

9 Rename a field, caption, or description

10 Insert Quick Start fields

11 Assign a field size or default value

Managing information is an integral part of operating a business. Information can come in a variety of forms, such as data about customers, including names, addresses, and telephone numbers; product data; or purchasing and sales data. Most companies manage data using system software. Microsoft Access is a database management system offered as part of the suite of applications in Microsoft Office 365. With Access, you can organize, store, maintain, retrieve, sort, and print all types of business data.

This chapter contains just a few ideas on how to manage data with Access. A properly designed and maintained database management system can help a company operate smoothly.

 Data Files

Before beginning chapter work, copy the AL1C1 folder to your storage medium and then make AL1C1 the active folder.

The online course includes additional review and assessment resources.

You will open a database and open and close objects in the database, including a table, a query, a form, and a report.

Tutorial

Opening a Blank
Database

Exploring a Database

A database is comprised of a series of objects (such as tables, queries, forms, and/or reports) used to enter, manage, view, and print data. Data in a database is organized into tables, which contain information for related items (such as customers, employees, orders, or products).

Quick Steps

Create New Database
1. Open Access.
2. Click *Blank desktop database* template.
3. Type database name.
4. Click Create button.

Create

To create a new database or open a previously created database, click the Windows Start button and then click the Access tile. (These steps may vary depending on your system configuration.) This displays the Access opening screen, as shown in Figure 1.1. At this screen, open a recently opened database, a blank database, a database from the Open backstage area, or a database based on a template.

To create a new blank database, click the *Blank database* template. At the Blank database window, type a name for the database in the *File Name* text box, and then click the Create button. To save the database in a particular location, click the Browse button at the right side of the *File Name* text box. At the File New Database dialog box, navigate to the location or folder, type the database name in the *File name* text box, and then click OK.

Tutorial

Opening an Existing
Database

Opening a Database

A previously saved database can be opened at the Open dialog box. To display this dialog box, display the Open backstage area and then click the *Browse* option.

Figure 1.1 Access Opening Screen

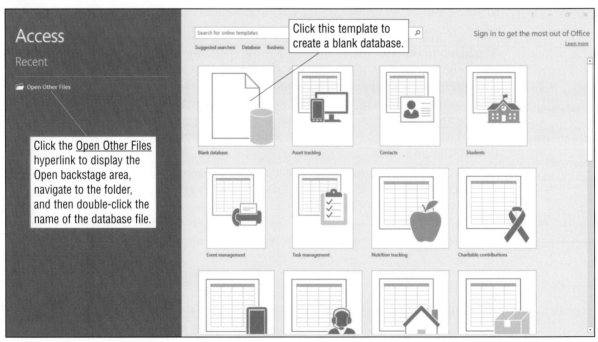

Display the Open backstage area by clicking the <u>Open Other Files</u> hyperlink at the Access opening screen. Or, click the File tab at the blank Access screen and then click the *Open* option. Other methods for displaying the Open backstage area include using the keyboard shortcut, Ctrl + O, or inserting an Open button on the Quick Access Toolbar.

At the Open backstage area, click the *Browse* option and the Open dialog box displays. At the Open dialog box, navigate to the desired location, such as the drive containing your storage medium, open the folder containing the database, and then double-click the database name in the Content pane. When a database is open, the Access screen looks similar to what is shown in Figure 1.2. Refer to Table 1.1 for descriptions of the Access screen elements.

Only one Access database can be open at a time. If a new database is opened in the current Access window, the existing database closes. However, multiple instances of Access can be opened and a database can be opened in each instance. In other applications in the Microsoft Office suite, a revised file must be saved after changes are made to the file. In an Access database, any changes made to data are saved automatically when moving to a new record, closing a table, or closing the database.

Enabling Content

A security warning message bar may appear below the ribbon if Access determines the file being opened did not originate from a trusted location on the computer and may contain viruses or other security hazards. This often occurs when an Access database is copied from another medium (such as an external drive or cloud storage location). Active content in the file is disabled until the Enable Content button is clicked. The message bar closes when the database is identified as a trusted source. Before making any changes to the database, the Enable Content button must be clicked.

Figure 1.2 Access Screen

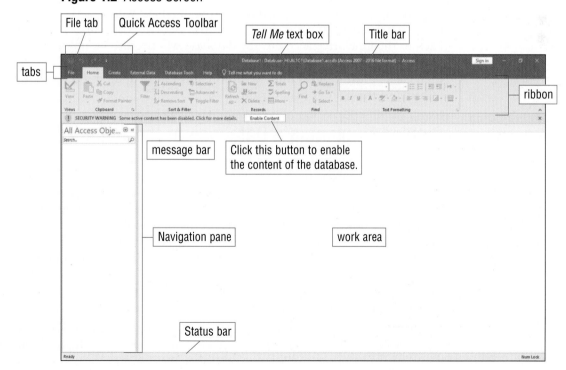

Table 1.1 Access Screen Elements

Feature	Description
File tab	when clicked, displays the backstage area that contains options for working with and managing databases
message bar	displays security alerts if the database being opened contains potentially unsafe content
Navigation pane	lists the names of objects within the database grouped by categories
Quick Access Toolbar	contains buttons for commonly used commands
ribbon	contains tabs with commands and buttons divided into groups
Status bar	displays messages, the current view, and view buttons
tabs	contain commands and features organized into groups
Tell Me text box	provides information as well as guidance on how to perform a function
Title bar	contains the database name followed by the program name
work area	displays opened objects

Managing the
Recent Option List

Opening a Database from the *Recent* Option List

At the Open backstage area with the *Recent* option selected, a list of the most recently opened databases displays. Access shows 50 of the most recently opened databases and groups them into categories such as *Today*, *Yesterday*, and perhaps another category such as *Last Week*. Click the database name in the *Recent* option list to open the database.

(Note: If opening a database from a OneDrive account, Access requires that a copy of the database be saved to a location such as the computer's hard drive or a USB flash drive. Any changes made to the database will be saved to the local copy of the database but not the database in the OneDrive account. To save a database back to the OneDrive account, the database will need to be uploaded by opening a web browser, going to OneDrive.com, logging in to the OneDrive account, and then clicking the Upload link. Microsoft constantly updates the OneDrive.com website, so these steps may vary.)

Pinning/Unpinning a Database at the *Recent* Option List

If a database is opened on a regular basis, consider pinning it to the *Recent* option list. To pin a database to the *Recent* option list at the Open backstage area, hover the mouse pointer over the database name and then click the small left-pointing push pin icon to the right of the database name. The left-pointing push pin changes to a down-pointing push pin and the pinned database is inserted into a new category named *Pinned*. The *Pinned* category appears at the top of the *Recent* option list. The next time the Open backstage area displays, the pinned database is in the *Pinned* category. A database can also be pinned to the Recent list at the Access opening screen. When a database is pinned, it is found at the top of the Recent list as well as the *Recent* option list at the Open backstage area.

To unpin a database from the Recent or *Recent* option list, click the pin icon to change it from a down-pointing push pin to a left-pointing push pin. More than one database can be pinned to a list. Another method for pinning and unpinning databases is to use the shortcut menu. Right-click a database name and then click the *Pin to list* or *Unpin from list* option.

Opening and Closing Objects

Tutorial

Opening and Closing an Object

The Navigation pane at the left side of the Access screen lists the objects contained in the database. Some common objects found in a database include tables, queries, forms, and reports. Refer to Table 1.2 for descriptions of these four types of objects.

Control what is shown in the pane by clicking the menu bar at the top of the Navigation pane and then clicking an option at the drop-down list or by clicking the button on the menu bar containing the down arrow. (The name of this button changes depending on what is selected.) For example, to display a list of all saved objects in the database, click the *Object Type* option at the drop-down list. This view lists the objects grouped by type: *Tables, Queries, Forms,* and *Reports.* To open an object, double-click the object in the Navigation pane. The object opens in the work area and a tab displays with the object name at the left side of the object. An object can also be opened with the shortcut menu by right-clicking the object in the Navigation pane and then clicking the *Open* option at the shortcut menu.

Shutter Bar Open/Close Button

💡 *Hint* Hide the Navigation pane by clicking the Shutter Bar Open/Close Button or by pressing the F11 function key.

To view more of an object, consider closing the Navigation pane by clicking the Shutter Bar Open/Close Button in the upper right corner of the Navigation pane or by pressing the F11 function key. Click the button or press the F11 function key again to reopen the Navigation pane. More than one object can be opened in the work area. Each object opens with a visible tab. Navigate to objects by clicking the object tab.

To close an object, click the Close button (X) for that object (depending on the version of Access, this button will be located on the object tab or at the right corner of the work area) or use the keyboard shortcut Ctrl + F4. The shortcut menu can also be used to close an object by right-clicking the object tab and then clicking *Close* at the shortcut menu. Close multiple objects by right-clicking an object tab and then clicking *Close All* at the shortcut menu.

Closing a Database

Tutorial

Closing a Database and Closing Access

Close

To close a database, click the File tab and then click the *Close* option. Close Access by clicking the Close button in the upper right corner of the screen or with the keyboard shortcut Alt + F4.

Table 1.2 Database Objects

Object Type	Description
table	Organizes data in fields (columns) and records (rows). A database must contain at least one table. The table is the base upon which other objects are created.
query	Displays data from a table or related tables that meets a conditional statement and/or performs calculations. For example, all records from a specific month can be shown or only those records containing a specific city.
form	Allows fields and records to be presented in a layout different from the datasheet. Used to facilitate data entry and maintenance.
report	Prints data from tables or queries.

1. Open Access by clicking the Windows Start button and then clicking the Access tile in the Start menu.
2. At the Access opening screen, click the <u>Open Other Files</u> hyperlink.
3. At the Open backstage area, click the *Browse* option.
4. At the Open dialog box, navigate to the AL1C1 folder on your storage medium and then double-click *1-SampleDatabase*. (This database contains data on orders, products, and suppliers for a specialty hiking and backpacking outfitters store named Pacific Trek.)
5. Click the Enable Content button in the message bar if a security warning message appears. (The message bar will display immediately below the ribbon.)
6. With the database open, click the Navigation pane menu bar and then click *Object Type* in the *Navigate To Category* section of the drop-down list. Click the Navigation pane menu bar again and then click *All Access Objects* in the *Filter By Group* section. (This option lists the objects grouped by type: *Tables*, *Queries*, *Forms*, and *Reports*.)

7. Double-click *Suppliers* in the Tables group in the Navigation pane. This opens the Suppliers table in the work area, as shown in Figure 1.3.
8. Close the Suppliers table by clicking the Close button for the table. (Note that the Close button for an object will be located on the object tab or in the corner of the work area, as shown in the image below. Be careful not to confuse it with the Close button for the Access window, which closes the application.)

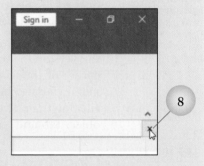

9. Double-click *OrdersOver$500* in the Queries group in the Navigation pane. A query displays data that meets a conditional statement. This query displays orders that meet the criterion of being more than $500.
10. Close the query by clicking the Close button for the query object (located on the object tab or in the right corner of the work area).
11. Right-click *SuppliersNotVancouver* in the Queries group in the Navigation pane and then click *Open* at the shortcut menu. This query displays information about suppliers but excludes those in Vancouver.
12. Right-click the SuppliersNotVancouver tab in the work area and then click *Close* at the shortcut menu.

13. Double-click *Orders* in the Forms group in the Navigation pane. This opens an order form. A form is used to view and edit data in a table one record at a time.
14. Double-click *Orders* in the Reports group in the Navigation pane. The Orders form is still open and the report opens in the work area over the Orders form. The Orders report displays information about orders and order amounts.

15. Close the Navigation pane by clicking the Shutter Bar Open/Close Button in the upper right corner of the pane.
16. After viewing the report, click the Shutter Bar Open/Close Button again to open the Navigation pane.
17. Right-click the Orders tab and then click *Close All* at the shortcut menu. (This closes both open objects.)
18. Close the database by clicking the File tab and then clicking the *Close* option.
19. Close Access by clicking the Close button in the upper right corner of the screen.

Figure 1.3 Suppliers Table

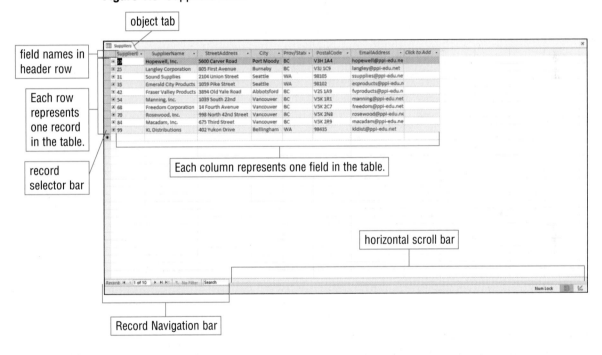

Activity 2 Manage Tables in a Database **7 Parts**

Pacific Trek is an outfitting store specializing in hiking and backpacking gear. Information about the store, including suppliers and products, is contained in a database. You will open the database and then insert and delete records; insert, move, and delete fields; preview and print tables; rename and delete a table; and create two new tables for the database. Additionally, you will modify field columns, change page layout, insert Quick Start fields, and modify field properties.

Managing Tables

In a new database, tables are the first objects created, since all other database objects rely on a table as the source for their data. Managing the tables in the database is important for keeping the database up to date and may include inserting or deleting records, inserting or deleting fields, renaming fields, creating a hard copy of the table by printing the table, and renaming and deleting tables.

Adding and Deleting Records in a Table

Adding and Deleting Records

When a table is opened, it displays in Datasheet view in the work area. The Datasheet view displays the contents of a table in a column and row format similar to an Excel worksheet. Columns contain field data, with the field names in the header row at the top of the table, and rows contain records. A Record Navigation bar displays at the bottom of the screen just above the Status bar and contains buttons to navigate in the table. Figure 1.4 identifies the buttons and other elements on the Record Navigation bar.

New

Quick Steps

Add New Record
1. Open table.
2. Click New button on Home tab.
3. Type data.
OR
1. Open table.
2. Click New (blank) record button on Record Navigation bar.
3. Type data.

To add a new record to the open table, make sure the Home tab is active and then click the New button in the Records group. This moves the insertion point to the first field in the blank row at the bottom of the table and the *Current Record* box on the Record Navigation bar indicates what record is being created (or edited). A new record can also be added by clicking the New (blank) record button on the Record Navigation bar.

When working in a table, press the Tab key to make the next field in the current record active or press Shift + Tab to make the previous field in the current record active. Make a field active with the mouse by clicking in the field. When typing data for the first field in the record, another row of cells is automatically inserted below the current row and a pencil icon appears in the record selector bar at the beginning of the current record. The pencil icon indicates that the record is being edited and that the changes to the data have not been saved. When data is entered in the last field and the insertion point is moved out of the field, the pencil icon is removed, indicating that the data has been saved.

 Delete

Quick Steps

Delete Record
1. Click field in record.
2. Click Delete button arrow.
3. Click *Delete Record*.
4. Click Yes.

To delete a record, click in one of the fields in the record, make sure the Home tab is active, click the Delete button arrow, and then click *Delete Record* at the drop-down list. At the message asking to confirm the deletion, click Yes. Click in a field in a record and the Delete button is dimmed unless specific data is selected.

Data entered in a table is automatically saved; however, changes to the layout of a table are not automatically saved. For example, if a column is deleted in a table, a deletion confirmation message will display when the table is closed.

Figure 1.4 Record Navigation Bar

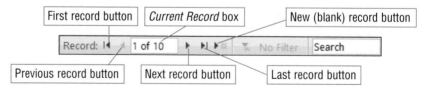

1. Open Access.
2. At the Access opening screen, click the <u>Open Other Files</u> hyperlink.
3. At the Open backstage area, click the *Browse* option.
4. At the Open dialog box, navigate to the AL1C1 folder on your storage medium and then double-click **1-PacTrek**.
5. Click the Enable Content button in the message bar if a security warning message appears. (The message bar will display immediately below the ribbon.)
6. With the database open, make sure the Navigation pane lists object types. (If it does not, click the Navigation pane menu bar and then click *Object Type* at the drop-down list.)
7. Double-click *Suppliers* in the Tables group in the Navigation pane. (This opens the table in Datasheet view.)
8. With the Suppliers table open and the Home tab active, add a new record by completing the following steps:

a. Click the New button in the Records group on the Home tab. (This moves the insertion point to the first field in the blank record at the bottom of the table and the *Current Record* box in the Record Navigation bar indicates what record you are creating or editing.)

b. Type 38. (This inserts *38* in the field immediately below *99*.)
c. Press the Tab key (to make the next field in the current record active) and then type Hadley Company.
d. Press the Tab key and then type 5845 Jefferson Street.
e. Press the Tab key and then type Seattle.
f. Press the Tab key and then type WA.
g. Press the Tab key and then type 98107.
h. Press the Tab key and then type hcompany@ppi-edu.net.
i. Press the Tab key and then type Jurene Miller.

SupplierID	SupplierNan	StreetAddre	City	Prov/State	PostalCode	EmailAddi	Contact	Click to Add
10	Hopewell, Inc.	5600 Carver Ro	Port Moody	BC	V3H 1A4	hopewell@¡	Jacob Hopewel	
25	Langley Corpor	805 First Avent	Burnaby	BC	V3J 1C9	langley@ppi	Mandy Shin	
31	Sound Supplie:	2104 Union Stri	Seattle	WA	98105	ssupplies@p	Regan Levine	
35	Emerald City Pi	1059 Pike Stree	Seattle	WA	98102	ecproducts@	Howard Greer	
42	Fraser Valley P	3894 Old Yale F	Abbotsford	BC	V2S 1A9	fvproducts@	Layla Adams	
54	Manning, Inc.	1039 South 22n	Vancouver	BC	V5K 1R1	manning@p¡	Jack Silversteir	
68	Freedom Corpi	14 Fourth Aver	Vancouver	BC	V5K 2C7	freedom@p¡	Opal Northwoc	
70	Rosewood, Inc	998 North 42no	Vancouver	BC	V5K 2N8	rosewood@¡	Clint Rivas	
84	Macadam, Inc.	675 Third Stree	Vancouver	BC	V5K 2R9	macadam@¡	Hans Reiner	
99	KL Distribution	402 Yukon Driv	Bellingham	WA	98435	kldist@ppi-e	Noland Dannis	
38	Hadley Compai	5845 Jefferson	Seattle	WA	98107	hcompany@	Jurene Miller	

9. Close the Suppliers table by clicking the table's Close button. (The text you entered was automatically saved by Access.)
10. Open the Products table by double-clicking *Products* in the Tables group in the Navigation pane. (This opens the table in Datasheet view.)
11. Insert two new records by completing the following steps:
a. Click the New button in the Records group and then enter the data for a new record as shown in Figure 1.5. (See the record that begins with *901-S*.)

b. After typing the last field entry in the record for product number 901-S, press the Tab key. This moves the insertion point to the blank field below *901-S*.

c. Type the new record as shown in Figure 1.5. (See the record that begins with *917-S*.)

12. With the Products table open, delete a record by completing the following steps:

a. Click in the field containing the data *780-2*.

b. Click the Delete button arrow in the Records group (notice that the button is dimmed) and then click *Delete Record* at the drop-down list.

c. At the message asking if you want to delete the record, click Yes.

13. Close the Products table by clicking the Close button for the table.

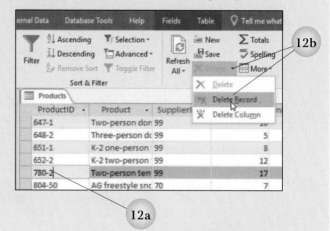

Check Your Work

Figure 1.5 Activity 2a, Step 11

ProductID ▾	Product ▾	SupplierID ▾	UnitsInStock ▾	UnitsOnOrde ▾	ReorderLevel ▾	Click to Add
602-XR	Binoculars, 8 x 42	35	3	5	5	
602-XT	Binoculars, 10.5 x 45	35	5	0	4	
602-XX	Binoculars, 10 x 50	35	7	0	5	
647-1	Two-person dome tent	99	10	15	15	
648-2	Three-person dome tent	99	5	0	10	
651-1	K-2 one-person tent	99	8	0	10	
652-2	K-2 two-person tent	99	12	0	10	
804-50	AG freestyle snowboard, X50	70	7	0	10	
804-60	AG freestyle snowboard, X60	70	8	0	5	
897-L	Lang blunt snowboard	70	8	0	7	
897-W	Lang blunt snowboard, wide	70	4	0	3	
901-S	Solar battery pack	38	16	0	15	
917-S	Silo portable power pack	38	8	0	10	

Step 11

Tutorial

Managing Fields in Datasheet View

Quick Steps

Insert New Field
1. Open table.
2. Click in first cell below *Click to Add*.
3. Type data.

Inserting, Moving, and Deleting Fields

When managing a database, an additional field may need to be added to a table. For example, a field column may need to be added for contact information, one for cell phone numbers, or for the number of items in stock. To insert a new field column in a table, open the table in Datasheet view and then click in the first cell below *Click to Add* in the header row. Type the data in the cell, press the Down Arrow key, and then type the data for the second row. Continue in this manner until all data has been entered for the new field. In addition to pressing the Down Arrow key, move the insertion point down to the next cell by clicking in the cell or by pressing the Tab key.

Quick Steps

Move Field
1. Select column.
2. Position mouse pointer on field name in header row.
3. Click and hold down left mouse button.
4. Drag to new location.
5. Release mouse button.

Delete Field
1. Click in field.
2. Click Delete button arrow.
3. Click *Delete Column*.
4. Click Yes.

A new field column is added to the right of existing field columns. Move a field column by positioning the mouse pointer on the field name in the header row until the pointer displays as a down-pointing black arrow and then click the left mouse button. This selects the entire field column. With the field column selected, position the mouse pointer on the field name; click and hold down the left mouse button; drag to the left or right until a thick, black vertical line displays in the desired location; and then release the mouse button. The thick, black vertical line indicates where the field column will be positioned when the mouse button is released. In addition, the pointer displays with the outline of a gray box attached to it, indicating that a move operation is being performed.

Delete a field column in a manner similar to deleting a row. Click in one of the fields in the column, make sure the Home tab is active, click the Delete button arrow, and then click *Delete Column* at the drop-down list. At the message asking to confirm the deletion, click Yes.

Activity 2b Inserting, Moving, and Deleting Fields   Part 2 of 7

1. With **1-PacTrek** open, add a new field to the Suppliers table by completing the following steps:
   a. Double-click *Suppliers* in the Tables group in the Navigation pane.
   b. Click in the field immediately below *Click to Add* in the header row.
   c. Type (604) 555-3843 and then press the Down Arrow key.
   d. Type the remaining telephone numbers as shown below and at the right.

   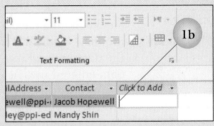

2. Move the *Field1* field column so it is positioned immediately left of the *EmailAddress* field column by completing the following steps:
   a. Position the mouse pointer on the *Field1* field name in the header row until the pointer displays as a down-pointing black arrow and then click the left mouse button. (This selects the field column.)
   b. Position the mouse pointer on the *Field1* field name. Click and hold down the left mouse button; drag to the left until the thick, black vertical line displays immediately left of the *EmailAddress* field column; and then release the mouse button.

3. Delete the *Contact* field by completing the following steps:
   a. Position the mouse pointer on the *Contact* field name in the header row until the pointer displays as a down-pointing black arrow and then click the left mouse button. (This selects the field column.)
   b. Click the Delete button arrow in the Records group and then click *Delete Column* at the drop-down list.
   c. At the message asking if you want to permanently delete the selected field(s) click Yes.

   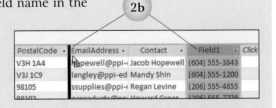

4. Close the Suppliers table. At the message asking if you want to save the changes to the layout of the table, click Yes.

Check Your Work

Tutorial

Hiding, Unhiding,
Freezing, and
Unfreezing Field
Columns

 More

Hiding, Unhiding, Freezing, and Unfreezing Field Columns

A field column in a table can be hidden if the column is not needed for data entry or editing purposes or to make viewing easier for two nonadjacent field columns containing data to be compared. To hide a field column, click in any field in the column, click the More button in the Records group on the Home tab, and then click *Hide Fields* at the drop-down list. Hide adjacent field columns by selecting the columns, clicking the More button in the Records group, and then clicking *Hide Fields* at the drop-down list. To unhide field columns, click the More button and then click *Unhide Fields*. At the Unhide Columns dialog box, insert a check mark in the check boxes for those field columns that should remain visible.

Another method for comparing field columns side by side is to freeze a column. Freezing a field column is also helpful when not all of the columns of data are visible at one time. To freeze a field column, click in any field in the column, click the More button, and then click *Freeze Fields* at the drop-down list. To freeze adjacent field columns, select the columns first, click the More button, and then click *Freeze Fields* at the drop-down list. To unfreeze all field columns in a table, click the More button and then click *Unfreeze All Fields* at the drop-down list.

 Tutorial

Adjusting Field
Column Width

Q̆uick Steps

Adjust Field Column Width
1. Click More button.
2. Click *Field Width*.
3. Type measurement.
4. Click OK.
OR
1. Click More button.
2. Click *Field Width*.
3. Click Best Fit button.
OR
Double-click column boundary line.
OR
Select field columns and then double-click column boundary line.
OR
Drag column boundary line to new position.

Adjusting Field Column Width

Field columns have a default width of 13.111 characters, which is the approximate number of characters that will be visible in a field in the column. Depending on the data entered in a field in a field column, not all of the data will be visible. Or data entered in a field may take up only a portion of the field. Change a field's column width with options at the Column Width dialog box. Display the dialog box by clicking the More button in the Records group on the Home tab and then clicking the *Field Width* option. Enter a measurement in the *Column Width* measurement box or click the Best Fit button to adjust the column width to accommodate the longest entry.

The width of a field column also can be adjusted to accommodate the longest entry by positioning the mouse pointer on the column boundary line at the right side of the column in the header row until the pointer turns into a left-and-right-pointing arrow with a vertical line in the middle and then double-clicking the left mouse button. Adjust the widths of adjacent field columns by selecting the columns first and then double-clicking one of the selected column boundary lines in the header row. To select adjacent field columns, position the mouse pointer on the first field name to be selected in the header row until the pointer turns into a down-pointing black arrow, click and hold down the left mouse button, drag to the last field name to be selected, and then release the mouse button. With the field columns selected, double-click one of the field column boundary lines.

Another method for adjusting the width of a field column is to drag a boundary line to the desired position. To do this, position the mouse pointer on the column boundary line (until the pointer turns into a left-and-right-pointing arrow with a vertical line in the middle), click and hold down the left mouse button, drag until the field column is the desired width, and then release the mouse button.

1. With **1-PacTrek** open, open the Suppliers table.
2. Hide the *PostalCode* field column by clicking the *PostalCode* field name in the header row, clicking the More button in the Records group on the Home tab, and then clicking *Hide Fields* at the drop-down list.

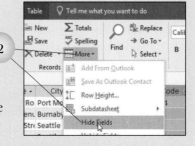

3. Unhide the field column by clicking the More button and then clicking *Unhide Fields* at the drop-down list. At the Unhide Columns dialog box, click the *PostalCode* check box to insert a check mark, and then click the Close button.

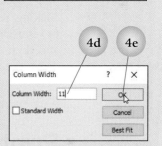

4. Adjust the width of the *SupplierID* field column by completing the following steps:
   a. Click the *SupplierID* field name in the header row.
   b. Click the More button in the Records group on the Home tab.
   c. Click the *Field Width* option.
   d. Type 11 in the *Column Width* measurement box.
   e. Click OK.
5. Adjust the *SupplierName* field column by completing the following steps:
   a. Click the *SupplierName* field name in the header row.
   b. Click the More button in the Records group.
   c. Click the *Field Width* option.
   d. Click the Best Fit button.
6. Select the remaining columns and adjust the column widths by completing the following steps:
   a. Position the mouse pointer on the *StreetAddress* field name in the header row until the pointer turns into a down-pointing black arrow, click and hold down the left mouse button, drag to the *EmailAddress* field name, and then release the mouse button.
   b. With the columns selected, double-click one of the column boundary lines in the header row.
   c. Click in any field in the table to deselect the field columns.
7. Increase the width of the *EmailAddress* field column by positioning the mouse pointer on the column boundary line in the header row at the right side of the *EmailAddress* field column until it turns into a left-and-right-pointing arrow with a vertical line in the middle, clicking and holding down the left mouse button, dragging all the way to the right side of the screen, and then releasing the mouse button. (Check the horizontal scroll bar at the bottom of the table and notice that the scroll bar contains a scroll box.)
8. Position the mouse pointer on the scroll box on the horizontal scroll bar and then drag to the left until the *SupplierID* field is visible.

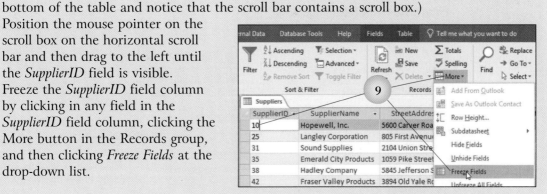

9. Freeze the *SupplierID* field column by clicking in any field in the *SupplierID* field column, clicking the More button in the Records group, and then clicking *Freeze Fields* at the drop-down list.

10. Using the mouse, drag the scroll box along the horizontal scroll to the right and left to see that the *SupplierID* field column remains visible on the screen.
11. Unfreeze the field column by clicking the More button in the Records group and then clicking *Unfreeze All Fields* at the drop-down list.
12. Double-click the column boundary line in the header row at the right side of the *EmailAddress* field column.
13. Close the Suppliers table and click Yes at the message that asks if you want to save the changes to the layout.
14. Open the Products table and then complete steps similar to those in Step 6 to select and then adjust the field column widths.
15. Close the Products table and click Yes at the message that asks if you want to save the changes to the layout.

 Check Your Work

 Tutorial

Renaming and Deleting Objects

Quick Steps

Rename Table
1. Right-click table name in Navigation pane.
2. Click *Rename*.
3. Type new name.
4. Press Enter key.

Delete Table
1. Right-click table name in Navigation pane.
2. Click *Delete*.
3. Click Yes, if necessary.

Renaming and Deleting a Table

Managing tables might include actions such as renaming and deleting a table. Rename a table by right-clicking the table name in the Navigation pane, clicking *Rename* at the shortcut menu, typing the new name, and then pressing the Enter key. Delete a table from a database by clicking the table name in the Navigation pane, clicking the Delete button in the Records group on the Home tab, and then clicking Yes at the message asking to confirm the deletion. Another method is to right-click the table in the Navigation pane, click *Delete* at the shortcut menu, and then click Yes at the message. If a table is deleted from the computer's hard drive, the message asking to confirm the deletion does not display. This is because Access automatically sends the deleted table to the Recycle Bin, where it can be retrieved if necessary.

 Tutorial

Previewing and Printing a Table

Quick Steps

Print Table
1. Click File tab.
2. Click *Print*.
3. Click *Quick Print*.
OR
1. Click File tab.
2. Click *Print*.
3. Click *Print*.
4. Click OK.

Printing Tables

Click the File tab and then click the *Print* option to display the Print backstage area as shown in Figure 1.6. Click the *Quick Print* option to send the table directly to the printer without making any changes to the printer setup or the table formatting. Click the *Print* option to display the Print dialog box, with options for specifying the printer, page range, and specific records to print. Click OK to close the dialog box and send the table to the printer. By default, Access prints a table on letter-size paper in portrait orientation.

Figure 1.6 Print Backstage Area

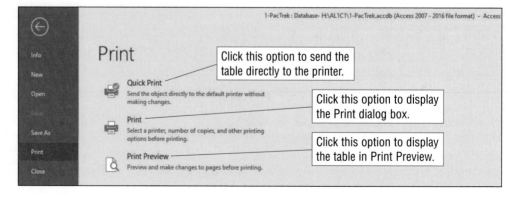

Previewing a Table

Quick Steps

Preview Table
1. Click File tab.
2. Click *Print* option.
3. Click *Print Preview* option.

Print Preview

Print

Close Print Preview

Before printing a table, consider displaying the table in Print Preview to see how the table will print on the page. To display a table in Print Preview, as shown in Figure 1.7, click the *Print Preview* option at the Print backstage area.

Use options in the Zoom group on the Print Preview tab to increase or decrease the size of the page display. The size of the page display can also be changed using the Zoom slider bar at the right side of the Status bar. If a table spans more than one page, use buttons on the Navigation bar to move to the next or previous page.

Print a table from Print Preview by clicking the Print button at the left side of the Print Preview tab. Click the Close Print Preview button to close Print Preview and continue working in the table without printing it.

Changing Page Size and Margins

Size

Margins

By default, Access prints a table in standard letter size (8.5 inches wide and 11 inches tall). Click the Size button in the Page Size group on the Print Preview tab and a drop-down list displays with options for changing the page size to legal, executive, envelope, and so on. Access uses default top, bottom, left, and right margins of 1 inch. Change these default margins by clicking the Margins button in the Page Size group and then clicking one of the predesigned margin options.

Figure 1.7 Print Preview

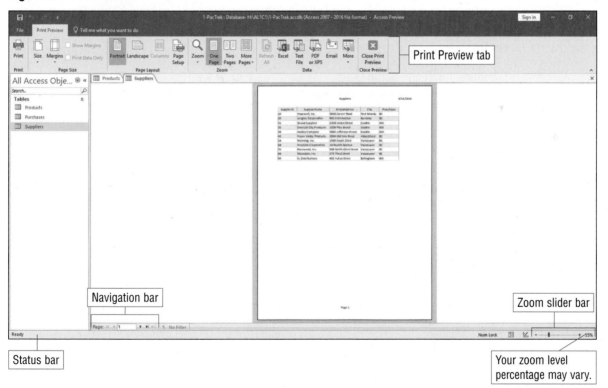

Changing Page Layout

The Print Preview tab contains the Page Layout group with buttons for controlling how data is printed on the page. By default, Access prints a table in portrait orientation, which prints the text on the page so that it is taller than it is wide (like a page in this textbook). If a table contains a number of columns, changing to landscape orientation allows more columns to fit on a page. Landscape orientation rotates the printout to be wider than it is tall. To change from the default portrait orientation to landscape orientation, click the Landscape button in the Page Layout group on the Print Preview tab.

Click the Page Setup button in the Page Layout group and the Page Setup dialog box displays as shown in Figure 1.8. At the Page Setup dialog box with the Print Options tab selected, notice that the default margins are 1 inch. Change these defaults by typing different numbers in the margin measurement boxes. By default, the table prints at the top center of the page and the current date prints in the upper right corner of the page. In addition, the word *Page* followed by the page number prints at the bottom of the page. To specify that the name of the table, date, and page number should not print, remove the check mark from the *Print Headings* option at the Page Setup dialog box with the Print Options tab selected.

Click the Page tab at the Page Setup dialog box and the dialog box displays as shown in Figure 1.9. Change the orientation with options in the *Orientation* section and change the paper size with options in the *Paper* section. Click the *Size* option box arrow and a drop-down list displays with paper sizes similar to the options available at the *Size* button drop-down list in the Page Size group on the Print Preview tab. Specify the printer with options in the *Printer for (table name)* section of the dialog box.

Landscape

Page Setup

Quick Steps

Display Page Setup Dialog Box
1. Click File tab.
2. Click *Print* option.
3. Click *Print Preview* option.
4. Click Page Setup button.

Figure 1.8 Page Setup Dialog Box with Print Options Tab Selected

Enter measurements in these measurement boxes to change the page margins.

Remove the check mark from this check box to specify that the table name, date, and page number should not print.

Figure 1.9 Page Setup Dialog Box with Page Tab Selected

Click this option to change the page orientation to landscape.

Change the paper size with this option.

1. With **1-PacTrek** open, open the Suppliers table.
2. Preview and then print the Suppliers table in landscape orientation by completing the following steps:

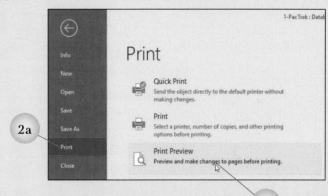

   a. Click the File tab and then click the *Print* option.
   b. At the Print backstage area, click the *Print Preview* option.
   c. Click the Two Pages button in the Zoom group on the Print Preview tab. (This displays two pages of the table.)
   d. Click the Zoom button arrow in the Zoom group on the Print Preview tab and then click *75%* at the drop-down list.

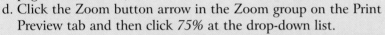

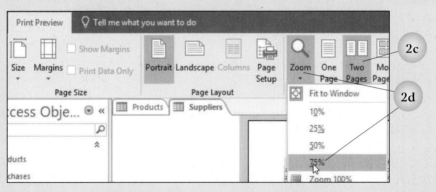

   e. Position the mouse pointer on the Zoom slider bar button at the right side of the Status bar, click and hold down the left mouse button, drag to the right until *100%* displays at the right of the Zoom slider bar, and then release the mouse button.

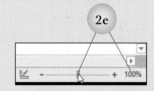

   f. Return the display to a full page by clicking the One Page button in the Zoom group on the Print Preview tab.
   g. Click the Margins button in the Page Size group on the Print Preview tab and then click the *Narrow* option at the drop-down list. (Notice how the data will print on the page with the narrow margins.)
   h. Change the margins back to the default by clicking the Margins button in the Page Size group and then clicking the *Normal* option at the drop-down list.
   i. Change to landscape orientation by clicking the Landscape button in the Page Layout group.
   j. Print the table by clicking the Print button at the left side of the Print Preview tab and then clicking OK at the Print dialog box.

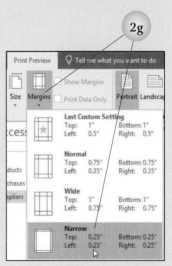

3. Close the Suppliers table.

4. Open the Products table and then print the table by completing the following steps:
   a. Click the File tab and then click the *Print* option.
   b. At the Print backstage area, click the *Print Preview* option.
   c. Click the Page Setup button in the Page Layout group on the Print Preview tab. (This displays the Page Setup dialog box with the Print Options tab selected.)
   d. At the Page Setup dialog box, click the Page tab.
   e. Click the *Landscape* option.

   f. Click the Print Options tab.
   g. Select the current measurement in the *Top* measurement box and then type 0.5.
   h. Select the current measurement in the *Bottom* measurement box and then type 0.5.
   i. Select the current measurement in the *Left* measurement box and then type 1.5.
   j. Click OK to close the dialog box.
   k. Click the Print button on the Print Preview tab and then click OK at the Print dialog box. (This table will print on two pages.)

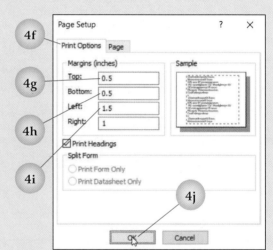

5. Close the Products table.
6. Rename the Purchases table by right-clicking *Purchases* in the Navigation pane, clicking *Rename* at the shortcut menu, typing Orders, and then pressing the Enter key.
7. Delete the Orders table by right-clicking *Orders* in the Tables group in the Navigation pane and then clicking *Delete* at the shortcut menu. If a message displays asking if you want to permanently delete the table, click Yes.

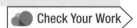

Designing a Table

Tables are the first objects created in a new database and all other objects in a database rely on tables for data. Designing a database involves planning the number of tables needed and the fields that will be included in each table. Each table in a database should contain information about only one subject. For example, the Suppliers table in the 1-PacTrek database contains data only about suppliers and the Products table contains data only about products.

Database designers often create a visual representation of the database's structure in a diagram similar to the one shown in Figure 1.10. Each table is represented by a box with the table name at the top. Within each box, the fields that will be stored in the table are listed with the field names that will be used when the table is created.

Notice that one field in each table has an asterisk next to its name. The field with the asterisk is called the *primary key field*. A primary key field holds data that uniquely identifies each record in a table and is usually an identification number. The lines drawn between each table in Figure 1.10 are called *join lines,* and they represent links established between tables (called *relationships*) so that data can be extracted from one or more tables. The join lines point to a common field name included in each table that is to be linked. (Joining tables is covered in Chapter 2.) A database with related tables is called a *relational database*.

The join line in the database diagram connects the *SupplierID* field in the Suppliers table with the *SupplierID* field in the Products table and another join line connects the *SupplierID* field in the Suppliers table with the *SupplierID* field in the Orders table. A join line connects the *ProductID* field in the Products table with the *ProductID* field in the Orders table.

♀**Hint** Join tables to minimize or eliminate data duplication.

Consider certain design principles when designing a database. The first principle is to reduce redundant (duplicate) data. Redundant data increases the amount of data entry required, increases the chances for errors and inconsistencies, and takes up additional storage space. The Products table contains a *SupplierID* field and that field reduces the duplicate data needed in the table by keeping the data in one table instead of two and then joining the tables by a common field.

Figure 1.10 Database Diagram

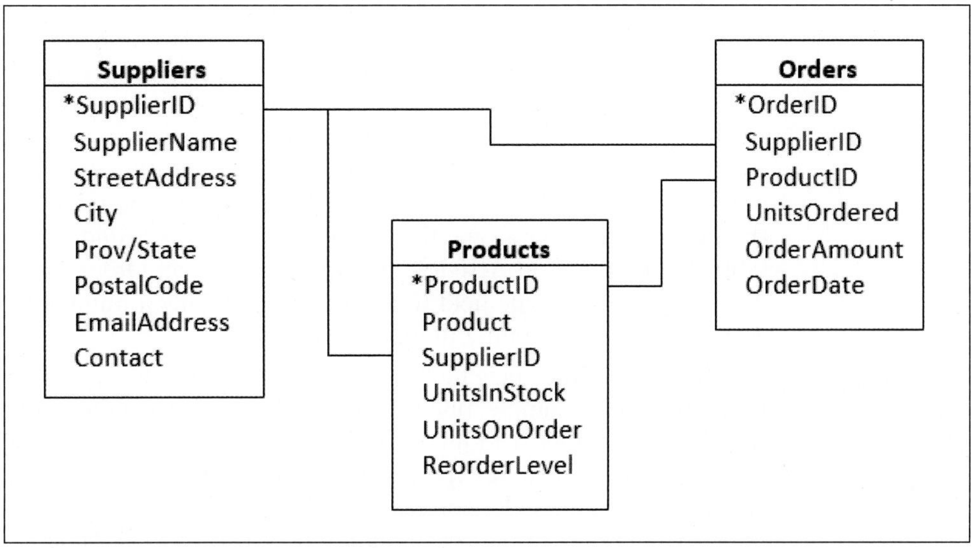

For example, rather than typing the supplier information in the Suppliers table *and* the Products table, type the information once in the Suppliers table and then join the tables with the connecting field *SupplierID*. If information is needed on suppliers as well as specific information about products, the information can be drawn into one object, such as a query or report using data from both tables. When creating the Orders table, the *SupplierID* field and the *ProductID* field will be used rather than typing all the information for the suppliers and the product description. Typing a two-letter unique identifier number for a supplier greatly reduces the amount of typing required to create the Orders table. Inserting the *ProductID* field in the Orders table eliminates the need to type the product description for each order; instead, a unique five-, six-, or seven-digit identifier number is typed.

Creating a Table

Creating a Table in Datasheet View

Creating a new table generally involves determining fields, assigning a data type to each field, modifying properties, designating the primary key field, and naming the table. This process is referred to as defining the table structure.

The first step in creating a table is to determine the fields. A field, commonly called a column, is one piece of information about a person, place, or item. Each field contains data about one aspect of the table subject, such as a company name or product number. All fields for one unit, such as a customer or product, are considered a record. For example, in the Suppliers table in the 1-PacTrek database, a record is all of the information pertaining to one supplier. A collection of records becomes a table.

Hint A database table contains fields that can describe a person, customer, client, object, place, idea, or event.

When designing a table, determine the fields to be included based on how the information will be used. When organizing fields be sure to consider not only the current needs for the data but also any future needs. For example, a company may need to keep track of customer names, addresses, and telephone numbers for current mailing lists. In the future, the company may want to promote a new product to customers who purchase a specific type of product. For this information to be available at a later date, a field that identifies product type must be included in the database. When organizing fields, consider all potential needs for the data but also try to keep the fields logical and manageable.

 Table

A table can be created in Datasheet view or Design view. To create a table in Datasheet view, open the database (or create a new database), click the Create tab, and then click the Table button in the Tables group. This inserts a blank table in the work area with the tab labeled *Table1*, as shown in Figure 1.11. Notice the column with the field name *ID* has been created automatically. Access creates *ID* as an AutoNumber data type field in which the field value is assigned automatically by Access as each record is entered in the table. In many tables, this AutoNumber data type field is used to create the unique identifier for the table. For example, in Activity 2e, you will create an Orders table and use the *ID* AutoNumber data type field to assign automatically a number to each order, since each order must contain a unique number.

Hint Assign a data type to each field that determines the type of information that can be entered into the field.

When creating a new field (column), determine the type of data to be inserted in the field. For example, one field might contain text such as a name or product description, another field might contain an amount of money, and another might contain a date. The data type defines the type of information Access will allow to be entered into the field. For example, Access will not allow alphabetic characters to be entered into a field with a data type set to Date & Time.

Figure 1.11 Blank Table in Datasheet View

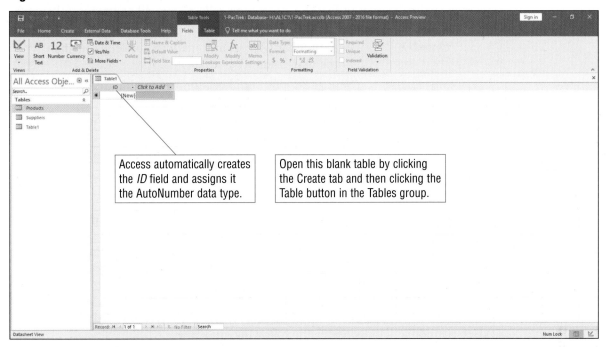

Access automatically creates the *ID* field and assigns it the AutoNumber data type.

Open this blank table by clicking the Create tab and then clicking the Table button in the Tables group.

 More Fields

The Add & Delete group on the Table Tools Fields tab contains five buttons for assigning data types plus a More Fields button. Descriptions of the five data types assigned by the buttons are provided in Table 1.3.

Table 1.3 Data Types

Button	Description
Short Text	Alphanumeric data up to 255 characters in length—for example, a name, an address, or a value such as a telephone number or social security number that is used as an identifier and not for calculating.
Number	Positive or negative values that can be used in calculations; not to be used for values that will calculate monetary amounts (see Currency).
Currency	Values that involve money; Access will not round off during calculations.
Date & Time	Used to ensure dates and times are entered and sorted properly.
Yes/No	Data in the field will be either *Yes* or *No*, *True* or *False*, *On* or *Off*.

In Activity 2e, you will create the Orders table, as shown in Figure 1.12. Looking at the diagram in Figure 1.10 (on page 21), you will assign the following data types to the columns:

OrderID:	AutoNumber (Access automatically assigns this data type to the first column)
SupplierID:	Short Text (the supplier numbers are identifiers, not numbers for calculating)
ProductID:	Short Text (the product numbers are identifiers, not numbers for calculating)
UnitsOrdered:	Number (the unit numbers are values for calculating)
Amount:	Currency
OrderDate:	Date & Time

Click a data type button and Access inserts a field to the right of the *ID* field and selects the field heading *Field1*. Type a name for the field; press the Enter key; and Access selects the next field heading, named *Click to Add*, and displays a drop-down list of data types. This drop-down list contains the same five data types as the buttons in the Add & Delete group as well as additional data types. Click the data type at the drop-down list, type the field name, and then press the Enter key. Continue in this manner until all field names have been entered for the table. When naming a field, consider the following guidelines:

- Each field must have a unique name.
- The name should describe the contents of the field.
- A field name can contain up to 64 characters.
- A field name can contain letters and numbers. Some symbols are permitted but others are excluded, so avoid using symbols other than the underscore (to separate words) and the number symbol (to indicate an identifier number).

💡 **Hint** Avoid using spaces in field names.

- Do not use spaces in field names. Although a space is an accepted character, most database designers avoid using spaces in field names and object names. Use field compound words for field names or the underscore character as a word separator. For example, a field name for a person's last name could be named *LastName*, *Last_Name*, or *LName*.
- Abbreviate field names so that they are as short as possible but still easily understood. For example, a field such as *CompanyName* could be shortened to *CoName* and a field such as *EmailAddress* could be shortened to *Email*.

Activity 2e Creating a Table and Entering Data   Part 5 of 7

1. With **1-PacTrek** open, create a new table and specify data types and field names by completing the following steps:
   a. Click the Create tab.
   b. Click the Table button in the Tables group.
   c. Click the Short Text button in the Add & Delete group.

   d. With the *Field1* field name selected, type SupplierID and then press the Enter key. (This displays a drop-down list of data types below the *Click to Add* heading.)

e. Click the *Short Text* option at the drop-down list.

f. Type ProductID in the next field name and then press the Enter key.

g. Click *Number* at the drop-down list, type UnitsOrdered in the next field name, and then press the Enter key.

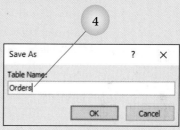

h. Click *Currency* at the drop-down list, type OrderAmount for the next field, and then press the Enter key.

i. Click *Date & Time* at the drop-down list and then type OrderDate. (Do not press the Enter key since this is the last field in the table.)

2. Enter the first record in the table, as shown in Figure 1.12, by completing the following steps:

a. Click two times in the first field below the *SupplierID* field name. (The first time you click the mouse button, the row is selected. Clicking the second time makes only the field below *SupplierID* active.)

b. Type the data in the fields as shown in Figure 1.12. Press the Tab key to move to the next field or press Shift + Tab to move to the previous field. Access will automatically insert the next number in the sequence in the first field column (the *ID* field column). When typing the money amounts in the *OrderAmount* field column, you do not need to type the dollar symbol or the comma. Access will automatically insert them when you make the next field active. Make sure to proofread the data after you type it to ensure it is accurate. ***Note: Since the ID field has the Autonumber data type applied, Access automatically inserts (New) in the row after the last record. Even though (New) appears in the field, the row does not contain a new record and the information in the row does not print.***

3. When the 14 records have been entered, click the Save button on the Quick Access Toolbar.

4. At the Save As dialog box, type Orders and then press the Enter key. (This saves the table with the name *Orders*.)

5. Close the Orders table by clicking the Close button for the table.

Check Your Work

Figure 1.12 Activity 2e

ID	SupplierID	ProductID	UnitsOrdered	OrderAmount	OrderDate	Cli
1	54	101-S3	10	$1,137.50	1/5/2021	
2	68	209-L	25	$173.75	1/5/2021	
3	68	209-XL	25	$180.00	1/5/2021	
4	68	209-XXL	20	$0.00	1/5/2021	
5	68	210-M	15	$0.00	1/5/2021	
6	68	210-L	25	$0.00	1/5/2021	
7	31	299-M2	10	$0.00	1/19/2021	
8	31	299-M3	10	$0.00	1/19/2021	
9	31	299-M5	10	$0.00	1/19/2021	
10	31	299-W1	8	$0.00	1/19/2021	
11	31	299-W3	10	$0.00	1/19/2021	
12	31	299-W4	10	$0.00	1/19/2021	
13	31	299-W5	10	$0.00	1/19/2021	
14	35	602-XR	5	$0.00	1/19/2021	
*	(New)		0	$0.00		

Renaming a Field

Click a data type button or click a data type at the data type drop-down list and the default field name (such as *Field1*) is automatically selected. With the default field name selected, type a name for the field. To change a field name, right-click the name, click *Rename Field* at the shortcut menu (which selects the current field name), and then type the new name.

Modifying Field
Properties in
Datasheet View

Name &
Caption

Inserting a Name, Caption, and Description

When creating a table that others will use, consider providing additional information so users understand the fields in the table and what should be entered in each one. Along with the field name, provide a caption and description for each field with options at the Enter Field Properties dialog box, shown in Figure 1.13. Display this dialog box by clicking the Name & Caption button in the Properties group on the Table Tools Fields tab.

At the Enter Field Properties dialog box, type the name for the field in the *Name* text box. To provide a more descriptive name for the field, type the descriptive name in the *Caption* text box. The text typed will become the field caption but the actual field name will still be part of the table structure. Creating a caption is useful if the field name is abbreviated or to show spaces between words in a field name. The field name is what Access uses for the table and the caption is what users see.

The *Description* text box is another source for providing information about the field to others using the database. Type information in the text box that specifies what should be entered in the field. The text typed in the *Description* text box displays at the left side of the Status bar when a field in a column is active. For example, type *Enter the total amount of the order* in the *Description* text box for the *OrderAmount* field and that text will display at the left side of the Status bar when a field in the column is active.

Figure 1.13 Enter Field Properties Dialog Box

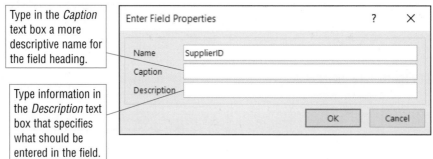

Type in the *Caption* text box a more descriptive name for the field heading.

Type information in the *Description* text box that specifies what should be entered in the field.

1. With **1-PacTrek** open, open the Orders table.
2. Access automatically named the first field *ID*. You want to make the field name more descriptive, so you decide to rename it. To do this, right-click the *ID* field name and then click *Rename Field* at the drop-down list.
3. Type OrderID and then press the Enter key.
4. To provide more information for others using the table, you decide to add information for the *SupplierID* field by creating a caption and description. To do this, complete the following steps:
   a. Click the *SupplierID* field name in the header row. (This selects the entire column.)
   b. Click the Table Tools Fields tab.
   c. Click the Name & Caption button in the Properties group. (At the Enter Field Properties dialog box, notice that *SupplierID* is already inserted in the *Name* text box.)
   d. At the Enter Field Properties dialog box, click in the *Caption* text box and then type Supplier Number.
   e. Click in the *Description* text box and then type Supplier identification number.
   f. Click OK to close the dialog box. (Notice that the field name now reads *Supplier Number*.)
5. Click the *ProductID* field name in the header row and then complete steps similar to those in Steps 4c through 4f to create the caption *Product Number* and the description *Product identification number*.
6. Click the *OrderAmount* field name in the header row and then complete steps similar to those in Steps 4c through 4f to create the caption *Order Amount* and the description *Total amount of order*.
7. Click the Save button on the Quick Access Toolbar to save the changes to the Orders table.
8. Close the Orders table.

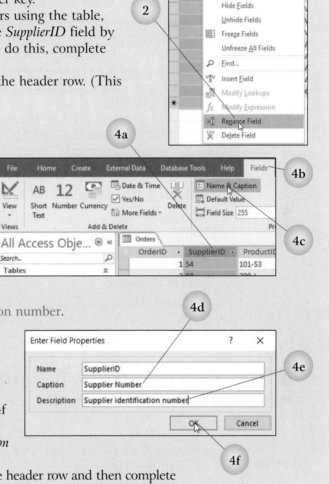

Check Your Work

Assigning a Default Value

The Properties group on the Table Tools Fields tab contains additional buttons for defining field properties in a table. If most records in a table are likely to contain the same field value in a column, consider inserting that value by default. Do this by clicking the Default Value button in the Properties group. At the Expression Builder dialog box, type the default value and then click OK.

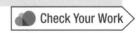 Default Value

For example, in Activity 2g, you will create a new table in the 1-PacTrek database containing information on customers, most of whom live in Vancouver, British Columbia. You will create a default value of *Vancouver* for the *City* field and a default value of *BC* for the *Prov/State* field. You can replace the default value with different text, so if a customer lives in Abbotsford instead of Vancouver, simply type *Abbotsford* in the *City* field instead.

Assigning a Field Size

The default field size property varies depending on the data type. For example, the Short Text data type assigns a maximum length of 255 characters that can be entered in the field. This number can be decreased depending on what data will be entered in the field. The field size can also be changed to control how much data is entered to help reduce errors. For example, if the two-letter state abbreviation is to be inserted in a state field, a field size of 2 characters can be assigned to the field. If someone entering data into the table tries to type more than two letters, Access will not accept the additional text. To change the field size, click in the *Field Size* text box in the Properties group on the Table Tools Fields tab and then type the number.

Changing the AutoNumber Data Type Field

Access automatically applies the AutoNumber data type to the first field in a table and assigns a unique number to each record in the table. In many cases, letting Access automatically assign a number to a record is a good idea. Some situations may arise, however, when the unique value in the first field should be something other than a number.

To change the AutoNumber data type for the first field, click the *Data Type* option box arrow in the Formatting group on the Table Tools Fields tab and then click the data type at the drop-down list.

Creating a Table Using Quick Start Fields

 Short Text

Inserting Quick Start Fields

The Add & Delete group on the Table Tools Fields tab contains buttons for specifying data types. The Short Text button was used to specify the data type for the *SupplierID* field when creating the Orders table. The field drop-down list was also used to choose a data type. In addition to these two methods, a data type can be assigned by clicking the More Fields button in the Add & Delete group on the Table Tools Fields tab. Click this button and a drop-down list displays with data types grouped into categories such as *Basic Types*, *Number*, *Large Number*, *Date and Time*, *Yes/No*, and *Quick Start*.

The options in the *Quick Start* category not only define a data type but also assign a field name. Additionally, with options in the *Quick Start* category, a group of related fields can be added in one step. For example, click the *Name* option in the *Quick Start* category and Access inserts the *LastName* field in one column and the *FirstName* field in the next column. Both fields are automatically assigned a Short Text data type. Click the *Address* option in the *Quick Start* category and Access inserts five fields, including *Address*, *City*, *StateProvince*, *ZIPPostal*, and *CountryRegion*—all with the Short Text data type assigned.

Activity 2g Creating a Customers Table, Inserting Quick Start Fields and Modifying Field Properties

1. The owners of Pacific Trek have decided to publish a semiannual product catalog and have asked customers who want to receive the catalog to fill out a form and include on the form whether or not they want to receive notices of upcoming sales in addition to the catalog. Create a table to store the data for customers by completing the following steps:

   a. With **1-PacTrek** open, click the Create tab.

   b. Click the Table button in the Tables group.

   c. With the *Click to Add* field active, click the More Fields button in the Add & Delete group on the Table Tools Fields tab.

   d. Scroll down the drop-down list and then click *Name* in the *Quick Start* category. (This inserts the captions *Last Name* and *First Name* in the table. The actual field names are *LastName* and *FirstName*.)

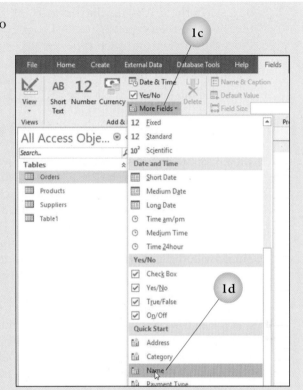

   e. Click *Click to Add* immediately right of the *First Name* field name in the header row. (Although the data type drop-down list displays, you are going to use the More Fields button rather than the drop-down list to create the next fields.)

   f. Click the More Fields button, scroll down the drop-down list, and then click *Address* in the *Quick Start* category. (This inserts five more fields in the table.)

   g. Scroll to the right in the table until *Click to Add* is visible in the header row following the *Country Region* field name. (You can scroll in the table using the horizontal scroll bar to the right of the Record Navigation bar.)

   h. Click *Click to Add* and then click *Yes/No* at the drop-down list.

   i. With the name *Field1* selected, type Mailers. (When entering records in the table, you will insert a check mark in the field check box if a customer wants to receive sales promotion mailers. If a customer does not want to receive the mailers, you will leave the check box blank.)

2. Rename and create a caption and description for the *ID* field by completing the following steps:

   a. Scroll to the beginning of the table and then click the *ID* field name in the header row.

   b. Click the Name & Caption button in the Properties group on the Table Tools Fields tab.

c. At the Enter Field Properties dialog box, select the text *ID* in the *Name* text box and then type CustomerID.

d. Press the Tab key and then type Customer Number in the *Caption* text box.

e. Press the Tab key and then type Access will automatically assign the record the next number in the sequence.

f. Click OK to close the Enter Field Properties dialog box. (Notice the description at the left side of the Status bar.)

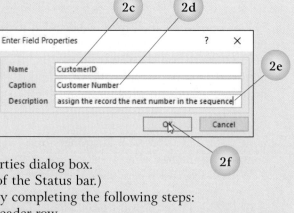

3. Add a description to the *Last Name* field by completing the following steps:

a. Click the *Last Name* field name in the header row.

b. Click the Name & Caption button in the Properties group.

c. At the Enter Field Properties dialog box, notice that Access named the field *LastName* but provided the caption *Last Name*. You do not want to change the name and caption so press the Tab key two times to make the *Description* text box active and then type Customer last name.

d. Click OK to close the dialog box.

4. You know that a customer's last name will not likely exceed 30 characters, so you decide to limit the field size. To do this, click in the *Field Size* text box in the Properties group (this selects *255*), type 30, and then press the Enter key.

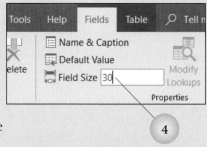

5. Click the *First Name* field name in the header row and then complete steps similar to those in Steps 3 and 4 to add the description *Customer first name* and change the field size to 30 characters.

6. Since most of Pacific Trek's customers live in the city of Vancouver, you decide to make it the default field value. To do this, complete the following steps:

a. Click the *City* field name in the header row.

b. Click the Default Value button in the Properties group.

c. At the Expression Builder dialog box, type Vancouver.

d. Click OK to close the dialog box.

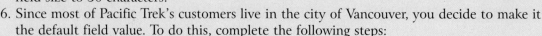

7. Change the *State Province* field name and insert a default value by completing the following steps:

a. Right-click the *State Province* field name in the header row and then click *Rename Field* at the shortcut menu.

b. Type Province.

c. Click the Default Value button.

d. Type BC in the Expression Builder dialog box and then click OK.

8. Click the *ZIP Postal* field name in the header row and then limit the field size to 7 characters by clicking in the *Field Size* text box (which selects *255*), typing 7, and then pressing the Enter key.

9. Since most customers want to be sent the sales promotional mailers, you decide to insert a check mark as the default value in the check boxes in the *Yes/No* column. To do this, complete the following steps:
   a. Click the *Mailers* field name in the header row.
   b. Click the Default Value button in the Properties group.
   c. At the Expression Builder dialog box, press the Backspace key two times to delete *No* and then type Yes.
   d. Click OK to close the dialog box.
10. Delete the *Country Region* field by clicking the *Country Region* field name in the header row and then clicking the Delete button in the Add & Delete group.
11. Save the table by completing the following steps:
    a. Click the Save button on the Quick Access Toolbar.
    b. At the Save As dialog box, type Customers and then press the Enter key.
12. Enter the six records in the table shown in Figure 1.14. To remove a check mark in the *Mailers* column, press the spacebar.
13. Adjust the field column widths to accommodate the longest entry in each column by completing the following steps:
    a. Position the mouse pointer on the *Customer Number* field name in the header row until the pointer turns into a down-pointing black arrow, click and hold down the left mouse button, drag to the *Mailers* field name, and then release the mouse button.
    b. With the columns selected, double-click one of the column boundary lines in the header row.
14. Click the Save button on the Quick Access Toolbar to save the Customers table.
15. Print the Customers table by completing the following steps:
    a. Click the File tab and then click the *Print* option.
    b. At the Print backstage area, click the *Print Preview* option.
    c. Click the Landscape button in the Page Layout group on the Print Preview tab.
    d. Click the Print button at the left side of the Print Preview tab.
    e. At the Print dialog box, click OK.
16. Close the Customers table.
17. Open the Orders table.
18. Adjust the field column widths to accommodate the longest entry in each column.
19. Click the Save button to save the Orders table.
20. Print the table in landscape orientation (refer to Step 15) and then close the table.
21. Close **1-PacTrek**.

 Check Your Work

Figure 1.14 Activity 2g

Customer Number	Last Name	First Name	Address	City	State Province	ZIP Postal	Mailers
1	Blakely	Mathias	7433 224th Ave E	Vancouver	BC	V5K 2M7	☑
2	Donato	Antonio	18225 Victoria Dr	Vancouver	BC	V5K 1H4	☐
3	Girard	Stephanie	430 Deer Lake Pl	Burnaby	BC	V3J 1E4	☑
4	Hernandez	Angelica	1233 E 59th Ave	Vancouver	BC	V5K 3H3	☑
5	Ives-Keller	Shane	9055 Gilber Rd	Richmond	BC	V6Y 1B2	☐
6	Kim	Keung	730 West Broadway	Vancouver	BC	V5K 5B2	☑
* (New)				Vancouver	BC		☑

Chapter Summary

- Microsoft Access is a database management system software program that can organize, store, maintain, retrieve, sort, and print all types of business data.

- In Access, open an existing database by clicking the <u>Open Other Files</u> hyperlink at the Access opening screen. At the Open backstage area, click the *Browse* option. At the Open dialog box, navigate to the location of the database and then double-click the database.

- Some common objects found in a database include tables, queries, forms, and reports.

- The Navigation pane displays at the left side of the Access screen and lists the objects that are contained in the database.

- Open a database object by double-clicking the object in the Navigation pane or by right-clicking the object and then clicking *Open* at the shortcut menu.

- Close an object by clicking the object's Close button or right-clicking the object tab and then clicking *Close* at the shortcut menu.

- When a table is open, the Record Navigation bar displays at the bottom of the screen and contains buttons for viewing records in the table.

- Insert a new record in a table by clicking the New button in the Records group on the Home tab or by clicking the New (blank) record button in the Record Navigation bar.

- Delete a record by clicking in a field in the record to be deleted, clicking the Delete button arrow on the Home tab, and then clicking *Delete Record* at the drop-down list.

- Add a field column to a table by clicking in the first field below *Click to Add* and then typing the data.

- Move a field column by selecting the column and then using the mouse to drag a thick, black, vertical line (representing the column) to the desired location.

- Delete a field column by clicking in any field in the column, clicking the Delete button arrow, and then clicking *Delete Column* at the drop-down list.

- Data entered in a table is automatically saved while changes to the layout of a table are not automatically saved.

- Hide, unhide, freeze, and unfreeze field columns with options at the More button drop-down list. Display this list by clicking the More button in the Records group on the Home tab.

- Adjust the width of a field column with options at the Column Width dialog box. Display the dialog box by clicking the More button and then clicking *Field Width* at the drop-down list. Enter a column measurement in the *Column Width* measurement box or click the Best Fit button.

- Adjust the width of a field column by dragging the column boundary line in the header row. Or adjust the width of a column (or selected columns) to accommodate the longest entry by double-clicking the column boundary line.

- Rename a table by right-clicking the table name in the Navigation pane, clicking *Rename*, and then typing the new name. Delete a table by right-clicking the table name in the Navigation pane and then clicking *Delete*.

- Print a table by clicking the File tab, clicking the *Print* option, and then clicking the *Quick Print* option. Preview a table before printing by clicking the *Print Preview* option at the Print backstage area.

- Use buttons and options on the Print Preview tab to change the page size, orientation, and margins.
- The first principle in database design is to reduce redundant data, because redundant data increases the amount of data entry required and the potential for errors, as well as takes up additional storage space.
- A data type defines the type of data Access will allow in the field. Assign a data type to a field with buttons in the Add & Delete group on the Table Tools Fields tab, by clicking an option from the field drop-down list, or with options at the More Fields button drop-down list.
- Rename a field by right-clicking the field name, clicking *Rename Field* at the shortcut menu, and then typing the new name.
- Type a name, caption, and description for a field with options at the Enter Field Properties dialog box.
- Use options in the *Quick Start* category in the More Fields button drop-down list to define a data type and assign a field name to a group of related fields.
- Insert a default value in a field with the Default Value button and assign a field size with the *Field Size* text box in the Properties group on the Table Tools Fields tab.
- Use the *Data Type* option box in the Formatting group on the Table Tools Fields tab to change the AutoNumber data type for the first column in a table.

Commands Review

FEATURE	RIBBON TAB, GROUP/OPTION	BUTTON, OPTION	KEYBOARD SHORTCUT
close Access		☒	Alt + F4
close database	File, *Close*		
create table	Create, Tables	▦	
Currency data type	Table Tools Fields, Add & Delete	▣	
Date & Time data type	Table Tools Fields, Add & Delete	▦	
delete column	Home, Records	☒, *Delete Column*	
delete record	Home, Records	☒, *Delete Record*	
Enter Field Properties dialog box	Table Tools Fields, Properties	▤	
Expression Builder dialog box	Table Tools Fields, Properties	▦	
freeze column	Home, Records	▦, *Freeze Fields*	
hide column	Home, Records	▦, *Hide Fields*	

FEATURE	RIBBON TAB, GROUP/OPTION	BUTTON, OPTION	KEYBOARD SHORTCUT
landscape orientation	File, *Print*		
new record	Home, Records		Ctrl + +
next field			Tab
Number data type	Table Tools Fields, Add & Delete	12	
page margins	File, *Print*		
Page Setup dialog box	File, *Print*		
page size	File, *Print*		
portrait orientation	File, *Print*		
previous field			Shift + Tab
Print backstage area	File, *Print*		
Print dialog box	File, *Print*	*Print*	Ctrl + P
Print Preview	File, *Print*	*Print Preview*	
Short Text data type	Table Tools Fields, Add & Delete	AB	
unfreeze column	Home, Records	, *Unfreeze Fields*	
unhide column	Home, Records	, *Unhide Fields*	
Yes/No data type	Table Tools Fields, Add & Delete	☑	

Microsoft®

Access®

Creating Relationships between Tables

Performance Objectives

Upon successful completion of Chapter 2, you will be able to:

1   Define a primary key field in a table

2   Create a one-to-many relationship

3   Specify referential integrity

4   Print, edit, and delete relationships

5   Edit, insert, and delete records

6   Create a one-to-one relationship

7   View and edit a subdatasheet

Access is a relational database program you can use to create tables that are related or connected within the same database. When a relationship is established between tables, you can view and edit records in related tables with a subdatasheet. In this chapter, you will learn how to identify a primary key field in a table that is unique to that table, how to join tables by creating a relationship between them, and how to view and edit subdatasheets.

Data Files

Before beginning chapter work, copy the AL1C2 folder to your storage medium and then make AL1C2 the active folder.

The online course includes additional training and assessment resources.

Activity 1 **Establish Relationships between Tables** **4 Parts**

You will specify the primary key fields in a table, establish one-to-many relationships between tables, specify referential integrity, and print the relationships. You will also insert, edit, and delete records and edit and delete a relationship.

Relating Tables

Hint Defining a relationship between tables is one of the most powerful features of a relational database management system.

Generally, a database management system fits into one of two categories: a file management system (also sometimes referred to as a *flat file database*) or a relational database management system. A flat file management system stores all data in a single directory and cannot contain multiple tables. This type of management system is a simple way to store data but becomes inefficient as more data is added. In a relational database management system, like Access, relationships are defined between sets of data, allowing greater flexibility in manipulating data and eliminating data redundancy (entering the same data in more than one place).

In Activity 1, you will define relationships between tables in the Pacific Trek database. Because the tables in the database will be related, information on the same product does not need to be repeated in multiple tables. If you used a flat file management system to maintain product information, you would need to repeat that product description each time.

Determining Relationships

Taking time to plan a database is extremely important. Creating a database with related tables takes even more consideration. Determine how to break down the required data and what tables to create to eliminate redundancies. One idea to help determine what tables are necessary in a database is to think of the word *about*. For example, the Pacific Trek store needs a table *about* products, another *about* suppliers, and another *about* orders. A table should be about only one subject, such as products, suppliers, or orders.

Along with determining the necessary tables for a database, determine the relationship between those tables. The ability to relate, or "join," tables is what makes Access a relational database management system. As explained in Chapter 1, database designers often create a visual representation of the database's structure in a diagram. The database diagram for the Pacific Trek database is shown in Figure 2.1. (Some of the fields in tables have been modified slightly from the database used in Chapter 1.)

Setting the Primary Key Field

Setting the Primary Key Field

A database table can contain two different types of key fields: a primary key field and a foreign key field. In the database diagram in Figure 2.1, notice that one field in each table contains an asterisk. The asterisk indicates a primary key field, which is a field that holds data that uniquely identifies each record in a table. For example, the *SupplierID* field in the Suppliers table contains a unique supplier number for each record in the table and the *ProductID* field in the Products table contains a unique product number for each product. A table can have only one primary key field and it is the field by which the table is sorted whenever the table is opened.

Figure 2.1 Pacific Trek Database Diagram

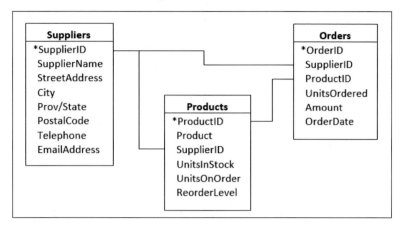

Quick Steps

Define Primary Key Field
1. Open table.
2. Click View button.
3. Click field.
4. Click Primary Key button.
5. Click Save button.

Hint You must enter a value in the primary key field in every record.

Hint Access uses a primary key field to associate data from multiple tables.

 Primary Key

When a new record is added to a table, Access checks to ensure that there is no existing record with the same data in the primary key field. If there is, Access displays an error message indicating there are duplicate values and does not allow the record to be saved. When adding a new record to a table, Access expects a value in each record in the table. This is is referred to as *entity integrity*. If a value is not entered in a field, Access enters a null value, but a null value cannot be given to a primary key field. Access will not allow a database to be closed that contains a primary key field with a null value.

By default, Access includes the *ID* field as the first field in a table, assigns the AutoNumber data type, and identifies the field as the primary key field. The AutoNumber data type assigns the first record a field value of *1* and each new record is assigned the next sequential number. Use this default field as the primary key field or define a different primary key field. To determine what field is the primary key field or to define a primary key field, display the table in Design view. To do this, open the table and then click the View button at the left side of the Home tab. A table also can be opened in Design view by clicking the View button arrow and then clicking *Design View* at the drop-down list. To add or remove a primary key designation from a field, click the field in the *Field Name* column and then click the Primary Key button in the Tools group on the Table Tools Design tab. A key icon is inserted in the field selector bar (the blank column to the left of the field names) for the field. Figure 2.2 shows the Products table in Design view with the *ProductID* field identified as the primary key field.

Typically, a primary key field in one table becomes the foreign key field in a related table. For example, the primary key field *SupplierID* in the Suppliers table is considered the foreign key field in the Orders table. In the Suppliers table, each entry in the *SupplierID* field must be unique since it is the primary key field but the same supplier number may appear more than once in the *SupplierID* field in the Orders table (for instance, in a situation when more than one product is ordered from the same supplier).

Data in the foreign key field must match data in the primary key field of the related table. For example, any supplier number entered in the *SupplierID* field in the Orders table must also be contained in the Suppliers table. In other words, an order would not be placed by a supplier that does not exist in the Suppliers table. Figure 2.3 identifies the primary and foreign key fields in the tables in the Pacific Trek database. Primary key fields are identified with *(PK)* and foreign key fields are identified with *(FK)* in the figure.

Figure 2.2 Products Table in Design View

A key icon in the field selector bar specifies the primary key field.

Figure 2.3 Pacific Trek Database Diagram with Primary and Foreign Key Fields Identified

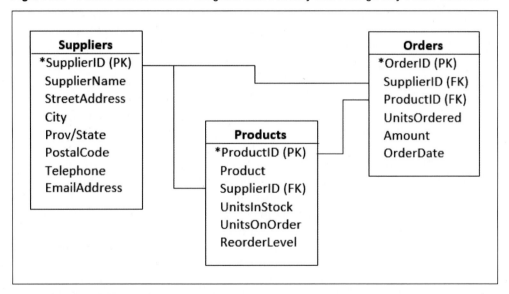

Activity 1a Defining a Primary Key Field

Part 1 of 4

1. Open Access.
2. At the Access opening screen, click the <u>Open Other Files</u> hyperlink at the left side of the screen.
3. At the Open backstage area, click the *Browse* option.
4. At the Open dialog box, navigate to your AL1C2 folder and then double-click the database *2-PacTrek*.
5. Click the Enable Content button in the message bar if the security warning message appears. (The message bar will display immediately below the ribbon.)
6. Open the Products table.

7. View the primary key field by completing the following steps:

a. Click the View button at the left side of the Home tab. (This displays the table in Design view.) **7a**

b. In Design view, notice the *Field Name, Data Type*, and *Description* columns and the information for each field. The first field, *ProductID*, is the primary key field and is identified by the key icon in the field selector bar.

c. Click the View button to return to Datasheet view.

d. Close the Products table.

8. Open the Suppliers table, click the View button to display the table in Design view, and then notice the *SupplierID* field is defined as the primary key field.

9. Click the View button to return to Datasheet view and then close the table.

10. Open the Orders table. (The first field in the Orders table has been changed from the AutoNumber data type field automatically assigned by Access to a Short Text data type field.)

11. Define the *OrderID* field as the primary key field by completing the following steps:

a. Click the View button at the left side of the Home tab.

b. With the table in Design view and the *OrderID* field selected in the *Field Name* column, click the Primary Key button in the Tools group on the Table Tools Design tab.

11b

c. Click the Save button on the Quick Access Toolbar.

d. Click the View button to return the table to Datasheet view.

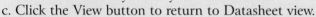

12. Move the *OrderDate* field by completing the following steps:

a. Click the *OrderDate* field name in the header row. (This selects the column.)

b. Position the mouse pointer on the field name in the header row; click and hold down the left mouse button; drag to the left until the thick, black vertical line displays immediately left of the *ProductID* field; and then release the mouse button.

12b

13. Adjust the column widths.

14. Save and then close the Orders table.

 Check Your Work

Creating a One-to-
Many Relationship

Creating a One-to-Many Relationship

In Access, one table can be related to another. Establishing this relationship is is generally referred to as performing a join. When tables with a common field are joined, data can be extracted from both tables as if they were one large table. Relating tables helps to ensure the integrity of the data by avoiding the need to enter the same data in multiple tables. For example, in Activity 1b, a relationship will be established between the Suppliers table and the Products table. The relationship will ensure that a supplier number cannot be entered in the Products table without first being entered in the Suppliers table. This type of relationship is called a *one-to-many relationship*, which means that one record in the Suppliers table will match zero, one, or many records in the Products table.

In a one-to-many relationship, the table containing the "one" is referred to as the *primary table* and the table containing the "many" is referred to as the *related table*. Access follows a set of rules that provide referential integrity, which enforces consistency between related tables. These rules are enforced when data is updated in related tables. The referential integrity rules ensure that a record added to a related table has a matching record in the primary table.

To create a one-to-many relationship, open the database containing the tables to be related. Click the Database Tools tab and then click the Relationships button in the Relationships group. This displays the Show Table dialog box, as shown in Figure 2.4. (If the Show Table dialog box is not visible, click the Show Table button in the Relationships group.) At the Show Table dialog box, each table that will be related must be added to the Relationships window. To do this, click the first table name to be included and then click the Add button (or double-click the table name). This inserts the fields of the table in a table field list box. Continue in this manner until all necessary tables (in table field list boxes) have been added to the Relationships window and then click the Close button.

At the Relationships window, such as the one shown in Figure 2.5, use the mouse to drag the common field from the primary table's table field list box (the "one") to the related table's table field list box (the "many"). This displays the Edit Relationships dialog box, as shown in Figure 2.6. At the Edit Relationships dialog box, check to make sure the correct field name is in the *Table/Query* and *Related Table/Query* list boxes and the relationship type at the bottom of the dialog box is *One-To-Many*.

 Relationships

Quick Steps

Create One-to-Many Relationship
1. Click Database Tools tab.
2. Click Relationships button.
3. Add tables.
4. Drag "one" field from primary table to "many" field in related table.
5. At Edit Relationships dialog box, click Create button.
6. Click Save button.

Figure 2.4 Show Table Dialog Box

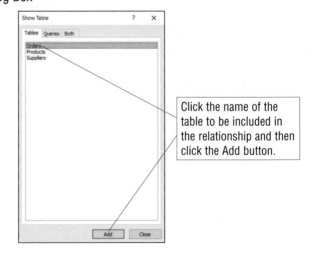

Click the name of the table to be included in the relationship and then click the Add button.

Figure 2.5 Relationships Window

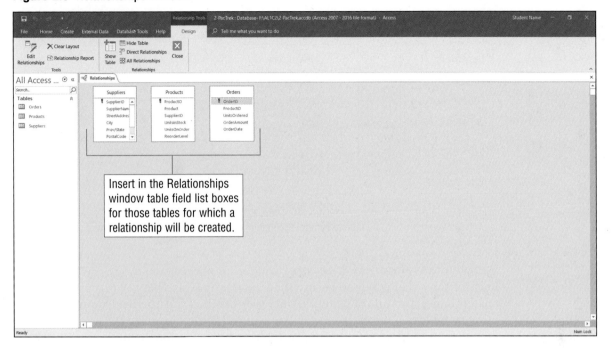

Insert in the Relationships window table field list boxes for those tables for which a relationship will be created.

Figure 2.6 Edit Relationships Dialog Box

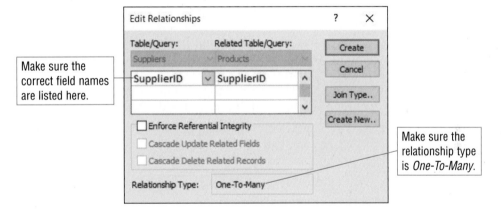

Make sure the correct field names are listed here.

Make sure the relationship type is *One-To-Many*.

Specify the relationship options by choosing *Enforce Referential Integrity*, as well as *Cascade Update Related Fields* and/or *Cascade Delete Related Records*, and then click the Create button. This closes the Edit Relationships dialog box and opens the Relationships window showing the relationship between the tables.

Relationships are indicated with a black line connecting the table field list boxes, as shown in Figure 2.7. The number 1 at one end of the line signifies the "one" side of the relationship and the infinity symbol (∞) at the other end signifies the "many" side of the relationship. The black line, called the *join line*, is thick at both ends if the *Enforce Referential Integrity* option is chosen. If this option is not chosen, the line is thin at both ends. Click the Save button on the Quick Access Toolbar to save the relationship. After saving, click the X to close the Relationships window.

Figure 2.7 One-to-Many Relationship

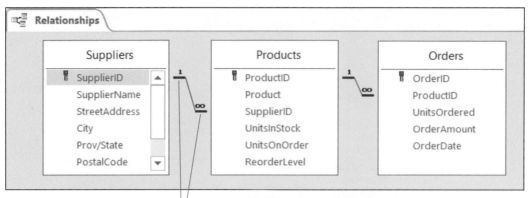

This is an example of a one-to-many relationship, where 1 identifies the "one" side of the relationship and the infinity symbol (∞) identifies the "many" side.

💡 **Hint** Referential integrity ensures that a record exists in the "one" table before the record can be entered in the "many" table.

Specifying Referential Integrity Choose *Enforce Referential Integrity* at the Edit Relationships dialog box to ensure that the relationships between records in related tables are valid. Referential integrity can be set if the field from the primary table is a primary key field and the related fields have the same data type. When referential integrity is established, a value for the primary key field must first be entered in the primary table before it can be entered in the related table.

If only *Enforce Referential Integrity* is selected and the related table contains a record, a primary key field value in the primary table cannot be changed. A record in the primary table cannot be deleted if its key value equals a foreign key field in the related table. Select *Cascade Update Related Fields* and, if changes are made to the primary key field value in the primary table, Access will automatically update the matching value in the related table. Choose *Cascade Delete Related Records* and, if a record is deleted in the primary table, Access will delete any related records in the related table.

In Activity 1b, you will create one-to-many relationships between tables in the 2-PacTrek database, as shown in Figure 2.8.

Figure 2.8 Relationships in the 2-PacTrek Database

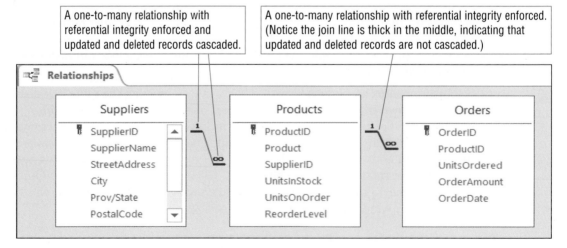

A one-to-many relationship with referential integrity enforced and updated and deleted records cascaded.

A one-to-many relationship with referential integrity enforced. (Notice the join line is thick in the middle, indicating that updated and deleted records are not cascaded.)

Tutorial

Creating a
Relationship Report

Relationship
Report

Creating and Printing a Relationship Report Once all relationships have been created in a database, printing a hard copy of the relationships is a good idea. The documentation is a quick reference of all of the table names, fields within each table, and relationships between tables. To create and print a relationship report, display the Relationships window and then click the Relationship Report button in the Tools group. This opens a relationship report in Print Preview. Click the Print button in the Print group on the Print Preview tab and then click OK at the Print dialog box. After printing the relationship report, click the Close button to close the relationship report.

Quick Steps

Create Relationship Report
1. Click Database Tools tab.
2. Click Relationships button.
3. Click Relationship Report button.

Activity 1b Relating Tables

Part 2 of 4

1. With **2-PacTrek** open, click the Database Tools tab and then click the Relationships button in the Relationships group. (The Show Table dialog box should display in the Relationships window. If it does not display, click the Show Table button in the Relationships group on the Relationship Tools Design tab.)

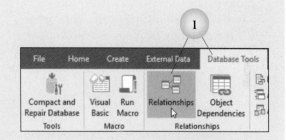

2. At the Show Table dialog box with the Tables tab selected, add the Suppliers, Products, and Orders tables to the Relationships window by completing the following steps:
   a. Click *Suppliers* in the list box and then click the Add button.

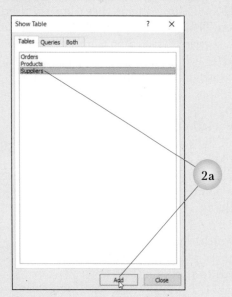

   b. Click *Products* in the list box and then click the Add button.
   c. Double-click *Orders* in the list box.
3. Click the Close button to close the Show Table dialog box.

4. At the Relationships window, drag the *SupplierID* field from the Suppliers table field list box to the Products table field list box by completing the following steps:

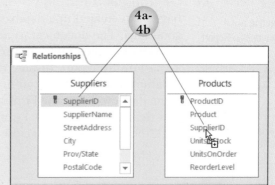

a. Position the mouse pointer on the *SupplierID* field in the Suppliers table field list box.

b. Click and hold down the left mouse button, drag the mouse pointer (with a field icon attached) to the *SupplierID* field in the Products table field list box, and then release the mouse button. (This displays the Edit Relationships dialog box.)

5. At the Edit Relationships dialog box, make sure *SupplierID* displays in the *Table/Query* and *Related Table/Query* list boxes and the relationship type at the bottom of the dialog box displays as *One-To-Many*.

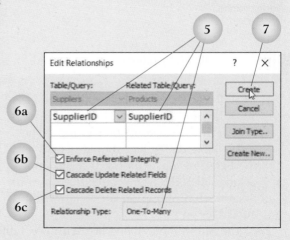

6. Enforce the referential integrity of the relationship by completing the following steps:

a. Click the *Enforce Referential Integrity* check box to insert a check mark. (This makes the other two options available.)

b. Click the *Cascade Update Related Fields* check box to insert a check mark.

c. Click the *Cascade Delete Related Records* check box to insert a check mark.

7. Click the Create button. (This closes the Edit Relationships dialog box and displays the Relationships window, showing a black join line [thick on the ends and thin in the middle] connecting the *SupplierID* field in the Suppliers table field list box to the *SupplierID* field in the Products table field list box. A *1* appears on the join line on the Suppliers table side and an infinity symbol [∞] appears on the join line on the Products table side.)

8. Click the Save button on the Quick Access Toolbar to save the relationship.

9. Create a one-to-many relationship between the Products table and the Orders table with the *ProductID* field by completing the following steps:

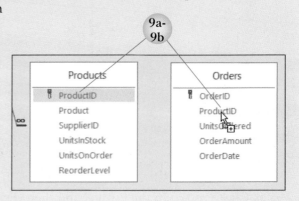

a. Position the mouse pointer on the *ProductID* field in the Products table field list box.

b. Click and hold down the left mouse button, drag the mouse pointer (with a field icon attached) to the *ProductID* field in the Orders table field list box, and then release the mouse button.

c. At the Edit Relationships dialog box, make sure *ProductID* displays in the *Table/Query* and *Related Table/Query* list boxes and the relationship type displays as *One-To-Many*.

d. Click the *Enforce Referential Integrity* check box to insert a check mark. (Do not insert check marks in the other two check boxes.)

e. Click the Create button.

10. Click the Save button on the Quick Access Toolbar to save the relationships.

11. Print the relationships by completing the following steps:

a. At the Relationships window, click the Relationship Report button in the Tools group. This opens the relationship report in Print Preview. (If a security notice displays, click the Open button.)

b. Click the Print button in the Print group at the left side of the Print Preview tab.

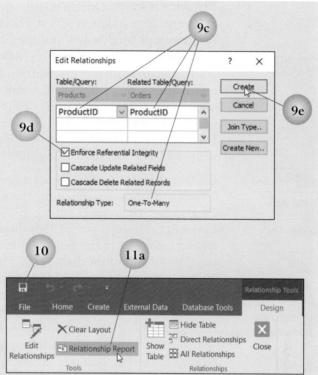

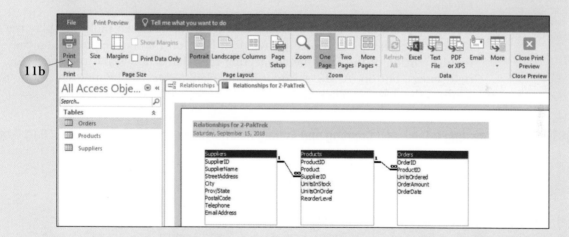

c. Click OK at the Print dialog box.

d. Close the report by clicking the report's Close button.

e. At the message asking if you want to save changes to the design of the report, click No.

12. Close the Relationships window by clicking the Close button for the window.

Check Your Work

Show Table

Showing Tables Once a relationship is established between tables and the Relationships window is closed, clicking the Relationships button displays the Relationships window without the Show Table dialog box. To display the Show Table dialog box, click the Show Table button in the Relationships group.

Pacific Trek offers a discount on one product each week. Keep track of this information by creating a Discounts table that includes the discount item for each week of the first three months of the year. A new record will be added to this field each week when the discount item is chosen. (In Activity 1c, you will create the Discounts table shown in Figure 2.9 on page 49 and then relate the Products table with the Discounts table using the *ProductID* field.)

Tutorial

Editing and Deleting a Relationship

Edit Relationships

Quick Steps

Edit Relationship
1. Click Database Tools tab.
2. Click Relationships button.
3. Click Edit Relationships button.
4. Make changes at Edit Relationships dialog box.
5. Click OK.

Editing a Relationship A relationship between tables can be edited or deleted. To edit a relationship, open the database containing the tables with the relationship, click the Database Tools tab, and then click the Relationships button in the Relationships group. This displays the Relationships window with the related tables. Click the Edit Relationships button in the Tools group to display the Edit Relationships dialog box. The dialog box will be similar to the one shown in Figure 2.6 on page 41. Identify the relationship to be edited by clicking the *Table/Query* option box arrow and then clicking the table name containing the "one" field. Click the *Related Table/Query* option box arrow and then click the table name containing the "many" field.

To edit a specific relationship, position the mouse pointer on the middle portion of the join line that connects the related tables and then click the right mouse button. At the shortcut menu, click the *Edit Relationship* option. This displays the Edit Relationships dialog box with the specific related field in both list boxes.

Deleting a Relationship To delete a relationship between tables, display the related tables in the Relationships window. Position the mouse pointer on the middle portion of the join line connecting the related tables and then click the right mouse button. At the shortcut menu, click *Delete*. At the message asking to confirm the deletion, click Yes.

Activity 1c Creating a Table and Creating and Editing Relationships   Part 3 of 4

1. With **2-PacTrek** open, create the Discounts table shown in Figure 2.9 on page 49 by completing the following steps:
   a. Click the Create tab.
   b. Click the Table button in the Tables group.
   c. Click the Short Text button in the Add & Delete group. (This creates and then selects *Field1* at the right of the *ID* column.)

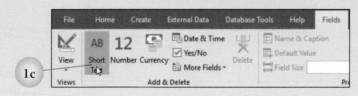

d. Type ProductID and then press the Enter key.

e. Click the *Short Text* option at the drop-down list and then type DiscountAmount.

f. Click the *ID* field name (in the first column), click the *Data Type* option box arrow in the Formatting group, and then click *Date/Time* at the drop-down list.

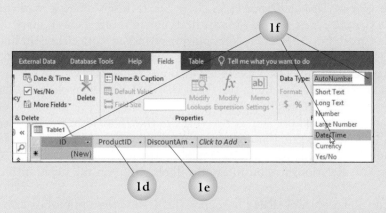

g. Right-click the *ID* field name, click *Rename Field* at the shortcut menu, type Week, and then press the Enter key.

h. Type the 13 records in the Discounts table shown in Figure 2.9 on page 49.

2. After typing the records, save the table by completing the following steps:

a. Click the Save button on the Quick Access Toolbar.

b. At the Save As dialog box, type Discounts and then press the Enter key.

3. Close the Discounts table.

4. Display the Relationships window and add the Discounts table to the window by completing the following steps:

a. Click the Database Tools tab and then click the Relationships button in the Relationships group.

b. Display the Show Table dialog box by clicking the Show Table button in the Relationships group.

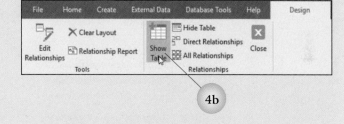

c. At the Show Table dialog box, double-click the *Discounts* table.

d. Click the Close button to close the Show Table dialog box.

5. At the Relationships window, create a one-to-many relationship between the Products table and the Discounts table with the *ProductID* field by completing the following steps:

a. Drag the *ProductID* field from the Products table field list box to the *ProductID* field in the Discounts table field list box.

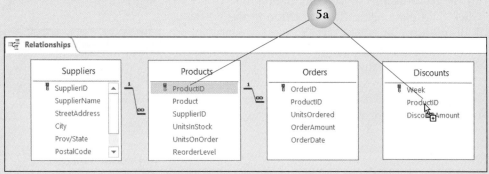

b. At the Edit Relationships dialog box, make sure *ProductID* displays in the *Table/Query* and *Related Table/Query* list boxes and the relationship type at the bottom of the dialog box displays as *One-To-Many*.

c. Click the *Enforce Referential Integrity* check box to insert a check mark.

d. Click the *Cascade Update Related Fields* check box to insert a check mark.

e. Click the *Cascade Delete Related Records* check box to insert a check mark.

f. Click the Create button. (At the Relationships window, notice the join line between the Products table and the Discounts table. If a message occurs stating that the relationship cannot be created, click OK. Open the Discounts table, check to make sure the product numbers are entered correctly in the *ProductID* field, and then close the Discounts table. Try again to create the relationship.)

6. Edit the one-to-many relationship between the *ProductID* field in the Products table and the Orders table and specify that you want to cascade updated and related fields and cascade and delete related records by completing the following steps:

a. Click the Edit Relationships button in the Tools group on the Relationship Tools Design tab.

b. At the Edit Relationships dialog box, click the *Table/Query* option box arrow and then click *Products* at the drop-down list.

c. Click the *Related Table/Query* option box arrow and then click *Orders* at the drop-down list.

d. Click the *Cascade Update Related Fields* check box to insert a check mark.

e. Click the *Cascade Delete Related Records* check box to insert a check mark.

f. Click OK.

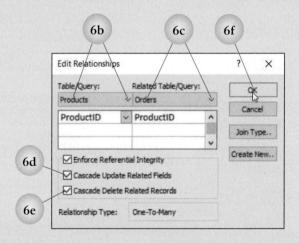

7. Click the Save button on the Quick Access Toolbar to save the relationship.

8. Print the relationships by completing the following steps:

a. Click the Relationship Report button in the Tools group.

b. Click the Print button in the Print group.

c. Click OK at the Print dialog box.

d. Close the report by clicking the Close button.

e. At the message asking if you want to save changes to the design of the report, click No.

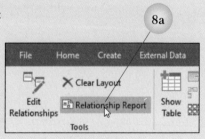

9. Delete the relationship between the Products table and the Discounts table by completing the following steps:
   a. Position the mouse pointer on the thin portion of the join line connecting the *ProductID* field in the Products table field list box with the *ProductID* field in the Discounts table field list box and then click the right mouse button.
   b. Click the *Delete* option at the shortcut menu.

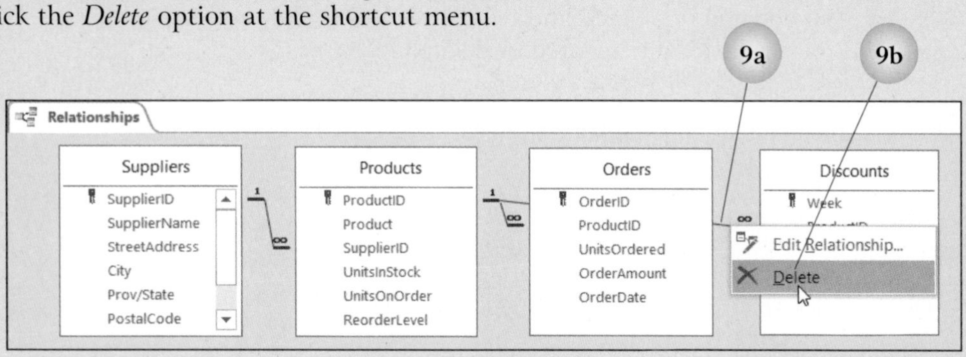

   c. At the message asking if you are sure you want to permanently delete the selected relationship from your database, click Yes.
10. Click the Save button on the Quick Access Toolbar to save the relationship.
11. Print the relationships by completing the following steps:
    a. Click the Relationship Tools Design tab and then click the Relationship Report button.
    b. Click the Print button.
    c. Click OK at the Print dialog box.
    d. Close the report by clicking the Close button.
    e. At the message asking if you want to save changes to the design of the report, click No.
12. Close the Relationships window.

Check Your Work

Figure 2.9 Discounts Table in Datasheet View

Week	ProductID	DiscountAmc	Click to Add
1/7/2021	155-45	20%	
1/14/2021	652-2	15%	
1/21/2021	443-1A	20%	
1/28/2021	202-CW	15%	
2/4/2021	804-60	10%	
2/11/2021	652-2	15%	
2/18/2021	101-S1B	5%	
2/25/2021	560-TL	20%	
3/4/2021	652-2	20%	
3/11/2021	602-XX	15%	
3/18/2021	100-05	10%	
3/25/2021	652-2	15%	
4/1/2021	202-CW	15%	

Inserting and Deleting Records in Related Tables In the relationship established in Activity 1b, a record must first be added to the Suppliers table before a related record can be added to the Products table. This is because the *Enforce Referential Integrity* option was selected at the Edit Relationships dialog box. Because the two options, *Cascade Update Related Fields* and *Cascade Delete Related Records*, were also selected, records in the Suppliers table (the primary table) can be updated or deleted and related records in the Products table (the related table) are automatically updated or deleted.

Activity 1d Editing, Inserting, and Deleting Records

Part 4 of 4

1. With **2-PacTrek** open, open the Suppliers table.
2. Change two supplier numbers in the Suppliers table (Access will automatically change them in the Products table and the Orders table) by completing the following steps:
   a. Double-click the field value *15* in the *SupplierID* field.
   b. Type *33*.
   c. Double-click the field value *42* in the *SupplierID* field.
   d. Type *51*.
   e. Click the Save button on the Quick Access Toolbar.
   f. Close the Suppliers table.
   g. Open the Products table and notice that supplier number *15* changed to *33* and supplier number *42* changed to *51*.
   h. Close the Products table.
3. Open the Suppliers table and then add the following records:

SupplierID	16	SupplierID	28
SupplierName	Olympic Suppliers	SupplierName	Gorman Company
StreetAddress	1773 50th Avenue	StreetAddress	543 26th Street
City	Seattle	City	Vancouver
Prov/State	WA	Prov/State	BC
PostalCode	98101	PostalCode	V5K 3C5
Telephone	(206) 555-9488	Telephone	(778) 555-4550
EmailAddress	olysuppliers@ppi-edu.net	EmailAddress	gormanco@ppi-edu.net

4. Delete the record for supplier number 38 (Hadley Company). At the message stating that relationships that specify cascading deletes are about to cause records in this table and related tables to be deleted, click Yes.

5. Display the table in Print Preview, change to landscape orientation, and then print the table.

6. Close the Suppliers table.

7. Open the Products table and then add the following records to the table:

ProductID	701-BK	ProductID	703-SP
Product	Basic first aid kit	Product	Medical survival pack
SupplierID	33	SupplierID	33
UnitsInStock	8	UnitsInStock	8
UnitsOnOrder	0	UnitsOnOrder	0
ReorderLevel	5	ReorderLevel	5
ProductID	185-10	ProductID	185-50
Product	Trail water filter	Product	Trail filter replacement cartridge
SupplierID	51	SupplierID	51
UnitsInStock	4	UnitsInStock	14
UnitsOnOrder	10	UnitsOnOrder	0
ReorderLevel	10	ReorderLevel	10

8. Display the Products table in Print Preview, change to landscape orientation, change the top and bottom margins to 0.5 inch and then print the table. (The table will print on three pages.)

9. Close the Products table.

10. Open the Orders table and then add the following record:

OrderID	1033
OrderDate	2/15/2021
ProductID	185-10
UnitsOrdered	10
Amount	$310.90

11. Print and then close the Orders table.

12. Close **2-PacTrek**.

 Check Your Work

---

Activity 2 Create Relationships and Display Subdatasheets in a Database **2 Parts**

You will open a company database and then create one-to-many relationships between tables, as well as a one-to-one relationship. You will also display and edit subdatasheets.

 Tutorial

Creating a One-to-
One Relationship

Creating a One-to-One Relationship

A one-to-one relationship can be created between tables in which each record in the first table matches only one record in the second table and one record in the second table matches only one record in the first table. (One-to-one relationships exist between primary key fields.) A one-to-one relationship is not as common as a one-to-many relationship, since the type of information used to create the relationship can be stored in one table. A one-to-one relationship is generally used to break a large table with many fields into two smaller tables.

Hint The Relationships window shows any relationship you have defined between tables.

In Activity 2a, you will create a one-to-one relationship between the Employees table and the Benefits table in the Griffin database. Each record in the Employees table and each record in the Benefits table pertains to one employee. These two tables could be merged into one but the data in each table is easier to manage when separated. Figure 2.10 shows the relationships you will define between the tables in the Griffin database. The Benefits table and the Departments table have been moved down so you can more easily see the relationships.

Figure 2.10 Griffin Database Table Relationships

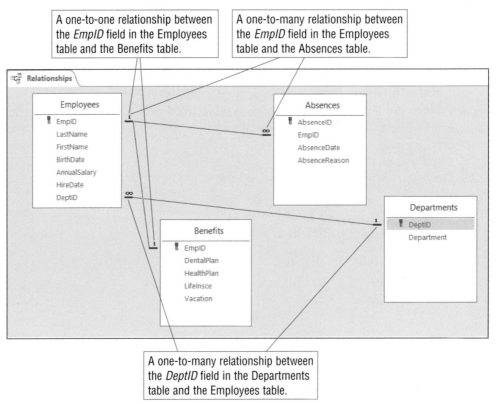

A one-to-one relationship between the *EmpID* field in the Employees table and the Benefits table.

A one-to-many relationship between the *EmpID* field in the Employees table and the Absences table.

A one-to-many relationship between the *DeptID* field in the Departments table and the Employees table.

Activity 2a   Creating One-to-Many and One-to-One Relationships

Part 1 of 2

1. Open **2-Griffin** and enable the content.
2. Click the Database Tools tab.
3. Click the Relationships button.
4. At the Show Table dialog box with the Tables tab selected, add all of the tables to the Relationships window by completing the following steps:
   a. Double-click *Employees* in the list box. (This inserts the table in the Relationships window.)
   b. Double-click *Benefits* in the list box.
   c. Double-click *Absences* in the list box.
   d. Double-click *Departments* in the list box.
   e. Click the Close button to close the Show Table dialog box.
5. At the Relationships window, create a one-to-many relationship with the *EmpID* field in the Employees table as the "one" and the *EmpID* field in the Absences table the "many" by completing the following steps:

a. Position the mouse pointer on the *EmpID* field in the Employees table field list box.
b. Click and hold down the left mouse button, drag the mouse pointer (with a field icon attached) to the *EmpID* field in the Absences table field list box, and then release the mouse button. (This displays the Edit Relationships dialog box.)

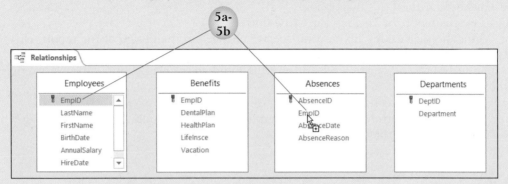

c. At the Edit Relationships dialog box, make sure *EmpID* displays in the *Table/Query* and *Related Table/Query* list boxes and the relationship type at the bottom of the dialog box displays as *One-To-Many*.
d. Click the *Enforce Referential Integrity* check box to insert a check mark.
e. Click the *Cascade Update Related Fields* check box to insert a check mark.
f. Click the *Cascade Delete Related Records* check box to insert a check mark.
g. Click the Create button. (A *1* appears at the Employees table field list box side and an infinity symbol [∞] appears at the Absences table field list box side of the black line.)
6. Complete steps similar to those in Step 5 to create a one-to-many relationship with the *DeptID* field in the Departments table the "one" and the *DeptID* field in the Employees table the "many." (You may need to scroll down the Employees table field list box to see the *DeptID* field.)
7. Create a one-to-one relationship with the *EmpID* field in the Employees table and the *EmpID* field in the Benefits table by completing the following steps:
a. Position the mouse pointer on the *EmpID* field in the Employees table field list box.
b. Click and hold down the left mouse button, drag the mouse pointer to the *EmpID* field in the Benefits table field list box, and then release the mouse button. (This displays the Edit Relationships dialog box; notice at the bottom of the dialog box that the relationship type displays as *One-To-One*.)
c. Click the *Enforce Referential Integrity* check box to insert a check mark.
d. Click the *Cascade Update Related Fields* check box to insert a check mark.
e. Click the *Cascade Delete Related Records* check box to insert a check mark.
f. Click the Create button. (Notice that a *1* appears at both the side of the Employees table field list box and at the side of the Benefits table field list box, indicating a one-to-one relationship.)

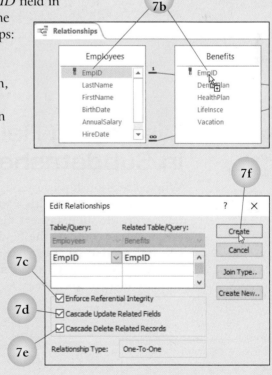

8. Click the Save button on the Quick Access Toolbar to save the relationships.
9. Print the relationships by completing the following steps:
   a. Click the Relationship Report button in the Tools group.
   b. Click the Print button in the Print group.
   c. Click OK at the Print dialog box.
   d. Close the report by clicking the Close button.
   e. At the message asking if you want to save changes to the design of the report, click No.
10. Close the Relationships window.
11. Add a record to and delete a record from the Employees and Benefits tables by completing the following steps:
    a. Open the Employees table.
    b. Click the New button in the Records group on the Home tab and then type the following data in the specified field:

EmpID	1096
LastName	Schwartz
FirstName	Bryan
BirthDate	5/21/1983
AnnualSalary	$45,000.00
HireDate	1/15/2010
DeptID	IT

    c. Delete the record for Trevor Sargent (employee number 1005). At the message stating that relationships that specify cascading deletes are about to cause records in this table and related tables to be deleted, click Yes.
    d. Print and then close the Employees table.
12. Open the Benefits table and notice that the record for Trevor Sargent is deleted but the new employee record you entered in the Employees table is not reflected in the Benefits table. Add a new record for Bryan Schwartz with the following information:

EmpID	1096
Dental Plan	(Press spacebar to remove check mark.)
Health Plan	(Leave check mark.)
Life Insurance	$100,000.00
Vacation	2 weeks

13. Print and then close the Benefits table.

Viewing a Subdatasheet

Displaying Related Records in Subdatasheets

When a relationship is established between tables, records in related tables can be viewed and edited with a subdatasheet. Figure 2.11 shows the Employees table with the subdatasheet displayed for employee Kate Navarro. The subdatasheet contains the fields in the Benefits table related to Kate Navarro. Use this subdatasheet to view and edit information in both the Employees table and Absences table. Changes made to fields in a subdatasheet affect the table and any related tables.

Quick Steps
Display Subdatasheet
1. Open table.
2. Click expand
   indicator at left side
   of record.
3. Click table at Insert
   Subdatasheet dialog
   box.
4. Click OK.

Access automatically inserts a plus symbol (referred to as an *expand indicator*) before each record in a table that is joined to another table by a one-to-many relationship. Click the expand indicator and, if the table is related to only one other table, a subdatasheet containing fields from the related table displays below the record, as shown in Figure 2.11. To close the subdatasheet, click the minus symbol (referred to as a *collapse indicator*) preceding the record. (The plus symbol turns into the minus symbol when a subdatasheet displays.)

If a table has more than one relationship defined, clicking the expand indicator will display the Insert Subdatasheet dialog box, as shown in Figure 2.12. At this dialog box, click the table in the Tables list box and then click OK. The Insert Subdatasheet dialog box can also be displayed by clicking the More button in the Records group on the Home tab, pointing to *Subdatasheet*, and then clicking *Subdatasheet*. Display subdatasheets for all records by clicking the More button, pointing to *Subdatasheet*, and then clicking *Expand All*. Close all subdatasheets by clicking the More button, pointing to *Subdatasheet*, and then clicking *Collapse All*.

If a table is related to two or more tables, specify the subdatasheet at the Insert Subdatasheet dialog box. To display a different subdatasheet, remove the subdatasheet first, before selecting the next subdatasheet. Do this by clicking the More button, pointing to *Subdatasheet*, and then clicking *Remove*.

Figure 2.11 Table with Subdatasheet Displayed in Datasheet View

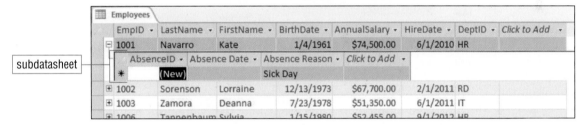

subdatasheet

Figure 2.12 Insert Subdatasheet Dialog Box

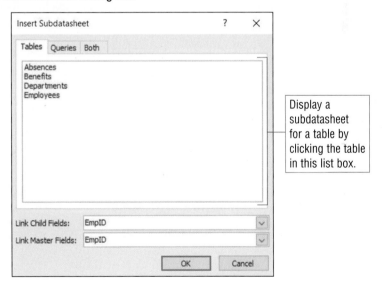

Display a subdatasheet for a table by clicking the table in this list box.

1. With **2-Griffin** open, open the Employees table.
2. Display a subdatasheet by clicking the expand indicator (plus symbol) at the left side of the first row (the row for Kate Navarro).
3. Close the subdatasheet by clicking the collapse indicator (minus symbol) at the left side of the record for Kate Navarro.
4. Display subdatasheets for all of the records by clicking the More button in the Records group, pointing to *Subdatasheet*, and then clicking *Expand All*.

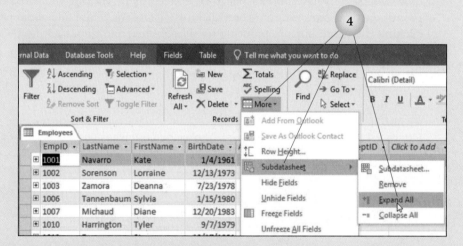

5. Remove the display of all subdatasheets by clicking the More button, pointing to *Subdatasheet*, and then clicking *Collapse All*.
6. Remove the connection between the Employees table and Absences table by clicking the More button, pointing to *Subdatasheet*, and then clicking *Remove*. (Notice that the expand indicators [plus symbols] no longer display before each record.)

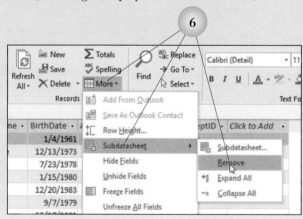

7. Suppose that the employee, Diane Michaud, has moved to a different department and has had an increase in salary. You would like to update her record in the tables. Display the Benefits subdatasheet and make changes to fields in the Employees table and Benefits table by completing the following steps:

a. Click the More button, point to *Subdatasheet*, and then click *Subdatasheet* at the side menu.

b. At the Insert Subdatasheet dialog box, click *Benefits* in the list box and then click OK.

c. Change the department ID for the record for *Diane Michaud (EmpID 1007)* from *DP* to *A*.

d. Change the salary from *$56,250.00* to *$57,500.00*.

e. Click the expand indicator (plus symbol) at the left side of the record for Diane Michaud.

f. Insert a check mark in the *Dental Plan* check box and change the vacation from 3 weeks to 4 weeks.

g. Click the collapse indicator (minus symbol) at the left side of the record for Diane Michaud.

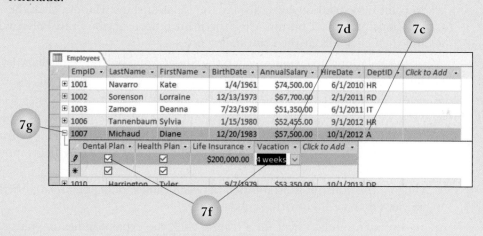

8. Click the Save button on the Quick Access Toolbar.
9. Print and then close the Employees table.
10. Open, print, and then close the Benefits table.
11. Close **2-Griffin**.

 Check Your Work

Chapter Summary

- Access is a relational database software program in which tables can be created that are related or connected.

- When planning a table, take time to determine how to break down the required data and what relationships need to be defined to eliminate data redundancies.

- Generally, one field in a table must be unique so that one record can be distinguished from another. A field with a unique value is considered a primary key field.

- A table can have only one primary key field and it is the field by which the table is sorted whenever it is opened.

- In a field defined as a primary key field, duplicate values are not allowed. Access also expects a value in each record in the primary key field.

- Typically, a primary key field in one table becomes the foreign key field in a related table. Data in a foreign key field must match data in the primary key field of the related tables.

- Tables are related by performing a join. When tables that have a common field are joined, data can be extracted from both tables as if they were one large table.

- A one-to-many relationship can be created between tables. In this relationship, a record must be added to the "one" table before it can be added to the "many" table.

- To print table relationships, display the Relationships window, click the Relationship Report button, click the Print button on the Print Preview tab, and then click OK at the Print dialog box.

- At the Relationships window, click the Show Table button to display the Show Table dialog box.

- A relationship between tables can be edited or deleted.

- A one-to-one relationship can be created between tables in which each record in the first table matches only one record in the related table. This type of relationship is generally used to break a large table with many fields into two smaller tables.

- When a relationship is established between tables, fields in a related table can be viewed and edited with a subdatasheet.

- To display a subdatasheet for a record, click the expand indicator (plus symbol) at the left of the record in Datasheet view. To display subdatasheets for all records, click the More button in the Records group on the Home tab, point to *Subdatasheet*, and then click *Expand All*.

- Display the Insert Subdatasheet dialog box by clicking the More button in the Reports group on the Home tab, pointing to *Subdatasheet*, and then clicking *Subdatasheet*.

- Turn off the display of a subdatasheet by clicking the collapse indicator (minus symbol) at the left of the record. To turn off the display of subdatasheets for all records, click the More button, point to *Subdatasheet*, and then click *Collapse All*.

Commands Review

FEATURE	RIBBON, GROUP	BUTTON, OPTION
Edit Relationships dialog box	Relationship Tools Design, Tools	
Insert Subdatasheet dialog box	Home, Records	, *Subdatasheet, Subdatasheet*
primary key field	Table Tools Design, Tools	
print relationships report	Relationship Tools Design, Tools	
Relationships window	Database Tools, Relationships	
Show Table dialog box	Relationship Tools Design, Relationships	

Microsoft®
Access®

Performing Queries

Performance Objectives

Upon successful completion of Chapter 3, you will be able to:

1 Design queries to extract specific data from tables

2 Modify queries

3 Design queries with *Or* and *And* criteria

4 Use the Simple Query Wizard to create and modify queries

5 Create and format a calculated field

6 Use aggregate functions in queries

7 Create a crosstab query, find duplicates, and find unmatched queries

One of the primary uses of a database is to extract the specific information needed to answer questions and make decisions. A company might need to know how much inventory is currently on hand, which products have been ordered, which accounts are past due, or which customers live in a particular city. You can extract specific information from a table or multiple tables by creating and running a query. You will learn how to perform a variety of queries on information in tables in this chapter.

 Data Files

Before beginning chapter work, copy the AL1C3 folder to your storage medium and then make AL1C3 the active folder.

The online course includes additional training and assessment resources.

You will design and run a number of queries including queries with fields from one table and queries with fields from more than one table. You will also use the Simple Query Wizard to design queries and create and format a calculated field.

 Tutorial

Creating a Query in Design View

💡 **Hint** The first step in designing a query is to choose the fields that you want to include in the query results datasheet.

Designing Queries

Being able to extract (pull out) specific data from a table is one of the most important functions of a database. Extracting data in Access is referred to as performing a query. The word *query* means "question" and to perform a query means to ask a question. Access provides several methods for designing a query. In this chapter, you will learn how to design your own query; use the Simple Query Wizard; create a calculated field; use aggregate functions in a query; and use the Crosstab Query Wizard, Find Duplicates Query Wizard, and Find Unmatched Query Wizard.

Designing a query consists of identifying the table or tables containing the data to be extracted, the field or fields from which the data will be drawn, and the criteria for selecting the data.

Specifying Tables and Fields for a Query

 Query Design

⏱ *Quick Steps*

Design Query
1. Click Create tab.
2. Click Query Design button.
3. Click table at Show Table dialog box.
4. Click Add button.
5. Add any additional tables.
6. In query design grid, click option box arrow in field in *Field* row.
7. Click field at drop-down list.
8. Insert criterion.
9. Click Run button.
10. Save query.

To design a query, open a database, click the Create tab, and then click the Query Design button in the Queries group. This displays a query window in the work area and also displays the Show Table dialog box, as shown in Figure 3.1.

Figure 3.1 Query Window with Show Table Dialog Box

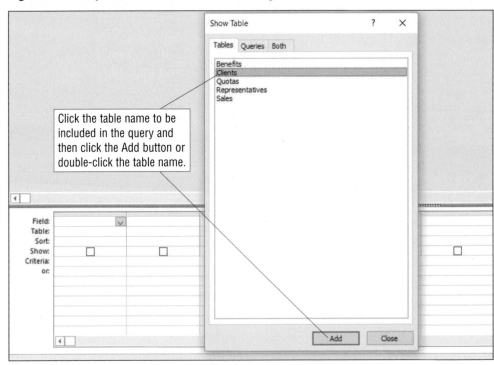

In the Show Table dialog box, click the name of the table to be included in the query and then click the Add button; or simply double-click the table name. This inserts a field list box for the table. Add any other tables required for the query. When all tables have been added, click the Close button. A sample query window is shown in Figure 3.2.

To insert a field in the query design grid, click the option box arrow in the field in the *Field* row and then click the specific field at the drop-down list. Or, double-click a field in a table field list box to insert it in the first available field in the *Field* row in the query design grid.

A third method for inserting a field in the query design grid is to drag a field from the table field list box to the desired location in the query design grid. To do this, position the mouse pointer on the field in the table field list box, click and hold down the left mouse button, drag to the specific field in the *Field* row in the query design grid, and then release the mouse button.

Tutorial

Adding a Criteria
Statement to a
Query

Quick Steps

**Establish Query
Criterion**

1. At query window, click in field in *Criteria* row in query design grid.
2. Type criterion and then press Enter key.
3. Click Run button.

! Run

Adding a Criteria Statement to a Query

Unless a criterion statement is added to a query in the *Criteria* row, the query will return (*return* is the term used for the query results) all records with the fields specified in the query design grid. While returning this information may be helpful, the information could easily be found in the table or tables. The value of performing a query is to extract specific information from a table or tables. To extract specific information, add a criterion statement to the query. To do this, click in the field in the *Criteria* row in the column containing the field name in the query design grid and then type the criterion. With the fields and criterion established, click the Run button in the Results group on the Query Tools Design tab. Access searches the specified tables for records that match the criterion and then displays those records in the query results datasheet.

Figure 3.2 Query Window

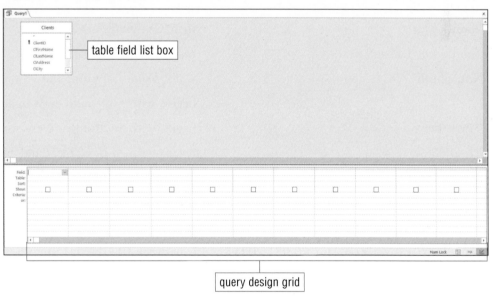

For example, to determine how many purchase orders were issued on a specific date, double-click the *PurchaseOrderID* field in the table field list box (which inserts *PurchaseOrderID* in the first field in the *Field* row in the query design grid) and then double-click the *OrderDate* field in the table field list box (which inserts *OrderDate* in the second field in the *Field* row in the query design grid). In this example, both fields are needed, so the purchase order ID is displayed along with the specific order date. After inserting the fields, add the criterion statement. The criterion statement for this example would be something like *#1/15/2021#*. After inserting the criterion, click the Run button in the Results group and the query returns only purchase orders for 1/15/2021 in the query results datasheet. If the criterion statement for the date was not added to the query, the query would return all purchase orders for all of the dates.

Access makes writing a criterion statement fairly simple by inserting the necessary symbols in the criterion. Type a city name, such as *Indianapolis*, in the *Criteria* row for a *City* field and then press the Enter key and Access changes the criterion to "*Indianapolis*". The quotation marks are inserted by Access and are necessary for the query to run properly. Let Access put the proper symbols around the criterion data in the field in the *Criteria* row or type the criterion with the symbols. Table 3.1 shows some examples of criteria statements, including what is typed and what is returned.

In the criteria examples, the asterisk is used as a wildcard character, which is a symbol that can be used to indicate any character. This is consistent with many other software applications. Two of the criteria examples in Table 3.1 use less-than and greater-than symbols. These symbols can be used for fields containing numbers, values, dates, amounts, and so forth. In the next several activities, you will design queries to extract specific information from different tables in databases.

Table 3.1 Criteria Examples

Typing This Criterion	Returns a Field Value Result That
"Smith"	matches *Smith*
"Smith" Or "Larson"	matches either *Smith* or *Larson*
Not "Smith"	is not *Smith* (anything but "Smith")
"s*"	begins with *S* or *s* and ends in anything
"*s"	begins with anything and ends in *S* or *s*
"[A-D]*"	begins with *A*, *B*, *C*, or *D* and ends in anything
#01/01/2021#	matches the date 01/01/2021
<#04/01/2021#	is less than (before) 04/01/2021
>#04/01/2021#	is greater than (after) 04/01/2021
Between #01/01/2021# And #03/31/2021#	is between 01/01/2021 and 03/31/2021

1. Open **3-Dearborn** from your AL1C3 folder and enable the content.
2. Create the following relationships and enforce referential integrity and cascade fields and records for each relationship:
   a. Create a one-to-many relationship with the *ClientID* field in the Clients table field list box the "one" and the *ClientID* field in the Sales table field list box the "many."
   b. Create a one-to-one relationship with the *RepID* field in the Representatives table field list box the "one" and the *RepID* field in the Benefits table field list box the "one."
   c. Create a one-to-many relationship with the *RepID* field in the Representatives table field list box the "one" and the *RepID* field in the Clients table field list box the "many."
   d. Create a one-to-many relationship with the *QuotaID* field in the Quotas table field list box the "one" and the *QuotaID* field in the Representatives table field list box the "many."
3. Click the Save button on the Quick Access Toolbar.
4. Print the relationships by completing the following steps:
   a. Click the Relationship Report button in the Tools group on the Relationship Tools Design tab.
   b. At the relationship report window, click the Landscape button in the Page Layout group on the Print Preview tab.
   c. Click the Print button at the left side of the Print Preview tab.
   d. At the Print dialog box, click OK.
5. Close the relationship report window without saving the report.
6. Close the Relationships window.
7. Extract records of those clients in Indianapolis by completing the following steps:

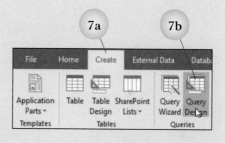

   a. Click the Create tab.
   b. Click the Query Design button in the Queries group.
   c. At the Show Table dialog box with the Tables tab selected, click *Clients* in the list box, click the Add button, and then click the Close button.
   d. Insert fields from the Clients table field list box into the *Field* row in the query design grid by completing the following steps:

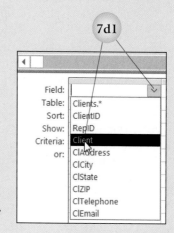

   1) Click the option box arrow at the right of the first field in the *Field* row in the query design grid and then click *Client* at the drop-down list.
   2) Click in the next field in the *Field* row (to the right of *Client*) in the query design grid, click the option box arrow, and then click *ClAddress* at the drop-down list.
   3) Click in the next field in the *Field* row (to the right of *ClAddress*), click the option box arrow, and then click *ClCity* at the drop-down list.
   4) Click in the next field in the *Field* row (to the right of *ClCity*), click the option box arrow, and then click *ClState* at the drop-down list.

5) Click in the next field in the *Field* row (to the right of *ClState*), click the option box arrow, and then click *ClZIP* at the drop-down list.

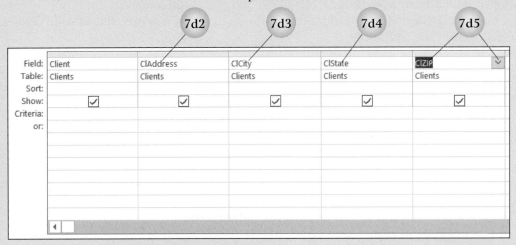

e. Insert the query criterion statement telling Access to display only those suppliers in Indianapolis by completing the following steps:
   1) Click in the *ClCity* field in the *Criteria* row in the query design grid. (This positions the insertion point in the field.)
   2) Type Indianapolis and then press the Enter key. (This changes the criterion to "*Indianapolis*".)

f. Return the results of the query by clicking the Run button in the Results group on the Query Tools Design tab. (This displays the results in the query results datasheet.)

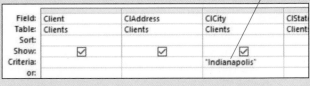

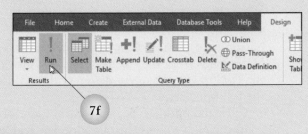

g. Save the results of the query by completing the following steps:
   1) Click the Save button on the Quick Access Toolbar.
   2) At the Save As dialog box, type Clients_Indianapolis and then press the Enter key or click OK.

h. Print the query results datasheet by clicking the File tab, clicking the *Print* option, and then clicking the *Quick Print* option.
i. Close the query.

8. Extract those records with quota identification numbers greater than 2 by completing the following steps:
   a. Click the Create tab and then click the Query Design button.
   b. Double-click *Representatives* in the Show Table dialog box and then click the Close button.
   c. In the query window, double-click *RepName*. (This inserts the field in the first field in the *Field* row in the query design grid.)
   d. Double-click *QuotaID*. (This inserts the field in the second field in the *Field* row in the query design grid.)

e. Insert the query criterion by completing the following steps:
   1) Click in the *QuotaID* field in the *Criteria* row in the query design grid.
   2) Type >2 and then press the Enter key. (Access will automatically insert quotation marks around *2* since the data type for the field is identified as *Short Text* [rather than *Number*].)

Field:	RepName	QuotaID
Table:	Representatives	Representatives
Sort:		
Show:	☑	☑
Criteria:		>"2"
or:		

8e2

f. Return the results of the query by clicking the Run button.
g. Save the query, typing QuotaID>2 as the query name.
h. Print and then close the query.

9. Extract those sales greater than $99,999 by completing the following steps:
   a. Click the Create tab and then click the Query Design button.
   b. Double-click *Sales* in the Show Table dialog box and then click the Close button.
   c. At the query window, double-click *ClientID*. (This inserts the field in the first field in the *Field* row in the query design grid.)
   d. Insert the *SalesAmount* field in the second field in the *Field* row.
   e. Insert the query criterion by completing the following steps:
      1) Click in the *SalesAmount* field in the *Criteria* row in the query design grid.
      2) Type >99999 and then press the Enter key. (Access will not insert quotation marks around *99999* since the field is identified as *Currency*.)

Field:	ClientID	SalesAmount
Table:	Sales	Sales
Sort:		
Show:	☑	☑
Criteria:		>99999
or:		

9e2

   f. Return the results of the query by clicking the Run button.
   g. Save the query, typing Sales>$99999 as the query name.
   h. Print and then close the query.

10. Extract records of those representatives with a telephone number that begins with the 765 area code by completing the following steps:
    a. Click the Create tab and then click the Query Design button.
    b. Double-click *Representatives* in the Show Table dialog box and then click the Close button.
    c. Insert the *RepName* field in the first field in the *Field* row.
    d. Insert the *RepTelephone* field in the second field in the *Field* row.
    e. Insert the query criterion by completing the following steps:
       1) Click in the *RepTelephone* field in the *Criteria* row.
       2) Type "(765*" and then press the Enter key. (You need to type the quotation marks in this criterion because the criterion contains a left parenthesis.)

Field:	RepName	RepTelephone
Table:	Representatives	Representatives
Sort:		
Show:	☑	☑
Criteria:		Like "(765*"
or:		

10e2

    f. Return the results of the query by clicking the Run button.
    g. Save the query, typing Reps_765AreaCode as the query name.
    h. Print and then close the query.

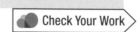
Check Your Work

Tutorial

Creating a Query in
Design View Using
Multiple Tables

In Activity 1a, each query was performed on fields from one table. Queries can also be performed on fields from multiple tables. In Activity 1b, queries will be performed on tables containing yes/no check boxes. When designing a query, both records that contain a check mark and records that do not contain a check mark can be extracted. To extract records that contain a check mark, click in the field in the *Criteria* row in the query design grid, type a *1*, press the Enter key, and Access changes the *1* to *True*. To extract records that do not contain a check mark, type *0* in the field in the *Criteria* row, press the Enter key, and Access changes the *0* to *False*.

The Zoom box can be used when entering a criterion in a query to provide a larger area for typing. To display the Zoom box, press Shift + F2 or right-click in the specific field in the *Criteria* row and then click *Zoom* at the shortcut menu. Type the criterion in the Zoom box and then click OK.

Activity 1b Performing Queries on Related Tables   Part 2 of 8

1. With **3-Dearborn** open, extract information on representatives hired between January 2014 and June 2014 and include the representatives' names by completing the following steps:
   a. Click the Create tab and then click the Query Design button.
   b. Double-click *Representatives* in the Show Table dialog box.
   c. Double-click *Benefits* in the Show Table dialog box and then click the Close button.
   d. At the query window, double-click *RepName* in the Representatives table field list box.
   e. Double-click *HireDate* in the Benefits table field list box.
   f. Insert the query criterion in the Zoom box by completing the following steps:
      1) Click in the *HireDate* field in the *Criteria* row.
      2) Press Shift + F2 to display the Zoom box.
      3) Type Between 1/1/2014 And 6/30/2014.
      4) Click OK.
   g. Return the results of the query by clicking the Run button.
   h. Save the query, typing Hires_2014JantoJun as the query name.
   i. Print and then close the query.

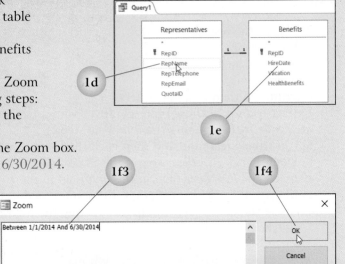

2. Extract records of those representatives who were hired in 2015 by completing the following steps:
   a. Click the Create tab and then click the Query Design button.
   b. Double-click *Representatives* in the Show Table dialog box.
   c. Double-click *Benefits* in the Show Table dialog box and then click the Close button.
   d. Double-click *RepID* in the Representatives table field list box.
   e. Double-click *RepName* in the Representatives table field list box.
   f. Double-click *HireDate* in the Benefits table field list box.

g. Insert the query criterion by completing the following steps:
   1) Click in the *HireDate* field in the *Criteria* row.
   2) Type *2015 and then press the Enter key.
h. Return the results of the query by clicking the Run button.
i. Save the query, typing Hires_2015 as the query name.
j. Print and then close the query.

Field:	RepID	RepName	HireDate
Table:	Representatives	Representatives	Benefits
Sort:			
Show:	☑	☑	☑
Criteria:			Like "*2015"
or:			

2g2

3. Suppose you need to determine sales for a company but you can only remember that the company name begins with *Blue*. Create a query that finds the company and identifies the sales by completing the following steps:
   a. Click the Create tab and then click the Query Design button.
   b. Double-click *Clients* in the Show Table dialog box.
   c. Double-click *Sales* in the Show Table dialog box and then click the Close button.
   d. Insert the *ClientID* field from the Clients table field list box in the first field in the *Field* row in the query design grid.
   e. Insert the *Client* field from the Clients table field list box in the second field in the *Field* row.
   f. Insert the *SalesAmount* field from the Sales table field list box in the third field in the *Field* row.
   g. Insert the query criterion by completing the following steps:
      1) Click in the *Client* field in the *Criteria* row.
      2) Type Blue* and then press the Enter key.

Field:	ClientID	Client	SalesAmount
Table:	Clients	Clients	Sales
Sort:			
Show:	☑	☑	☑
Criteria:		Like "Blue*"	
or:			

   h. Return the results of the query by clicking the Run button.
   i. Save the query, typing Sales_BlueRidge as the query name.
   j. Print and then close the query.

3g2

4. Close **3-Dearborn**.
5. Display the Open dialog box with your AL1C3 folder active.
6. Open **3-PacTrek** and enable the content.
7. Extract information on products ordered between February 1, 2021, and February 28, 2021, by completing the following steps:
   a. Click the Create tab and then click the Query Design button.
   b. Double-click *Products* in the Show Table dialog box.
   c. Double-click *Orders* in the Show Table dialog box and then click the Close button.
   d. Insert the *ProductID* field from the Products table field list box in the first field in the *Field* row.
   e. Insert the *Product* field from the Products table field list box in the second field in the *Field* row.
   f. Insert the *OrderDate* field from the Orders table field list box in the third field in the *Field* row.
   g. Insert the query criterion by completing the following steps:
      1) Click in the *OrderDate* field in the *Criteria* row.
      2) Type Between 2/1/2021 And 2/28/2021 and then press the Enter key.

Field:	ProductID	Product	OrderDate
Table:	Products	Products	Orders
Sort:			
Show:	☑	☑	☑
Criteria:			Between #2/1/2021#
or:			

7g2

h. Return the results of the query by clicking the Run button.

i. Save the query, typing Orders_February as the query name.

j. Print and then close the query.

8. Close **3-PacTrek**.

9. Open **3-CopperState** and enable the content.

10. Display the Relationships window and create the following additional relationships (enforce referential integrity and cascade fields and records):

a. Create a one-to-many relationship with the *AgentID* field in the Agents table field list box the "one" and the *AgentID* field in the Assignments table field list box the "many."

b. Create a one-to-many relationship with the *OfficeID* field in the Offices table field list box the "one" and the *OfficeID* field in the Assignments table field list box the "many."

c. Create a one-to-many relationship with the *OfficeID* field in the Offices table field list box the "one" and the *OfficeID* field in the Agents table field list box the "many."

11. Save and then print the relationships.

12. Close the relationship report without saving it and then close the Relationships window.

13. Extract records of clients that have uninsured motorist coverage by completing the following steps:

a. Click the Create tab and then click the Query Design button.

b. Double-click *Clients* in the Show Table dialog box.

c. Double-click *Coverage* in the Show Table dialog box and then click the Close button.

d. Insert the *ClientID* field from the Clients table field list box in the first field in the *Field* row.

e. Insert the *ClFirstName* field from the Clients table field list box in the second field in the *Field* row.

f. Insert the *ClLastName* field from the Clients table field list box in the third field in the *Field* row.

g. Insert the *UninsMotorist* field from the Coverage table field list box in the fourth field in the *Field* row. (You may need to scroll down the Coverage table field list box to see the *UninsMotorist* field.)

h. Insert the query criterion by clicking in the *UninsMotorist* field in the *Criteria* row, typing 1, and then pressing the Enter key. (Access changes the *1* to *True*.)

Field:	ClientID	ClFirstName	ClLastName	UninsMotorist
Table:	Clients	Clients	Clients	Coverage
Sort:				
Show:	☑	☑	☑	☑
Criteria:				True
or:				

13h

i. Click the Run button.

j. Save the query, typing Coverage_UninsuredMotorist as the query name.

k. Print and then close the query.

14. Extract records of claims in January over $999 by completing the following steps:

a. Click the Create tab and then click the Query Design button.

b. Double-click *Clients* in the Show Table dialog box.

c. Double-click *Claims* in the Show Table dialog box and then click the Close button.

d. Insert the *ClientID* field from the *Clients* table field list box in the first field in the *Field* row.

e. Insert the *ClFirstName* field from the Clients table field list box in the second field in the *Field* row.

f. Insert the *ClLastName* field from the Clients table field list box in the third field in the *Field* row.

g. Insert the *ClaimID* field from the Claims table field list box in the fourth field in the *Field* row.

h. Insert the *DateOfClaim* field from the Claims table field list box in the fifth field in the *Field* row.

i. Insert the *AmountOfClaim* field from the Claims table field list box in the sixth field in the *Field* row.

j. Click in the *DateOfClaim* field in the *Criteria* row, type Between 1/1/2021 And 1/31/2021, and then press the Enter key.

k. With the insertion point positioned in the *AmountOfClaim* field in the *Criteria* row, type >999 and then press the Enter key.

Field:	ClientID	ClFirstName	ClLastName	ClaimID	DateOfClaim	AmountOfClaim
Table:	Clients	Clients	Clients	Claims	Claims	Claims
Sort:						
Show:	☑	☑	☑	☑	☑	☑
Criteria:					Between #1/1/2021#	>999
or:						

14j   14k

l. Click the Run button.

m. Save the query, typing Claims_2021Jan>$999 as the query name.

n. Print and then close the query.

Check Your Work

 Tutorial

Sorting Data and Hiding Fields in Query Results

Sorting in a Query

When designing a query, the sort order of a field or fields can be specified. Click in a field in the *Sort* row and an option box arrow displays at the right side of the field. Click this option box arrow and a drop-down list displays with the choices *Ascending*, *Descending*, and *(not sorted)*. Click *Ascending* to sort from lowest to highest or click *Descending* to sort from highest to lowest.

Quick Steps

Sort Fields in Query
1. At query window, click in field in *Sort* row in query design grid.
2. Click option box arrow in field.
3. Click *Ascending* or *Descending*.

Hiding Fields in a Query

By default, each check box in the fields in the *Show* row in the query design grid contains a check mark, indicating that the column will be included in the query results datasheet. If a specific field is needed for the query but not needed when viewing the query results, remove the check mark from the field in the *Show* row to hide the field in the query results datasheet.

Arranging Fields in a Query

Use buttons in the Query Setup group on the Query Design Tools tab to insert a new field column in or delete an existing field column from the query design grid. To insert a field column, click in a field in the column that will display immediately to the right of the new column and then click the Insert Columns button in the Query Setup group on the Query Design Tools tab. To remove a column, click in a field in the column to be deleted and then click the Delete Columns button in the Query Setup group. Complete similar steps to insert or delete a row in the query design grid.

 Insert Columns

 Delete Columns

Columns in the query design grid can be rearranged by selecting the field column and then dragging the column to the desired position. To select a column in the query design grid, position the mouse pointer at the top of the column until the pointer turns into a small, black, down-pointing arrow and then click the left mouse button. Position the mouse pointer at the top of the selected column until the mouse displays as a pointer, click and hold down the left mouse button, drag to the desired position in the query design grid, and then release the mouse button. While the column is being dragged, a thick, black, vertical line displays identifying the location where the column will be inserted.

Activity 1c Performing Queries on Related Tables and Sorting in Field Values

Part 3 of 8

1. With **3-CopperState** open, extract information on clients with agents from the West Bell Road Glendale office and sort the information alphabetically by client last name by completing the following steps:
   a. Click the Create tab and then click the Query Design button.
   b. Double-click *Assignments* in the Show Table dialog box.
   c. Double-click *Clients* in the Show Table dialog box and then click the Close button.
   d. Insert the *OfficeID* field from the Assignments table field list box in the first field in the *Field* row.
   e. Insert the *AgentID* field from the Assignments table field list box in the second field in the *Field* row.
   f. Insert the *ClFirstName* field from the Clients table field list box in the third field in the *Field* row.
   g. Insert the *ClLastName* field from the Clients table field list box in the fourth field in the *Field* row.
   h. Click in the *OfficeID* field in the *Criteria* row, type GW, and then press the Enter key.
   i. Sort the *ClLastName* field column in ascending alphabetical order (A–Z) by completing the following steps:
      1) Click in the *ClLastName* field in the *Sort* row. (This displays an option box arrow in the field.)
      2) Click the option box arrow in the field in the *Sort* row and then click *Ascending*.

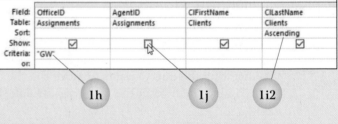

   j. Specify that you do not want the *AgentID* field to show in the query results by clicking in the check box in the *AgentID* field in the *Show* row to remove the check mark.
   k. Click the Run button.
   l. Save the query, typing Clients_GW as the query name.
   m. Print and then close the query.
2. Close **3-CopperState**.
3. Open **3-PacTrek**.
4. Extract information on orders less than $1,500 by completing the following steps:
   a. Click the Create tab and then click the Query Design button.
   b. Double-click *Products* in the Show Table dialog box.
   c. Double-click *Orders* in the Show Table dialog box and then click the Close button.
   d. Insert the *ProductID* field from the Products table field list box in the first field in the *Field* row.
   e. Insert the *SupplierID* field from the Products table field list box in the second field in the *Field* row.

f. Insert the *UnitsOrdered* field from the Orders table field list box in the third field in the *Field* row.

g. Insert the *OrderAmount* field from the Orders table field list box in the fourth field in the *Field* row. (You may need to scroll down the table field list box to display this field.)

h. Insert the query criterion by completing the following steps:
   1) Click in the *OrderAmount* field in the *Criteria* row.
   2) Type <1500 and then press the Enter key.

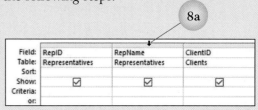

4h2

i. Sort the *OrderAmount* field column values from highest to lowest by completing the following steps:
   1) Click in the *OrderAmount* field in the *Sort* row. (This displays an option box arrow in the field.)
   2) Click the option box arrow in the field in the *Sort* row and then click *Descending*.

j. Return the results of the query by clicking the Run button.

4i2

k. Save the query, typing Orders<$1500 as the query name.

l. Print and then close the query.

5. Close **3-PacTrek**.

6. Open **3-Dearborn**.

7. Design a query by completing the following steps:
   a. Click the Create tab and then click the Query Design button.
   b. Double-click *Representatives* in the Show Table dialog box.
   c. Double-click *Clients* in the Show Table dialog box.
   d. Double-click *Sales* in the Show Table dialog box and then click the Close button.
   e. Insert the *RepID* field from the Representatives table field list box in the first field in the *Field* row.
   f. Insert the *RepName* field from the Representatives table field list box in the second field in the *Field* row.
   g. Insert the *ClientID* field from the Clients table field list box in the third field in the *Field* row.
   h. Insert the *SalesAmount* field from the Sales table field list box in the fourth field in the *Field* row.

8. Move the *RepName* field column by completing the following steps:
   a. Position the mouse pointer at the top of the *RepName* field column until the pointer turns into a small, black, down-pointing arrow and then click the left mouse button. (This selects the entire column.)

   8a

   b. Position the mouse pointer at the top of the selected column until the pointer turns into a white arrow.

   c. Click and hold down the left mouse button; drag to the right until a thick, black horizontal line displays between the *ClientID* field column and the *SalesAmount* field column; and then release the mouse button.

   8c

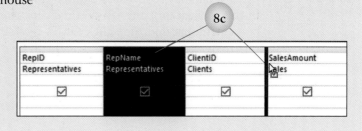

9. Delete the *RepID* field column by clicking in a field in the column and then clicking the Delete Columns button in the Query Setup group.
10. Insert a new field column and insert a new field in the column by completing the following steps:
    a. Click in a field in the *SalesAmount* field column and then click the Insert Columns button in the Query Setup group.
    b. Click the option box arrow in the *Field* row in the new field column and then click *Clients.Client* at the drop-down list.

11. Hide the *ClientID* field so it is not included in the query results by clicking the *Show* check box in the *ClientID* field in the *Show* row to remove the check mark.
12. Insert the query criterion that extracts information on sales over $100,000 by completing the following steps:
    a. Click in the *SalesAmount* field in the *Criteria* row.
    b. Type >100000 and then press the Enter key.
13. Sort the *SalesAmount* field column values from highest to lowest by completing the following steps:
    a. Click in the *SalesAmount* field in the *Sort* row.
    b. Click the option box arrow in the field in the *Sort* row and then click *Descending*.
14. Return the results of the query by clicking the Run button.
15. Save the query, typing Sales>$100000 as the query name.
16. Print and then close the query.

Check Your Work

Modifying a Query

A query can be modified and used for a new purpose. For example, if a query is designed to return sales of more than $100,000, the query can be used to find sales that are less than $100,000. Rather than design a new query, open the existing query, make any needed changes, and then run the query.

Quick Steps

Modify Query
1. Double-click query in Navigation pane.
2. Click View button.
3. Make changes to query.
4. Click Run button.
5. Click Save button.

Hint Save time designing a new query by modifying an existing query.

To modify an existing query, double-click the query in the Navigation pane. (This opens the query in Datasheet view.) Click the View button to display the query in Design view. A query can also be opened in Design view by right-clicking the query in the Navigation pane and then clicking *Design View* at the shortcut menu. Make changes to the query and then click the Run button in the Results group. Click the Save button on the Quick Access Toolbar to save the query with the same name. To save the query with a new name, click the File tab, click the *Save As* option, click the *Save Object As* option, and then click the Save As button. At the Save As dialog box, type a name for the query and then press the Enter key.

If a database contains a number of queries, the queries can be listed together in the Navigation pane. To do this, click the option box arrow in the Navigation pane menu bar and then click *Object Type* at the drop-down list. This groups objects in categories, such as *Tables* and *Queries*.

Tutorial

Renaming and
Deleting Objects

Renaming and Deleting a Query

If a query has been modified, consider renaming it. To do this, right-click the query name in the Navigation pane, click *Rename* at the shortcut menu, type the new name, and then press the Enter key. If a query is no longer needed in the database, delete it by clicking the query name in the Navigation pane, clicking the Delete button in the Records group on the Home tab, and then clicking the Yes button at the deletion message. Another method is to right-click the query in the Navigation pane, click *Delete* at the shortcut menu, and then click Yes at the deletion message. If a query is being deleted from a file on the computer's hard drive, the deletion message will not display. This is because Access automatically sends the deleted query to the Recycle Bin, where it can be retrieved if necessary.

Activity 1d Modifying and Deleting Queries

Part 4 of 8

1. With **3-Dearborn** open, find sales less than $100,000 by completing the following steps:
   a. Double-click *Sales>$100000* in the Queries group in the Navigation pane.
   b. Click the View button in the Views group to switch to Design view.
   c. Click in the field in the *Criteria* row containing the text *>100000* and then edit the text so it reads *<100000*.

Field:	RepName	Client	SalesAmount
Table:	Representatives	Clients	Sales
Sort:			Descending
Show:	☑	☑	☑
Criteria:			<100000
or:			

1c

   d. Click the Run button.
2. Save the query with a new name by completing the following steps:
   a. Click the File tab, click the *Save As* option, click the *Save Object As* option, and then click the Save As button.
   b. At the Save As dialog box, type Sales<$100000 and then click OK or press the Enter key.
   c. Print and then close the query.
3. Modify an existing query and find employees with three weeks of vacation by completing the following steps:
   a. Right-click *Hires_2014JantoJun* in the Queries group in the Navigation pane and then click *Design View* at the shortcut menu.
   b. Click in the *HireDate* field in the *Field* row.
   c. Click the option box arrow in the field and then click *Vacation* at the drop-down list.
   d. Select the current text in the *Vacation* field in the *Criteria* row, type 3 weeks, and then press the Enter key.
   e. Click the Run button.
   f. Save and then close the query.

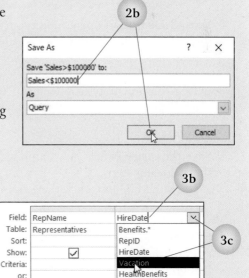

2b

Save As

Save 'Sales>$100000' to:

Sales<$100000

As

Query

OK   Cancel

3b

Field:	RepName	HireDate
Table:	Representatives	Benefits.*
Sort:		RepID
Show:	☑	HireDate
Criteria:		Vacation
or:		HealthBenefits

3c

4. Rename the query by completing the following steps:
   a. Right-click *Hires_2014JantoJun* in the Navigation pane and then click *Rename* at the shortcut menu.
   b. Type *Reps_3WeeksVacay* and then press the Enter key.
   c. Open, print, and then close the query.
5. Delete the *Sales>$99999* query by right-clicking the query name in the Navigation pane and then clicking *Delete* at the shortcut menu. If a deletion message displays, click Yes.

 Tutorial

Designing a Query
with *Or* Criteria

 Tutorial

Designing a Query
with *And* Criteria

💡**Hint** A query
can be designed that
combines *And* and *Or*
statements.

Designing Queries with *Or* and *And* Criteria

The query design grid contains an *or* row for designing a query that instructs Access to display records matching any of the criteria. Multiple criterion statements on different rows in a query become an *Or* statement, which means that any of the criterion can be met for a record to be included in the query results datasheet. For example, to display a list of employees with three weeks of vacation *or* four weeks of vacation, type *3 weeks* in the *Vacation* field in the *Criteria* row and then type *4 weeks* in the field immediately below *3 weeks* in the *or* row. Other examples include finding clients that live in *Muncie* or *Lafayette* and finding representatives with quotas of *1* or *2*.

When designing a query, criteria statements can be entered into more than one field in the *Criteria* row. Multiple criteria all entered in the same row become an *And* statement, for which each criterion must be met for Access to select the record. For example, a query can be designed to return clients in the Indianapolis area with sales greater than $100,000.

Activity 1e Designing Queries with *Or* and *And* Criteria

Part 5 of 8

1. With **3-Dearborn** open, modify an existing query and find employees with three weeks or four weeks of vacation by completing the following steps:
   a. Double-click the *Reps_3WeeksVacay* query in the Navigation pane.
   b. Click the View button in the Views group to switch to Design view.
   c. Click in the empty field below "*3 weeks*" in the *or* row, type *4 weeks*, and then press the Enter key.
   d. Click the Run button.
2. Save the query with a new name by completing the following steps:
   a. Click the File tab, click the *Save As* option, click the *Save Object As* option, and then click the Save As button.
   b. At the Save As dialog box, type *Reps_3or4WeeksVacay* and then press the Enter key.
   c. Print and then close the query.
3. Design a query that finds records of clients in the Indianapolis area with sales over $100,000 by completing the following steps:
   a. Click the Create tab and then click the Query Design button.

Field:	RepName	Vacation
Table:	Representatives	Benefits
Sort:		
Show:	☑	☑
Criteria:		"3 weeks"
or:		"4 weeks"

1c

b. Double-click *Clients* in the Show Table dialog box.

c. Double-click *Sales* in the Show Table dialog box and then click the Close button.

d. Insert the *Client* field from the Clients table field list box in the first field in the *Field* row.

e. Insert the *ClCity* field from the Clients table field list box in the second field in the *Field* row.

f. Insert the *SalesAmount* field from the Sales table field list box in the third field in the *Field* row.

g. Insert the query criteria by completing the following steps:

1) Click in the *ClCity* field in the *Criteria* row.

2) Type Indianapolis and then press the Enter key.

3) With the insertion point positioned in the *SalesAmount* field in the *Criteria* row, type >100000 and then press the Enter key.

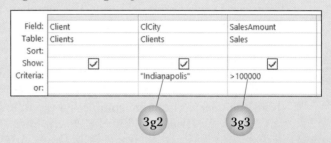

h. Click the Run button.

i. Save the query, typing Sales_Indianapolis>$100000 as the query name.

j. Print and then close the query.

4. Close **3-Dearborn**.

5. Open **3-PacTrek**.

6. Design a query that finds products available from supplier numbers 25, 31, and 42 by completing the following steps:

a. Click the Create tab and then click the Query Design button.

b. Double-click *Suppliers* in the Show Table dialog box.

c. Double-click *Products* in the Show Table dialog box and then click the Close button.

d. Insert the *SupplierID* field from the Suppliers table field list box in the first field in the *Field* row.

e. Insert the *SupplierName* field from the Suppliers table field list box in the second field in the *Field* row.

f. Insert the *Product* field from the Products table field list box in the third field in the *Field* row.

g. Insert the query criteria by completing the following steps:

1) Click in the *SupplierID* field in the *Criteria* row.

2) Type 25 and then press the Down Arrow key. (This makes the field below 25 active.)

3) Type 31 and then press the Down Arrow key. (This makes the field below 31 active.)

4) Type 42 and then press the Enter key.

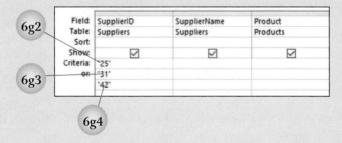

h. Click the Run button.

i. Save the query, typing Suppliers_25-31-42 as the query name.

j. Print and then close the query.

7. Design a query that finds the number of ski hats or gloves on order by completing the following steps:
   a. Click the Create tab and then click the Query Design button.
   b. Double-click *Orders* in the Show Table dialog box.
   c. Double-click *Suppliers* in the Show Table dialog box.
   d. Double-click *Products* in the Show Table dialog box and then click the Close button.
   e. Insert the *OrderID* field from the Orders table field list box in the first field in the *Field* row.
   f. Insert the *SupplierName* field from the Suppliers table field list box in the second field in the *Field* row.
   g. Insert the *Product* field from the Products table field list box in the third field in the *Field* row.
   h. Insert the *UnitsOrdered* field from the Orders table field list box in the fourth field in the *Field* row.
   i. Insert the query criteria by completing the following steps:
      1) Click in the *Product* field in the *Criteria* row.
      2) Type *ski hat* and then press the Down Arrow key. (You need to type the asterisk before and after *ski hat* so the query will find any product that includes the words *ski hat* in the description, no matter what text comes before or after the words. When you press the Down Arrow key, Access changes the criteria to *Like* "*ski hat*".)
      3) Type *gloves* and then press the Enter key.

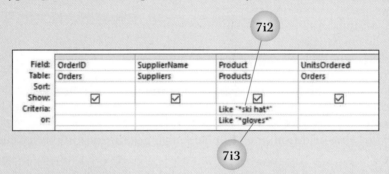

7i2

Field:	OrderID	SupplierName	Product	UnitsOrdered
Table:	Orders	Suppliers	Products	Orders
Sort:				
Show:	☑	☑	☑	☑
Criteria:			Like "*ski hat*"	
or:			Like "*gloves*"	

7i3

   j. Click the Run button.
   k. Save the query, typing Orders_HatsGloves as the query name.
   l. Print and then close the query.
8. Design a query that finds boots, sleeping bags, or backpacks and the suppliers that produce them by completing the following steps:
   a. Click the Create tab and then click the Query Design button.
   b. Double-click *Products* in the Show Table dialog box.
   c. Double-click *Suppliers* in the Show Table dialog box and then click the Close button.
   d. Insert the *ProductID* field from the Products table field list box in the first field in the *Field* row.
   e. Insert the *Product* field from the Products table field list box in the second field in the *Field* row.
   f. Insert the *SupplierName* field from the Suppliers table field list box in the third field in the *Field* row.
   g. Insert the query criteria by completing the following steps:
      1) Click in the *Product* field in the *Criteria* row.

2) Type *boots* and then press the Down Arrow key.
3) Type *sleeping bag* and then press the Down Arrow key.
4) Type *backpack* and then press the Enter key.

h. Click the Run button.

i. Save the query, typing Products_Equipment as the query name.

j. Print and then close the query.

9. Close **3-PacTrek**.

10. Open **3-CopperState**.

11. Design a query that finds clients who have only liability auto coverage by completing the following steps:

a. Click the Create tab and then click the Query Design button.

b. Double-click *Clients* in the Show Table dialog box.

c. Double-click *Coverage* in the Show Table dialog box and then click the Close button.

d. Insert the *ClientID* field from the Clients table field list box in the first field in the *Field* row.

e. Insert the *ClFirstName* field from the Clients table field list box in the second field in the *Field* row.

f. Insert the *ClLastName* field from the Clients table field list box in the third field in the *Field* row.

g. Insert the *Medical* field from the Coverage table field list box in the fourth field in the *Field* row.

h. Insert the *Liability* field from the Coverage table field list box in the fifth field in the *Field* row.

i. Insert the *Comprehensive* field from the Coverage table field list box in the sixth field in the *Field* row.

j. Insert the *UninsMotorist* field from the Coverage table field list box in the seventh field in the *Field* row. (You may need to scroll down the Coverage table field list box to see the *UninsMotorist* field.)

k. Insert the *Collision* field from the Coverage table field list box in the eighth field in the *Field* row. (You may need to scroll down the Coverage table field list box to see the *Collision* field.)

l. Insert the query criteria by completing the following steps:

1) Click in the *Medical* field in the *Criteria* row, type 0, and then press the Enter key. (Access changes the *0* to *False*.)

2) With the insertion point in the *Liability* field in the *Criteria* row, type 1 and then press the Enter key. (Access changes the *1* to *True*.)

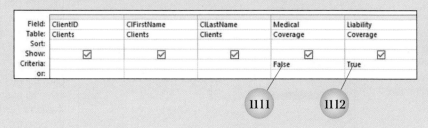

3) With the insertion point in the *Comprehensive* field in the *Criteria* row, type 0 and then press the Enter key.

4) With the insertion point in the *UninsMotorist* field in the *Criteria* row, type 0 and then press the Enter key.

5) With the insertion point in the *Collision* field in the *Criteria* row, type 0 and then press the Enter key.

  m. Click the Run button.

  n. Save the query, typing Coverage_LiabilityOnly as the query name.

  o. Print the query in landscape orientation.

  p. Close the query.

12. Close **3-CopperState**.

 Check Your Work

 Tutorial

Creating a Query Using the Simple Query Wizard

 Query Wizard

Quick Steps

Create Query Using Simple Query Wizard

1. Click Create tab.
2. Click Query Wizard button.
3. Make sure *Simple Query Wizard* is selected in list box and then click OK.
4. Follow query steps.
5. Click Finish button.

Creating a Query Using the Simple Query Wizard

The Simple Query Wizard provides steps for preparing a query. To use this wizard, open the database, click the Create tab, and then click the Query Wizard button in the Queries group. At the New Query dialog box, make sure *Simple Query Wizard* is selected in the list box and then click OK. At the first Simple Query Wizard dialog box, shown in Figure 3.3, specify the table(s) in the *Tables/Queries* option box. After specifying the table(s), insert the fields to be included in the query in the *Selected Fields* list box and then click the Next button.

At the second Simple Query Wizard dialog box, specify a detail or summary query and then click the Next button. At the third (and last) Simple Query Wizard dialog box, shown in Figure 3.4, type a name for the completed query or accept the name provided by the wizard. The third Simple Query Wizard dialog box also includes options for choosing to open the query to view the information or to modify the query design. To extract specific information, be sure to choose the *Modify the query design* option. After making any necessary changes, click the Finish button.

If the query design is not modified in the last Simple Query Wizard dialog box, the query displays all records for the fields identified in the first Simple Query Wizard dialog box.

Figure 3.3 First Simple Query Wizard Dialog Box

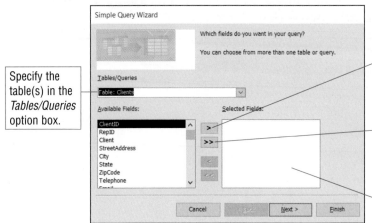

Specify the table(s) in the *Tables/Queries* option box.

Click the One Field button to insert the selected field in the *Available Fields* list box into the *Selected Fields* list box.

Click the All Fields button to insert all of the fields in the *Available Fields* list box into the *Selected Fields* list box.

Insert in the *Selected Fields* list box the fields to be included in the query.

Figure 3.4 Last Simple Query Wizard Dialog Box

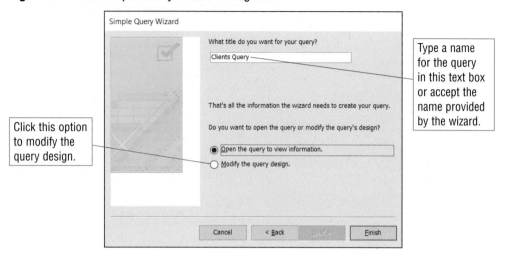

Type a name for the query in this text box or accept the name provided by the wizard.

Click this option to modify the query design.

Activity 1f Creating Queries Using the Simple Query Wizard

1. Open **3-Dearborn** and then use the Simple Query Wizard to create a query that returns client names along with sales by completing the following steps:

   a. Click the Create tab and then click the Query Wizard button in the Queries group.

   b. At the New Query dialog box, make sure *Simple Query Wizard* is selected in the list box and then click OK.

   c. At the first Simple Query Wizard dialog box, click the *Tables/Queries* option box arrow and then click *Table: Clients*.

   d. With *ClientID* selected in the *Available Fields* list box, click the One Field button (the button containing the greater-than symbol, >). This inserts the *ClientID* field in the *Selected Fields* list box.

   e. Click *Client* in the *Available Fields* list box and then click the One Field button.

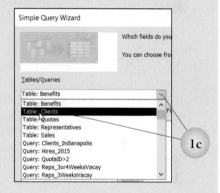

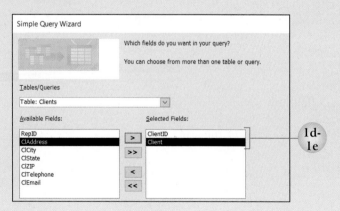

f. Click the *Tables/Queries* option box arrow and then click *Table: Sales*.

g. Click *SalesAmount* in the *Available Fields* list box and then click the One Field button.

h. Click the Next button.

i. At the second Simple Query Wizard dialog box, click the Next button.

j. At the last Simple Query Wizard dialog box, select the name in the *What title do you want for your query?* text box, type Clients_Sales, and then press the Enter key.

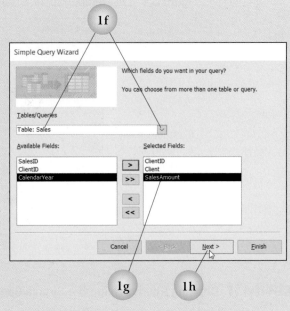

k. When the results of the query display, print the results.

2. Close the query.

3. Close **3-Dearborn**.

4. Open **3-PacTrek**.

5. Create a query that returns the products on order, order amounts, and supplier names by completing the following steps:

a. Click the Create tab and then click the Query Wizard button.

b. At the New Query dialog box, make sure *Simple Query Wizard* is selected in the list box and then click OK.

c. At the first Simple Query Wizard dialog box, click the *Tables/Queries* option box arrow and then click *Table: Suppliers*.

d. With *SupplierID* selected in the *Available Fields* list box, click the One Field button. (This inserts the *SupplierID* field in the *Selected Fields* list box.)

e. With *SupplierName* selected in the *Available Fields* list box, click the One Field button.

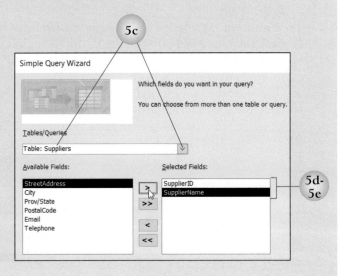

f. Click the *Tables/Queries* option box arrow and then click *Table: Orders*.
g. Click *ProductID* in the *Available Fields* list box and then click the One Field button.
h. Click *OrderAmount* in the *Available Fields* list box and then click the One Field button.
i. Click the Next button.

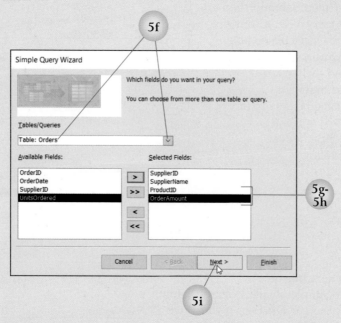

j. At the second Simple Query Wizard dialog box, click the Next button.
k. At the last Simple Query Wizard dialog box, select the text in the *What title do you want for your query?* text box, type Products_OrderAmounts, and then press the Enter key.

l. When the results of the query display, print the results.
m. Close the query.

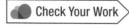 Check Your Work

To extract specific information when using the Simple Query Wizard, tell the wizard that the query design is to be modified. This displays the query window with the query design grid, where query criteria can be entered.

1. With **3-PacTrek** open, use the Simple Query Wizard to create a query that returns suppliers outside British Columbia by completing the following steps:
   a. Click the Create tab and then click the Query Wizard button.
   b. At the New Query dialog box, make sure *Simple Query Wizard* is selected and then click OK.
   c. At the first Simple Query Wizard dialog box, click the *Tables/Queries* option box arrow and then click *Table: Suppliers*.
   d. Insert the following fields in the *Selected Fields* list box:

   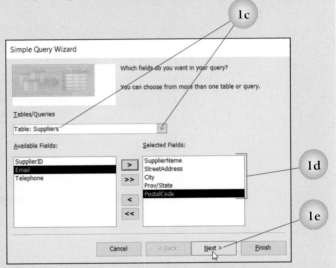

   > *SupplierName*
   > *StreetAddress*
   > *City*
   > *Prov/State*
   > *PostalCode*

   e. Click the Next button.
   f. At the last Simple Query Wizard dialog box, select the current text in the *What title do you want for your query?* text box and then type Suppliers_NotBC.
   g. Click the *Modify the query design* option and then click the Finish button.

   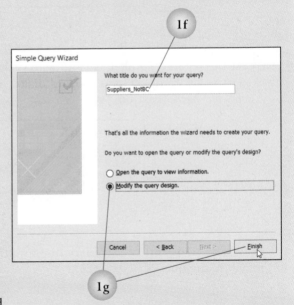

   h. At the query window, complete the following steps:
      1) Click in the *Prov/State* field in the *Criteria* row in the query design grid.
      2) Type Not BC and then press the Enter key.
   i. Specify that the fields are to be sorted in descending order by postal code by completing the following steps:
      1) Click in the *PostalCode* field in the *Sort* row.
      2) Click the option box arrow in the field and then click *Descending*.

   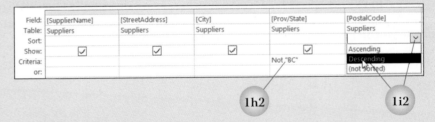

   j. Click the Run button. (This returns suppliers that are not in British Columbia and sorts the records by postal code in descending order.)
   k. Save, print, and then close the query.

2. Close **3-PacTrek**.
3. Open **3-Dearborn**.
4. Use the Simple Query Wizard to create a query that returns clients in Muncie by completing the following steps:
   a. Click the Create tab and then click the Query Wizard button.
   b. At the New Query dialog box, make sure *Simple Query Wizard* is selected and then click OK.
   c. At the first Simple Query Wizard dialog box, click the *Tables/Queries* option box arrow and then click *Table: Clients*. (You may need to scroll up the list to see this table.)
   d. Insert the following fields in the *Selected Fields* list box:

   > *Client*
   > *ClAddress*
   > *ClCity*
   > *ClState*
   > *ClZIP*

   e. Click the Next button.
   f. At the last Simple Query Wizard dialog box, select the current text in the *What title do you want for your query?* text box and then type Clients_Muncie.
   g. Click the *Modify the query design* option and then click the Finish button.
   h. At the query window, complete the following steps:
      1) Click in the *ClCity* field in the *Criteria* row in the query design grid.
      2) Type Muncie and then press the Enter key.

Field:	[Client]	[ClAddress]	[ClCity]	[ClState]	[ClZIP]
Table:	Clients	Clients	Clients	Clients	Clients
Sort:					
Show:	☑	☑	☑	☑	☑
Criteria:			"Muncie"		
or:					

4h2

   i. Click the Run button. (This displays clients in Muncie.)
   j. Save, print, and then close the query.
5. Close **3-Dearborn**.
6. Open **3-CopperState**.
7. Use the Simple Query Wizard to display clients that live in Phoenix with claims over $500 by completing the following steps:
   a. Click the Create tab and then click the Query Wizard button.
   b. At the New Query dialog box, make sure *Simple Query Wizard* is selected in the list box and then click OK.
   c. At the first Simple Query Wizard dialog box, click the *Tables/Queries* option box arrow and then click *Table: Clients*.
   d. Insert the following fields in the *Selected Fields* list box:

   > *ClientID*
   > *ClFirstName*
   > *ClLastName*
   > *ClAddress*
   > *ClCity*
   > *ClState*
   > *ClZIP*

   e. Click the *Tables/Queries* option box arrow and then click *Table: Claims*.

f. With *ClaimID* selected in the *Available Fields* list box, click the One Field button.
g. Click *AmountOfClaim* in the *Available Fields* list box and then click the One Field button.
h. Click the Next button.
i. At the second Simple Query Wizard dialog box, click the Next button.
j. At the last Simple Query Wizard dialog box, select the current text in the *What title do you want for your query?* text box and then type Claims_Phoenix>$500.
k. Click the *Modify the query design* option and then click the Finish button.
l. At the query window, complete the following steps:
   1) Click in the *ClCity* field in the *Criteria* row in the query design grid.
   2) Type Phoenix and then press the Enter key.
   3) Click in the *AmountOfClaim* field in the *Criteria* row. (You may need to scroll to the right to see this field.)
   4) Type >500 and then press the Enter key.

ClCity	ClState	ClZIP	ClaimID	AmountOfClaim
Clients	Clients	Clients	Claims	Claims
☑	☑	☑	☑	☑
"Phoenix"				>500

712   714

m. Click the Run button. (This displays clients in Phoenix with claims greater than $500.)
n. Save the query, print the query in landscape orientation, and then close the query.
8. Close **3-CopperState**.

Check Your Work

Tutorial

Performing Calculations in a Query

Performing Calculations in a Query

In a query, values from a field can be calculated by inserting a calculated field in a field in the *Field* row in the query design grid. To insert a calculated field, click in a field in the *Field* row, type a field name followed by a colon, and then type the equation. For example, to determine pension contributions as 3% of an employee's annual salary, type *PensionContribution:[AnnualSalary]*0.03* in the field in the *Field* row. Use brackets to specify field names and use mathematical operators to perform the equation. Some basic operators include the plus (+) for addition, the hyphen (-) for subtraction, the asterisk (*) for multiplication, and the forward slash (/) for division.

Builder

Type a calculation in a field in the *Field* row or in the Expression Builder dialog box. To display the Expression Builder dialog box, open the query in Design view, click in the field where the calculated field expression is to be inserted, and then click the Builder button in the Query Setup group on the Query Tools Design tab. Type field names in the Expression Builder and click OK and the equation is inserted in the field with the correct symbols. For example, type *AnnualSalary*0.03* in the Expression Builder and *Expr1: [AnnualSalary]*0.03* is inserted in the field in the *Field* row when OK is clicked. If a name is not typed for the field, Access creates the alias *Expr1* for the field name. Provide a specific name for the field, such as *PensionContribution*, by typing the name in the Expression Builder, followed by a colon, and then typing the expression.

Property
Sheet

If the results of the calculation should display as currency, apply numeric formatting and define the number of digits past the decimal point using the Property Sheet task pane. In Design view, click the Property Sheet button in the Show/Hide group on the Query Tools Design tab and the Property Sheet task pane displays at the right side of the screen. Click in the *Format* property box, click the option box arrow, and then click *Currency* at the drop-down list.

Activity 1h Creating and Formatting a Calculated Field in a Query

1. Open **3-MRInvestments** and enable the content.
2. Create a query that returns employer pension contributions at 3% of employees' annual salary by completing the following steps:
   a. Click the Create tab and then click the Query Design button.
   b. Double-click *Employees* in the Show Table dialog box and then click the Close button.
   c. Insert the *EmpID* field from the Employees table field list box in the first field in the *Field* row.
   d. Insert the *FirstName* field in the second field in the *Field* row.
   e. Insert the *LastName* field in the third field in the *Field* row.
   f. Insert the *AnnualSalary* field in the fourth field in the *Field* row.
   g. Click in the fifth field in the *Field* row.
   h. Type PensionContribution:[AnnualSalary]*0.03 and then press the Enter key.
   i. Click in the *PensionContribution* field and then click the Property Sheet button in the Show/Hide group.
   j. Click in the *Format* property box, click the option box arrow, and then click *Currency* at the drop-down list.
   k. Click the Close button in the upper right corner of the Property Sheet task pane.
   l. Click the Run button.
   m. Save the query, typing Benefits_PensionContri as the query name.
   n. Print and then close the query.

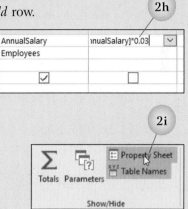

3. Modify *PensionContributionsQuery* and use the Expression Builder to write an equation finding the total amount of annual salary plus a 3% employer pension contribution by completing the following steps:
   a. Right-click *Benefits_PensionContri* in the Queries group in the Navigation pane and then click *Design View* at the shortcut menu.
   b. Click in the field in the *Field* row containing *PensionContribution:[AnnualSalary]*0.03*.
   c. Click the Builder button in the Query Setup group on the Query Tools Design tab.
   d. In the Expression Builder, select the existing expression *PensionContribution: [AnnualSalary]*0.03*.
   e. Type Salary&Pension:[AnnualSalary]*1.03 and then click OK.

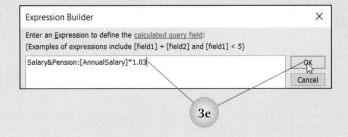

4. Click the Run button.
5. Save the query by completing the following steps:
   a. Click the File tab, click the *Save As* option, click the *Save Object As* option, and then click the Save As button.
   b. At the Save As dialog box, type Benefits_Salary&Pension and then click OK.
6. Print and then close the query.
7. Close **3-MRInvestments**.
8. Open **3-PacTrek**.
9. Create a query that returns orders and total order amounts by completing the following steps:
   a. Click the Create tab and then click the Query Design button.
   b. Double-click *Products* in the Show Table dialog box.
   c. Double-click *Orders* in the Show Table dialog box and then click the Close button.
   d. Insert the *Product* field from the Products table field list box in the first field in the *Field* row.
   e. Insert the *OrderID* field from the Orders table field list box in the second field in the *Field* row.
   f. Insert the *UnitsOrdered* field from the Orders table field list box in the third field in the *Field* row.
   g. Insert the *OrderAmount* field from the Orders table field list box in the fourth field in the *Field* row.
   h. Click in the fifth field in the *Field* row.
   i. Click the Builder button.
   j. Type Total:OrderAmount*UnitsOrdered in the Expression Builder and then click OK.

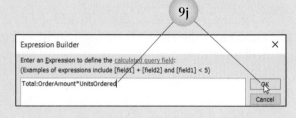

   k. Click the Run button.
   l. Adjust the width of the columns to fit the longest entries.
   m. Save the query, typing Orders_TotalAmount as the query name.
   n. Print and then close the query.
10. Close **3-PacTrek**.

> Check Your Work

Activity 2 Create Aggregate Functions and Crosstab Queries, Find Duplicates, and Find Unmatched Queries   6 Parts

You will create an aggregate functions query that determines the total, average, minimum, and maximum order amounts and then calculate total and average order amounts grouped by supplier. You will also use the Crosstab Query Wizard, Find Duplicates Query Wizard, and Find Unmatched Query Wizard to design queries.

Tutorial

Using Aggregate
Functions

Designing Queries with Aggregate Functions

An *aggregate function*—such as Sum, Avg, Min, Max, or Count—can be included in a query to calculate statistics from numeric field values of all the records in the table. When an aggregate function is used, Access displays one row in the query results datasheet with the formula result for the function used. For example, in a table with a numeric field containing annual salary amounts, the Sum function can be used to calculate the total of all salary amount values.

 Totals

To use aggregate functions, click the Totals button in the Show/Hide group on the Query Tools Design tab. Access adds a *Total* row to the query design grid with a drop-down list of functions. Access also inserts the words *Group By* in the field in the *Total* row. Click the option box arrow and then click an aggregate function at the drop-down list. In Activity 2a, Step 1, you will create a query in Design view and use aggregate functions to find the total of all sales, average sales, maximum and minimum sales, and total number of sales. The completed query is shown in Figure 3.5. Access automatically determines the column heading names.

Figure 3.5 Query Results for Activity 2a, Step 1

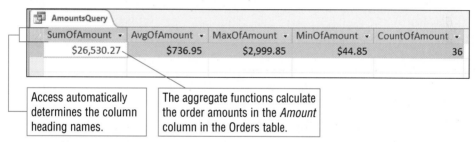

Access automatically determines the column heading names.

The aggregate functions calculate the order amounts in the *Amount* column in the Orders table.

Activity 2a Using Aggregate Functions

Part 1 of 6

1. Open **3-PacTrek** and then create a query with aggregate functions that determines total, average, maximum, and minimum order amounts, as well as the total number of orders, by completing the following steps:
   a. Click the Create tab and then click the Query Design button.
   b. At the Show Table dialog box, make sure *Orders* is selected in the list box, click the Add button, and then click the Close button.
   c. Insert the *OrderAmount* field in the first, second, third, fourth, and fifth fields in the *Field* row. (You may need to scroll down the Orders table field list box to see the *OrderAmount* field.)

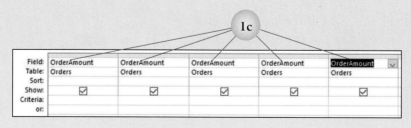

d. Click the Totals button in the Show/Hide group on the Query Tools Design tab. (This adds a *Total* row to the query design grid between *Table* and *Sort* with the default option of *Group By*.)

e. Specify a Sum function for the first field in the *Total* row by completing the following steps:
   1) Click in the first field in the *Total* row.
   2) Click the option box arrow in the field.
   3) Click *Sum* at the drop-down list.

f. Complete steps similar to those in Step 1e to insert *Avg* in the second field in the *Total* row.

g. Complete steps similar to those in Step 1e to insert *Max* in the third field in the *Total* row.

h. Complete steps similar to those in Step 1e to insert *Min* in the fourth field in the *Total* row.

i. Complete steps similar to those in Step 1e to insert *Count* in the fifth field in the *Total* row.

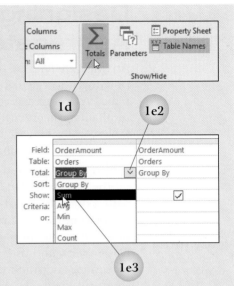

1d   1e2

1e3

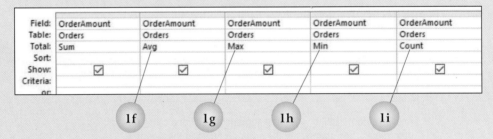

1f   1g   1h   1i

j. Click the Run button. (Notice the headings that Access assigns to the columns.)

k. Adjust the widths of the columns.

l. Save the query, typing Orders_Statistics as the query name.

m. Print and then close the query.

2. Close **3-PacTrek**.

3. Open **3-CopperState**.

4. Create a query with aggregate functions that determines total, average, maximum, and minimum claim amounts by completing the following steps:

   a. Click the Create tab and then click the Query Design button.

   b. At the Show Table dialog box, double-click *Claims*.

   c. Click the Close button to close the Show Table dialog box.

   d. Insert the *AmountOfClaim* field in the first, second, third, and fourth fields in the *Field* row.

   e. Click the Totals button in the Show/Hide group.

   f. Click in the first field in the *Total* row, click the option box arrow in the field, and then click *Sum* at the drop-down list.

   g. Click in the second field in the *Total* row, click the option box arrow, and then click *Avg* at the drop-down list.

4f

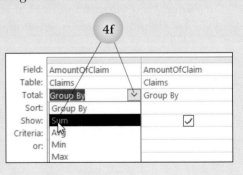

h. Click in the third field in the *Total* row, click the option box arrow, and then click *Max* at the drop-down list.

i. Click in the fourth field in the *Total* row, click the option box arrow, and then click *Min* at the drop-down list.

j. Click the Run button. (Notice the headings that Access chooses for the columns.)

k. Automatically adjust the widths of the columns.

l. Save the query, typing Claims_Amounts as the query name.

m. Print the query in landscape orientation and then close the query.

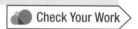 Check Your Work

Use the *Group By* field in the *Total* row to add a field to the query that groups records for statistical calculations. For example, to calculate the total of all orders for a specific supplier, add the *SupplierID* field to the query design grid with the *Total* row set to *Group By*. In Activity 2b, Step 1, you will create a query in Design view and use aggregate functions to find the total of all order amounts and the average order amounts grouped by supplier number.

Activity 2b Using Aggregate Functions and Grouping Records

Part 2 of 6

1. With **3-CopperState** open, determine the sum and average of client claims by completing the following steps:
   a. Click the Create tab and then click the Query Design button.
   b. At the Show Table dialog box, double-click *Clients* in the list box.
   c. Double-click *Claims* in the list box and then click the Close button.
   d. Insert the *ClientID* field from the Clients table field list box to the first field in the *Field* row.
   e. Insert the *AmountOfClaim* field from the Claims table field list box to the second field in the *Field* row.
   f. Insert the *AmountOfClaim* field from the Claims table field list box to the third field in the *Field* row.
   g. Click the Totals button in the Show/ Hide group.
   h. Click in the second field in the *Total* row, click the option box arrow, and then click *Sum* at the drop-down list.
   i. Click in the third field in the *Total* row, click the option box arrow, and then click *Avg* at the drop-down list.
   j. Make sure *Group By* displays in the first field in the *Total* row.
   k. Click the Run button.
   l. Adjust column widths.
   m. Save the query, typing Claims_AmountsSumAvg as the query name.
   n. Print and then close the query.

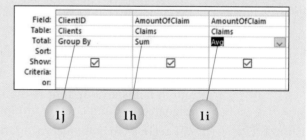

2. Close **3-CopperState**.
3. Open **3-PacTrek**.
4. Determine the total and average order amounts for each supplier by completing the following steps:
   a. Click the Create tab and then click the Query Design button.

b. At the Show Table dialog box, make sure *Orders* is selected in the list box and then click the Add button.

c. Click *Suppliers* in the list box, click the Add button, and then click the Close button.

d. Insert the *OrderAmount* field from the Orders table field list box to the first field in the *Field* row. (You may need to scroll down the Orders table field list box to see the *OrderAmount* field.)

e. Insert the *OrderAmount* field from the Orders table field list box to the second field in the *Field* row.

f. Insert the *SupplierID* field from the Suppliers table field list box to the third field in the *Field* row.

g. Insert the *SupplierName* field from the Suppliers table field list box to the fourth field in the *Field* row.

h. Click the Totals button.

i. Click in the first field in the *Total* row, click the option box arrow, and then click *Sum* at the drop-down list.

j. Click in the second field in the *Total* row, click the option box arrow, and then click *Avg* at the drop-down list.

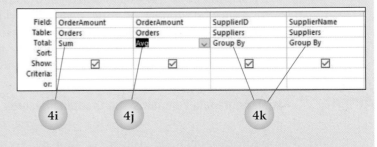

k. Make sure *Group By* displays in the third and fourth fields in the *Total* row.

l. Click the Run button.

m. Adjust column widths.

n. Save the query, typing Suppliers_Amounts as the query name.

o. Print and then close the query.

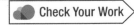
Check Your Work

Tutorial

Creating a Crosstab Query

Creating a Crosstab Query

A crosstab query calculates aggregate functions, such as Sum and Avg, in which field values are grouped by two fields. A wizard is included that provides the steps to create the query. The first field selected displays one row in the query results datasheet for each group. The second field selected displays one column in the query results datasheet for each group. A third field is specified that is the numeric field to be summarized. The cell at the intersection of each row and column holds a value that is the result of the specified aggregate function for the designated row and column group.

Quick Steps

Create Crosstab Query
1. Click Create tab.
2. Click Query Wizard button.
3. Double-click *Crosstab Query Wizard*.
4. Complete wizard steps.

Create a crosstab query from fields in one table. To include fields from more than one table, first create a query containing the fields, and then create the crosstab query. For example, in Activity 2c, Step 2, you will create a new query that contains fields from each of the three tables in the Pacific Trek database. Using this query, you will use the Crosstab Query Wizard to create a query that summarizes the order amounts by supplier name and product ordered. The results of that crosstab query are shown in Figure 3.6. The first column displays the supplier names, the second column displays the total amounts for each supplier, and the remaining columns display the amounts by suppliers for specific items.

Figure 3.6 Crosstab Query Results for Activity 2c, Step 2

Order amounts are grouped by supplier name and individual product.

SupplierName	Total Of Orde	Binoculars, 8	Cascade R4 je	Cascade R4 je	Cascade R4 je	Cascade R4 je	Deluxe map c	Eight-piece st
Bayside Supplies	$224.00							$99.75
Cascade Gear	$3,769.00		$1,285.00	$1,285.00	$599.50	$599.50		
Emerald City Products	$2,145.00	$2,145.00						
Fraser Valley Products	$3,892.75							
Freedom Corporation	$1,286.65							
Hopewell, Inc.	$348.60							
KL Distributions	$4,288.35							
Langley Corporation	$593.25							
Macadam, Inc.	$175.70						$129.75	
Manning, Inc.	$4,282.25							
Sound Supplies	$5,524.72							

Activity 2c Creating Crosstab Queries

Part 3 of 6

1. With **3-PacTrek** open, create a query containing fields from the three tables by completing the following steps:
   a. Click the Create tab and then click the Query Design button.
   b. At the Show Table dialog box with *Orders* selected in the list box, click the Add button.
   c. Double-click *Products* in the list box.
   d. Double-click *Suppliers* in the list box and then click the Close button.
   e. Insert the following fields to the specified fields in the *Field* row:
      1) From the Products table field list box, insert the *ProductID* field in the first field in the *Field* row.
      2) From the Products table field list box, insert the *Product* field in the second field in the *Field* row.
      3) From the Orders table field list box, insert the *UnitsOrdered* field in the third field in the *Field* row.
      4) From the Orders table field list box, insert the *OrderAmount* field in the fourth field in the *Field* row.
      5) From the Suppliers table field list box, insert the *SupplierName* field in the fifth field in the *Field* row.
      6) From the Orders table field list box, insert the *OrderDate* field in the sixth field in the *Field* row.

1e

Field:	ProductID	Product	UnitsOrdered	OrderAmount	SupplierName	OrderDate	
Table:	Products	Products	Orders	Orders	Suppliers	Orders	
Sort:							
Show:	☑	☑	☑	☑	☑	☑	
Criteria:							
or:							

   f. Click the Run button to run the query.
   g. Save the query, typing Items_Ordered as the query name.
   h. Close the query.

2. Create a crosstab query that summarizes the orders by supplier name and by product ordered by completing the following steps:

a. Click the Create tab and then click the Query Wizard button.

b. At the New Query dialog box, double-click *Crosstab Query Wizard* in the list box.

c. At the first Crosstab Query Wizard dialog box, click the *Queries* option in the *View* section and then click *Query: Items_Ordered* in the list box.

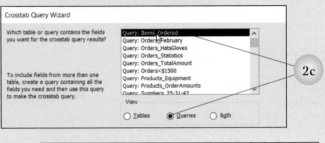

d. Click the Next button.

e. At the second Crosstab Query Wizard dialog box, click *SupplierName* in the *Available Fields* list box and then click the One Field button. (This inserts *SupplierName* in the *Selected Fields* list box and specifies that you want the *SupplierName* field for the row headings.)

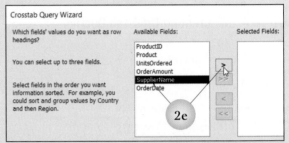

f. Click the Next button.

g. At the third Crosstab Query Wizard dialog box, click *Product* in the list box. (This specifies that you want the *Product* field for the column headings.)

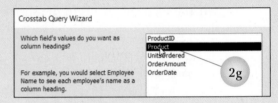

h. Click the Next button.

i. At the fourth Crosstab Query Wizard dialog box, click *OrderAmount* in the *Fields* list box and then click *Sum* in the *Functions* list box.

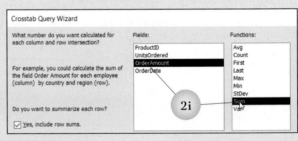

j. Click the Next button.

k. At the fifth Crosstab Query Wizard dialog box, select the current text in the *What do you want to name your query?* text box and then type Orders_Crosstab.

l. Click the Finish button.

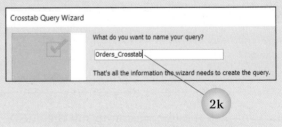

3. Display the query in Print Preview, change to landscape orientation, change the left margin to 0.4 inch and the right margin to 0.5 inch, and then print the query. (The query will print on four pages.)

4. Close the query.

5. Close **3-PacTrek**.

6. Open **3-CopperState**.

7. Create a crosstab query from fields in one table that summarizes clients' claims by completing the following steps:

   a. Click the Create tab and then click the Query Wizard button.

   b. At the New Query dialog box, double-click *Crosstab Query Wizard* in the list box.

   c. At the first Crosstab Query Wizard dialog box, click *Table: Claims* in the list box.

   d. Click the Next button.

   e. At the second Crosstab Query Wizard dialog box, click the One Field button. (This inserts the *ClaimID* field in the *Selected Fields* list box.)

   f. Click the Next button.

   g. At the third Crosstab Query Wizard dialog, make sure *ClientID* is selected in the list box and then click the Next button.

   h. At the fourth Crosstab Query Wizard dialog box, click *AmountOfClaim* in the *Fields* list box and click *Sum* in the *Functions* list box.

   i. Click the Next button.

   j. At the fifth Crosstab Query Wizard dialog box, click the Finish button.

8. Change to landscape orientation and then print the query. The query will print on two pages.

9. Close the query.

10. Close **3-CopperState**.

 Check Your Work

 Tutorial

Creating a Find
Duplicates Query

Quick Steps

**Create Find
Duplicates Query**

1. Click Create tab.
2. Click Query Wizard button.
3. Double-click *Find Duplicates Query Wizard*.
4. Complete wizard steps.

Creating a Find Duplicates Query

Use a find duplicates query to search a specified table or query for duplicate field values within a designated field or fields. Create this type of query, for example, if a record (such as a product record) may have been entered two times inadvertently (perhaps under two different product numbers). A find duplicates query has many applications. Here are a few other examples of how to use a find duplicates query:

- In an orders table, find records with the same customer number to identify loyal customers.

- In a customers table, find records with the same last name and mailing address so only one mailing will be sent to a household to save on printing and postage costs.

- In an employee expenses table, find records with the same employee number to determine which employee is submitting the most claims.

   Access provides the Find Duplicates Query Wizard to build the query based on the selections made in a series of dialog boxes. To use this wizard, open a database, click the Create tab, and then click the Query Wizard button. At the New Query dialog box, double-click *Find Duplicates Query Wizard* in the list box and then complete the steps provided by the wizard.

   In Activity 2d, you will assume that you have been asked to update the address for a supplier in the Pacific Trek database. Instead of updating the address, you create a new record. You will then use the Find Duplicates Query Wizard to find duplicate field values in the Suppliers table.

1. Open **3-PacTrek** and then open the Suppliers table.
2. Add the following record to the table:

   SupplierID   29
   SupplierName   Langley Corporation
   StreetAddress   1248 Larson Avenue
   City   Burnaby
   Prov/State   BC
   PostalCode   V5V 9K2
   Email   lc@ppi-edu.net
   Telephone   (604) 555-1200

3. Close the Suppliers table.
4. Use the Find Duplicates Query Wizard to find any duplicate supplier names by completing the following steps:
   a. Click the Create tab and then click the Query Wizard button.
   b. At the New Query dialog box, double-click *Find Duplicates Query Wizard*.
   c. At the first wizard dialog box, click *Table: Suppliers* in the list box.
   d. Click the Next button.
   e. At the second wizard dialog box, click *SupplierName* in the *Available fields* list box and then click the One Field button. (This moves the *SupplierName* field to the *Duplicate-value fields* list box.)
   f. Click the Next button.
   g. At the third wizard dialog box, click the All Fields button (the button containing the two greater-than symbols, >>). This moves all the fields to the *Additional query fields* list box. You are doing this because if you find a duplicate supplier name, you want to view all the fields to determine which record is accurate.
   h. Click the Next button.
   i. At the fourth (and last) wizard dialog box, type Suppliers_Duplicate in the *What do you want to name your query?* text box.

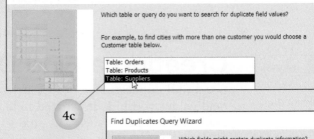

4c

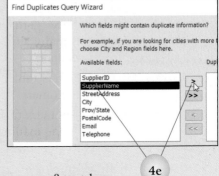

4e

4i

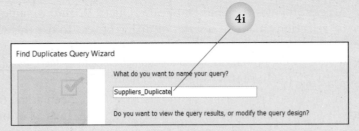

   j. Click the Finish button.
   k. Change to landscape orientation and then print the query.

5. As you look at the query results, you realize that an inaccurate record was entered for the Langley Corporation, so you decide to delete one of the records. To do this, complete the following steps:

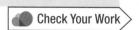

a. Select the row with a supplier ID of *29*.
b. Click the Home tab and then click the Delete button in the Records group.
c. At the message asking you to confirm the deletion, click Yes.
d. Close the query.

6. Change the street address for Langley Corporation by completing the following steps:
a. Open the Suppliers table in Datasheet view.
b. Change the address for Langley Corporation from *805 First Avenue* to *1248 Larson Avenue*. Do not make any changes to the other fields.
c. Close the Suppliers table.

> **Check Your Work**

The Find Duplicates Query can be used to find accidental duplications in database records, as shown in Activity 2d. This query is also useful for finding other types of information. In Activity 2e, the duplicate entries found by the wizard are not mistaken entries but rather indicate repeated use of the same supplier. This information could be used to negotiate for better prices or to ask for discounts.

Activity 2e Finding Duplicate Orders

Part 5 of 6

1. With **3-PacTrek** open, create a query with the following fields (in the order shown) from the specified tables:

   SupplierID   Suppliers table
   SupplierName   Suppliers table
   ProductID   Orders table
   Product   Products table

2. Run the query.
3. Save the query with the name *Suppliers_Orders* and then close the query.
4. Use the Find Duplicates Query Wizard to find the suppliers you order from the most by completing the following steps:
a. Click the Create tab and then click the Query Wizard button.
b. At the New Query dialog box, double-click *Find Duplicates Query Wizard*.
c. At the first wizard dialog box, click the *Queries* option in the *View* section and then click *Query: Suppliers_Orders*. (You may need to scroll down the list to display this query.)

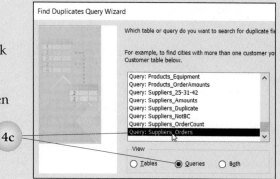

d. Click the Next button.
e. At the second wizard dialog box, click *SupplierName* in the *Available fields* list box and then click the One Field button.
f. Click the Next button.
g. At the third wizard dialog box, click the Next button.
h. At the fourth (and last) wizard dialog box, type Suppliers_OrderCount in the *What do you want to name your query?* text box.

4h

i. Click the Finish button.
j. Adjust the widths of the columns to fit the longest entries.
k. Save and then print the query.
5. Close the query.

> Check Your Work

 Tutorial

Creating a Find
Unmatched Query

Creating a Find Unmatched Query

Create a find unmatched query to compare two tables and produce a list of the records in one table that have no matching record in the other table. This type of query can be useful, for instance, to produce a list of customers who have never placed orders or a list of invoices that have no record of payment. Access provides the Find Unmatched Query Wizard to build the query.

Quick Steps

**Create Find
Unmatched Query**
1. Click Create tab.
2. Click Query Wizard button.
3. Double-click *Find Unmatched Query Wizard.*
4. Complete wizard steps.

In Activity 2f, you will use the Find Unmatched Query Wizard to find all of the products that have no units on order. This information is helpful in identifying which products are not selling and might need to be discontinued or returned. To use the Find Unmatched Query Wizard, click the Create tab and then click the Query Wizard button in the Queries group. At the New Query dialog box, double-click *Find Unmatched Query Wizard* in the list box and then follow the wizard steps.

Activity 2f **Creating a Find Unmatched Query**   Part 6 of 6

1. With **3-PacTrek** open, use the Find Unmatched Query Wizard to find all products that do not have units on order by completing the following steps:
   a. Click the Create tab and then click the Query Wizard button.
   b. At the New Query dialog box, double-click *Find Unmatched Query Wizard.*
   c. At the first wizard dialog box, click *Table: Products* in the list box. (This is the table containing the fields you want to see in the query results.)
   d. Click the Next button.
   e. At the second wizard dialog box, make sure *Table: Orders* is selected in the list box. (This is the table containing the related records.)
   f. Click the Next button.

1c

g. At the third wizard dialog box, make sure *ProductID* is selected in both the *Fields in 'Products'* list box and in the *Fields in 'Orders'* list box.

h. Click the Next button.

i. At the fourth wizard dialog box, click the All Fields button to move all of the fields from the *Available fields* list box to the *Selected fields* list box.

j. Click the Next button.

k. At the fifth wizard dialog box, click the Finish button. (Let the wizard determine the query name: *Products Without Matching Orders*.)

2. Print the query in landscape orientation and then close the query.

3. Close **3-PacTrek**.

Check Your Work

Chapter Summary

- One of the most important uses of a database is to select the information needed to answer questions and make decisions. Data can be extracted from an Access database by performing a query, which can be accomplished by designing a query or using a query wizard.

- Designing a query consists of identifying the table, the field or fields from which the data will be drawn, and the criterion or criteria for selecting the data.

- In designing a query, type the criterion (or criteria) statement for extracting the specific data. Access inserts any necessary symbols in the criterion when the Enter key is pressed.

- In a criterion, quotation marks surround field values and pound symbols (#) surround dates. Use the asterisk (*) as a wildcard character.

- A query can be performed on fields within one table or on fields from related tables.

- When designing a query, the sort order of a field or fields can be specified.

- An existing query can be modified and used for a new purpose rather than creating a new one from scratch.

- Enter a criterion in the *or* row in the query design grid to instruct Access to display records that match any of the criteria.

- Multiple criteria entered in the *Criteria* row in the query design grid become an *And* statement, where each criterion must be met for Access to select the record.

- The Simple Query Wizard provides the steps for preparing a query.

- A calculated field can be inserted in a field in the *Field* row when designing a query. If the results of the calculation should display as currency, apply numeric formatting and decimal places using the Format property box at the Property Sheet task pane. Display the task pane by clicking the Property Sheet button in the Show/Hide group on the Query Tools Design tab.

- Include an aggregate function (such as Sum, Avg, Min, Max, or Count) to calculate statistics from numeric field values. Click the Totals button in the Show/Hide group on the Query Tools Design tab to display the aggregate function list.
- Use a *Group By* field in the *Total* row to add a field to a query for grouping records for statistical calculations.
- Create a crosstab query to calculate aggregate functions (such as Sum and Avg), in which fields are grouped by two. Create a crosstab query from fields in one table. To include fields from more than one table, create a query first and then create the crosstab query.
- Use a find duplicates query to search a specified table for duplicate field values within a designated field or fields.
- Create a find unmatched query to compare two tables and produce a list of the records in one table that have no matching records in the other related table.

Commands Review

FEATURE	RIBBON TAB, GROUP	BUTTON, OPTION
add *Total* row to query design	Query Tools Design, Show/Hide	Σ
Crosstab Query Wizard	Create, Queries	, Crosstab Query Wizard
Expression Builder dialog box	Query Tools Design, Query Setup	
Find Duplicates Query Wizard	Create, Queries	, Find Duplicates Query Wizard
Find Unmatched Query Wizard	Create, Queries	, Find Unmatched Query Wizard
New Query dialog box	Create, Queries	
Property Sheet task pane	Query Tools Design, Show/Hide	
query results	Query Tools Design, Results	!
query window	Create, Queries	
Simple Query Wizard	Create, Queries	, Simple Query Wizard

Access®

Creating and Modifying Tables in Design View

Performance Objectives

Upon successful completion of Chapter 4, you will be able to:

1 Create a table in Design view

2 Assign a default value for a field

3 Use the Input Mask Wizard and the Lookup Wizard

4 Validate field entries

5 Insert, move, and delete fields in Design view

6 Insert a *Total* row

7 Sort records in a table

8 Print selected records in a table

9 Apply text formatting

10 Complete a spelling check

11 Find and replace data in records in a table

12 Use the Help and Tell Me features

In Chapter 1, you learned how to create a table in Datasheet view. A table can also be created in Design view, where the table's structure and properties are established before entering data. In this chapter, you will learn how to create a table in Design view and use the Input Mask Wizard and Lookup Wizards; insert, move, and delete fields in Design view; sort records; check spelling in a table; find and replace data; apply text formatting to a table; and use the Help and Tell Me features.

 Data Files

Before beginning chapter work, copy the AL1C4 folder to your storage medium and then make AL1C4 the active folder.

The online course includes additional training and assessment resources.

Tutorial

Creating a Table in Design View

Tutorial

Modifying Field Properties in Design View

 Table

 View

Quick Steps

Create Table in Design View
1. Open database.
2. Click Create tab.
3. Click Table button.
4. Click View button.
5. Type name for table.
6. Press Enter key or click OK.
7. Type field names, specify data types, and include descriptions.
8. Click Save button.

Creating a Table in Design View

In Datasheet view, a table is created by assigning each column a data type and typing the field name. Once the columns are defined, the data is entered into records. A table can also be created in Design view, where field properties can be set before entering data.

To display a table in Design view, open the database, click the Create tab, and then click the Table button. This opens a new blank table in Datasheet view. Display the table in Design view by clicking the View button at the left side of the Table Tools Fields tab in the Views group. Click the View button in a new table and Access displays the Save As dialog box. Type a name for the table and then press the Enter key or click OK. The Properties table in Design view in the SunProperties database is shown in Figure 4.1.

In Design view, each row in the top section of the work area represents one field in the table and is used to define the field name, the field data type, and a description. The *Field Properties* section in the lower half of the work area displays the properties for the active field. The properties vary depending on the active field. In the lower right corner of Design view, Help information is available for the active field or property in the Design window. In Figure 4.1, the *PropID* field name is active in Design view, so information on field names is available in the Help area.

Define each field in the table in the rows in the top section of Design view. When a table is created in Design view, Access automatically assigns the first field the name *ID* and assigns the AutoNumber data type. Leave this field name as *ID* or type a new name and accept the AutoNumber data type or change to a different data type. To create a new field in the table, click in a field in the *Field Name* column, type the field name, and then press the Tab key or Enter key. This makes the field in the *Data Type* column active. Click the option box arrow in the field in the *Data Type* column and then click the data type at the drop-down list. In Chapter 1, you created tables in Datasheet view and assigned data types of Short Text, Date/Time, Currency, and Yes/No. The drop-down list in the *Data Type* column includes these data types plus additional types, as described in Table 4.1.

Click the specific data type at the drop-down list and then press the Tab key and the field in the *Description* column becomes active. In the field, type a description that provides useful information to someone entering data in the table. For example, consider identifying the field's purpose or contents or providing instructional information for data entry. The description typed in the field in the *Description* column displays in the Status bar when the table's field is active in the table in Datasheet view.

Figure 4.1 Properties Table in Design View

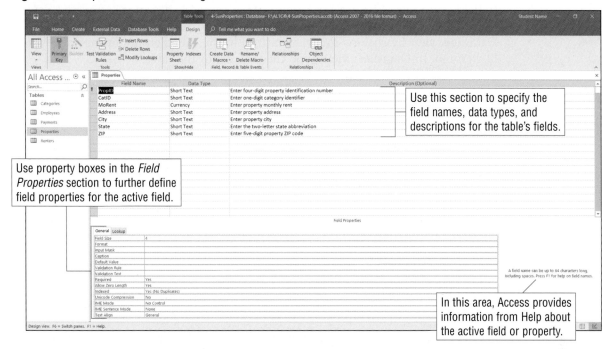

Use this section to specify the field names, data types, and descriptions for the table's fields.

Use property boxes in the *Field Properties* section to further define field properties for the active field.

In this area, Access provides information from Help about the active field or property.

Table 4.1 Data Types

Data Type	Description
Short Text	Used for alphanumeric data up to 255 characters in length—for example, a name, address, or value (such as a telephone number or social security number) that is used as an identifier and not for calculating.
Long Text	Used for alphanumeric data up to 65,535 characters in length.
Number	Used for positive and negative values that can be used in calculations. Do not use for values that will calculate monetary amounts (see Currency).
Date/Time	Used to ensure dates and times are entered and sorted properly.
Currency	Used for values that involve money. Access will not round off during calculations.
AutoNumber	Used to automatically number records sequentially (increments of 1); each new record is numbered as it is typed.
Yes/No	Used for values of *Yes* or *No*, *True* or *False*, or *On* or *Off*.
OLE Object	Used to embed or link objects created in other Office applications.
Hyperlink	Used to store a hyperlink, such as a URL.
Attachment	Used to add file attachments to a record such as a Word document or Excel workbook.
Calculated	Used to display the Expression Builder dialog box, where an expression is entered to calculate the value of the calculated column.
Lookup Wizard	Used to enter data in the field from another existing table or to display a list of values in a drop-down list from which the user chooses.

Save

When creating the table, continue typing field names, assigning data types to fields, and typing field descriptions. When the table design is completed, save the table by clicking the Save button on the Quick Access Toolbar. Return to Datasheet view by clicking the View button in the Views group on the Table Tools Design tab. In Datasheet view, type the records for the table.

Activity 1a Creating a Renters Table in Design View

1. Open Access and then open **4-SunProperties** from your AL1C4 folder.
2. Click the Enable Content button in the message bar.
3. View the Properties table in Design view by completing the following steps:
   a. Open the Properties table.
   b. Click the View button in the Views group on the Home tab.

   c. Click each field name and then look at the information in the *Field Properties* section.
   d. Click in various fields or properties in the work area and then read the information in the Help area in the lower right corner of Design view.
   e. Click the View button to return the table to Datasheet view.
   f. Close the Properties table.
4. Create a new table in Design view, as shown in Figure 4.2, by completing the following steps:
   a. Click the Create tab and then click the Table button in the Tables group.
   b. Click the View button in the Views group on the Table Tools Fields tab.
   c. At the Save As dialog box, type Renters and then press the Enter key.
   d. Type RenterID in the first field in the *Field Name* column and then press the Tab key.
   e. Change to the Short Text data type by clicking the option box arrow in the field in the *Data Type* column and then clicking *Short Text* at the drop-down list.
   f. Change the field size from the default of 255 characters to 3 characters by selecting *255* in the *Field Size* property box in the *Field Properties* section and then typing *3*.

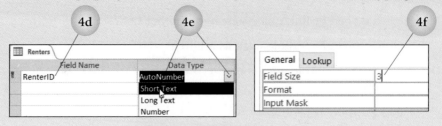

g. Click in the *Description* column for the *RenterID* field, type Enter three-digit renter identification number, and then press the Tab key.

Renters			
Field Name	Data Type		
RenterID	Short Text	Enter three-digit renter identification number	

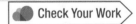
4g

h. Type RenFirstName in the second field in the *Field Name* column and then press the Tab key.
i. Select *255* in the *Field Size* property box in the *Field Properties* section and then type 20.
j. Click in the *Description* column for the *RenFirstName* field, type Enter renter's first name, and then press the Tab key.
k. Type RenLastName in the third field in the *Field Name* column and then press the Tab key.
l. Change the field size to 30 characters (at the *Field Size* property box).
m. Click in the *Description* column for the *RenLastName* field, type Enter renter's last name, and then press the Tab key.
n. Enter the remaining field names, data types, and descriptions as shown in Figure 4.2. (Change the field size to 4 characters for the *PropID* field, 5 characters for the *EmpID* field, and 3 characters for the *RenCreditScore* field.)
o. After all of the fields are entered, click the Save button on the Quick Access Toolbar.
p. Make sure the *RenterID* field is identified as the primary key field. (A key icon displays in the *RenterID* field selector bar.)
q. Click the View button to return the table to Datasheet view.
5. Enter the records in the Renters table as shown in Figure 4.3.
6. After all of the records are entered, adjust the column widths.
7. Save, print in landscape orientation, and then close the Renters table.

Check Your Work

Figure 4.2 Activity 1a Renters Table in Design View

Renters		
Field Name	Data Type	
RenterID	Short Text	Enter three-digit renter identification number
RenFirstName	Short Text	Enter renter's first name
RenLastName	Short Text	Enter renter's last name
PropID	Short Text	Enter four-digit property identification number
EmpID	Short Text	Enter five-digit employee identification number
RenCreditScore	Short Text	Enter renter's current credit score
LeaseBegDate	Date/Time	Enter beginning date of lease
LeaseEndDate	Date/Time	Enter ending date of lease

Figure 4.3 Activity 1a Renters Table in Datasheet View

RenterID	RenFirstName	RenLastName	PropID	EmpID	RenCreditScore	LeaseBegDate	LeaseEndDate	*Clic*
110	Greg	Hamilton	1029	04-14	624	1/1/2021	12/31/2021	
111	Julia	Perez	1013	07-20	711	1/1/2021	12/31/2021	
115	Dana	Rozinski	1026	02-59	538	2/1/2021	1/31/2022	
117	Miguel	Villegas	1007	07-20	695	2/1/2021	1/31/2022	
118	Mason	Ahn	1004	07-23	538	3/1/2021	2/28/2022	
119	Michelle	Bertram	1001	03-23	621	3/1/2021	2/28/2022	
121	Travis	Jorgenson	1010	04-14	590	3/1/2021	2/28/2022	
123	Richard	Terrell	1014	07-20	687	3/1/2021	2/28/2022	
125	Rose	Wagoner	1015	07-23	734	4/1/2021	3/31/2022	
127	William	Young	1023	05-31	478	4/1/2021	3/31/2022	
129	Susan	Lowrey	1002	04-14	634	4/1/2021	3/31/2022	
130	Ross	Molaski	1027	03-23	588	5/1/2021	4/30/2022	
131	Danielle	Rubio	1020	07-20	722	5/1/2021	4/30/2022	
133	Katie	Smith	1018	07-23	596	5/1/2021	4/30/2022	
134	Carl	Weston	1009	03-23	655	6/1/2021	5/31/2022	
135	Marty	Lobdell	1006	04-14	510	6/1/2021	5/31/2022	
136	Nadine	Paschal	1022	05-31	702	6/1/2021	5/31/2022	

Assigning a Default Value

Chapter 1 covered how to specify a default value for a field in a table in Datasheet view using the Default Value button in the Properties group on the Table Tools Fields tab. In addition to this method, a default value for a field can be specified in Design view with the *Default Value* property box in the *Field Properties* section. Click in the *Default Value* property box and then type the field value.

In Activity 1b, a health insurance field will be created with a Yes/No data type. Since most of the agents of Sun Properties have signed up for health insurance benefits, the default value for the field will be set to *Yes*. If a new field containing a default value is added to an existing table, the existing records do not reflect the default value. Only new records entered in the table reflect the default value.

Creating an Input Mask

Tutorial

Creating an Input Mask

Hint An input mask is a set of characters that control what can and cannot be entered in a field.

To maintain consistency and control data entered in a field, consider using the *Input Mask* property box to set a pattern for how data is entered in the field. For example, a pattern can be set for a zip code field that requires that the nine-digit zip code is entered in the field rather than the five-digit zip code. Or, a pattern can be set for a telephone field that requires that the three-digit area code is entered with the telephone number. Access includes an Input Mask Wizard that provides the steps for creating an input mask. The Input Mask is available for fields with a data type of Short Text or Date/Time.

 Build

Use the Input Mask Wizard when assigning a data type to a field. In Design view, click in the *Input Mask* property box in the *Field Properties* section and then run the Input Mask Wizard by clicking the Build button (contains three black dots) that appears at the right side of the *Input Mask* property box. This displays the first Input Mask Wizard dialog box, as shown in Figure 4.4. In the *Input Mask* list box, choose which input mask the data should look like and then click the Next button. At the second Input Mask Wizard dialog box, as shown in

Quick Steps

Use Input Mask Wizard

1. Open table in Design view.
2. Type text in *Field Name* column.
3. Press Tab key.
4. Change data type to *Short Text* or *Date/Time*.
5. Click Save button.
6. Click in *Input Mask* property box.
7. Click Build button.
8. Complete wizard steps.

Figure 4.5, specify the appearance of the input mask and the placeholder character and then click the Next button. At the third Input Mask Wizard dialog box, specify if the data should be stored with or without the symbol in the mask and then click the Next button. At the fourth dialog box, click the Finish button.

The input mask controls how data is entered into a field. In some situations, such as establishing an input mask to enter the date in the Medium Date format, what is entered will not match what Access displays. An input mask with the Medium Date data type format will require that the date be entered as *12-Sep-21* but, after the date is entered, Access will change the display to *09/12/2021*. Use the *Format* property box to match how Access displays the date with the input mask. Click in the *Format* property box, click the option box arrow in the property box and then click *Medium Date* at the drop-down list.

Figure 4.4 First Input Mask Wizard Dialog Box

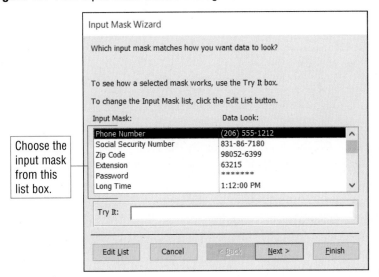

Choose the input mask from this list box.

Figure 4.5 Second Input Mask Wizard Dialog Box

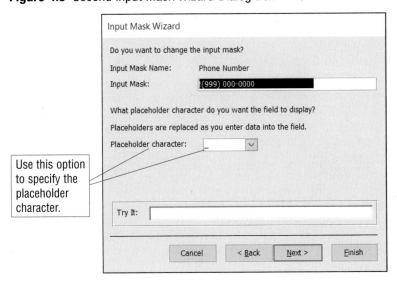

Use this option to specify the placeholder character.

1. With **4-SunProperties** open, create the Employees table in Design view as shown in Figure 4.6. Begin by clicking the Create tab and then clicking the Table button.
2. Click the View button to switch to Design view.
3. At the Save As dialog box, type Employees and then press the Enter key.
4. Type EmpID in the first field in the *Field Name* column and then press the Tab key.
5. Change to the Short Text data type by clicking the option box arrow in the *Data Type* column and then clicking *Short Text* at the drop-down list.
6. Change the field size from the default of 255 characters to 5 characters by selecting *255* in the *Field Size* property box in the *Field Properties* section and then typing 5.
7. Click in the *Description* column for the *EmpID* field, type Enter five-digit employee identification number, and then press the Tab key.

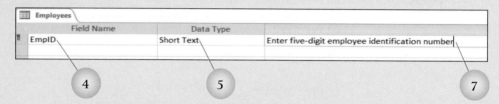

8. Type EmpFirstName in the second field in the *Field Name* column and then press the Tab key.
9. Select *255* in the *Field Size* property box in the *Field Properties* section and then type 20.
10. Click in the *Description* column for the *EmpFirstName* field, type Enter employee's first name, and then press the Tab key.
11. Complete steps similar to those in Steps 8 through 10 to create the *EmpLastName*, *EmpAddress*, and *EmpCity* fields as shown in Figure 4.6 (on page 111). Change the field size for the *EmpLastName* field and *EmpAddress* field to 30 characters and change the *EmpCity* field to 20 characters.
12. Create the *EmpState* field with a default value of *CA*, since all employees live in California, by completing the following steps:
    a. Type EmpState in the field below the *EmpCity* field in the *Field Name* column and then press the Tab key.
    b. Click in the *Default Value* property box in the *Field Properties* section and then type CA.
    c. Click in the *Description* column for the *EmpState* field, type CA automatically entered as state, and then press the Tab key.
13. Type EmpZIP and then press the Tab key.
14. Select *255* in the *Field Size* property box in the *Field Properties* section and then type 5.
15. Click in the *Description* column for the *EmpZIP* field, type Enter employee's five-digit ZIP code, and then press the Tab key.
16. Type EmpTelephone and then press the Tab key.
17. Create an input mask for the telephone number by completing the following steps:
    a. Click the Save button on the Quick Access Toolbar to save the table. (You must save the table before using the Input Mask Wizard.)

b. Click in the *Input Mask* property box in the *Field Properties* section.
c. Click the Build button (contains three black dots) at the right side of the *Input Mask* property box.

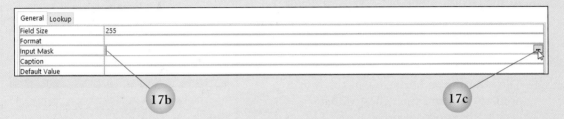

17b   17c

d. At the first Input Mask Wizard dialog box, make sure *Phone Number* is selected in the *Input Mask* list box and then click the Next button.
e. At the second Input Mask Wizard dialog box, click the *Placeholder character* option box arrow and then click # at the drop-down list.

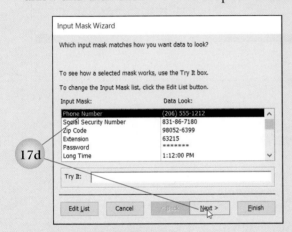

17d

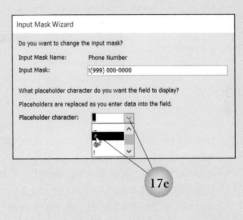

17e

f. Click the Next button.
g. At the third Input Mask Wizard dialog box, click the *With the symbols in the mask, like this* option.
h. Click the Next button.
i. At the fourth Input Mask Wizard dialog box, click the Finish button.

17g

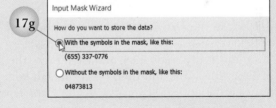

18. Click in the *Description* column for the *EmpTelephone* field, type Enter employee's telephone number, and then press the Tab key.
19. Type HireDate and then press the Tab key.
20. Click the option box arrow in the *Date Type* column and then click *Date/Time*.
21. Create an input mask for the date by completing the following steps:
    a. Click the Save button on the Quick Access Toolbar to save the table.
    b. Click in the *Input Mask* property box in the *Field Properties* section.
    c. Click the Build button (contains three black dots) at the right side of the *Input Mask* property box.

d. At the first Input Mask Wizard dialog box, click *Medium Date* in the *Input Mask* list box and then click the Next button.

e. At the second Input Mask Wizard dialog box, click the Next button.

f. At the third Input Mask Wizard dialog box, click the Finish button.

g. Click in the *Format* property box.

h. Click the option box arrow in the *Format* property box and then click *Medium Date* at the drop-down list.

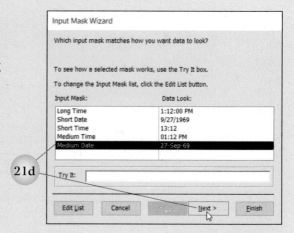

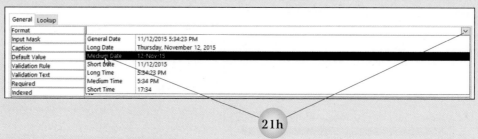

22. Click in the *Description* column for the *HireDate* field, type Enter employee's hire date and then press the Tab key.

23. Type HealthIns and then press the Tab key.

24. Click the option box arrow in the *Data Type* column and then click *Yes/No* at the drop-down list.

25. Click in the *Default Value* property box in the *Field Properties* section, delete the text *No*, and then type Yes.

26. Click in the *Description* column for the *HealthIns* field, type Leave check mark if employee is signed up for health insurance, and then press the Tab key.

27. Type DentalIns and then press the Tab key.

28. Click the option box arrow in the *Data Type* column and then click *Yes/No* at the drop-down list. (The text in the *Default Value* property box will remain as *No*.)

29. Click in the *Description* column for the *DentalIns* field, type Insert a check mark if employee is signed up for dental insurance, and then press the Tab key.

30. After all of the fields are entered, click the Save button on the Quick Access Toolbar.

31. Click the View button to return the table to Datasheet view.

32. Enter the records in the Employees table as shown in Figure 4.7.

33. After all of the records are entered, adjust the widths of the columns in the table.

34. Save, print, and then close the Employees table.

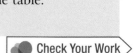

Figure 4.6 Activity 1b Employees Table in Design View

Field Name	Data Type	
EmpID	Short Text	Enter five-digit employee identification number
EmpFirstName	Short Text	Enter employee's first name
EmpLastName	Short Text	Enter employee's last name
EmpAddress	Short Text	Enter employee's address
EmpCity	Short Text	Enter employee's city
EmpState	Short Text	CA automatcially entered as state
EmpZIP	Short Text	Enter employee's five-digit ZIP code
EmpTelephone	Short Text	Enter employee's telephone number
HireDate	Date/Time	Enter employee's hire date
HealthIns	Yes/No	Leave check mark if employee is signed up for health insurance
DentalIns	Yes/No	Insert a check mark if employee is signed up for dental insurance

Figure 4.7 Activity 1b Employees Table in Datasheet View

EmpID	EmpFirstName	EmpLastName	EmpAddress	EmpCity	EmpState	EmpZIP	EmpTelephone	HireDate	HealthIns	DentalIns
02-59	Christina	Solomon	12241 East 51st	Citrus Heights	CA	95611	(916) 555-8844	01-Feb-08	✓	✓
03-23	Douglas	Ricci	903 Mission Road	Roseville	CA	95678	(916) 555-4125	01-Mar-09	✓	☐
03-55	Tatiana	Kasadev	6558 Orchard Drive	Citrus Heights	CA	95610	(916) 555-8534	15-Nov-11	✓	☐
04-14	Brian	West	12232 142nd Avenue East	Citrus Heights	CA	95611	(916) 555-0967	01-Apr-12	✓	✓
04-32	Kathleen	Addison	21229 19th Street	Citrus Heights	CA	95621	(916) 555-3408	01-Feb-13	✓	✓
05-20	Teresa	Villanueva	19453 North 42nd Street	Citrus Heights	CA	95611	(916) 555-2302	15-Jul-14	✓	✓
05-31	Marcia	Griswold	211 Haven Road	North Highlands	CA	95660	(916) 555-1449	01-May-15	☐	☐
06-24	Tiffany	Gentry	12312 North 20th	Roseville	CA	95661	(916) 555-0043	15-Apr-17	✓	✓
06-33	Joanna	Gallegos	6850 York Street	Roseville	CA	95747	(916) 555-7446	01-Jul-18	☐	☐
07-20	Jesse	Scholtz	3412 South 21st Street	Fair Oaks	CA	95628	(916) 555-4204	15-Feb-19	✓	☐
07-23	Eugene	Bond	530 Laurel Road	Orangevale	CA	95662	(916) 555-9412	01-Mar-20	✓	☐
*					CA				✓	☐

Tutorial

Applying a Validation Rule in Design View

Tutorial

Applying a Validation Rule in Datasheet View

Validating Field Entries

Use the *Validation Rule* property box in the *Field Properties* section in Design view to enter a statement containing a conditional test that is checked each time data is entered into a field. If data is entered that fails to satisfy the test, Access rejects the entry and returns an error message. Such errors can be reduced by entering a conditional statement in the *Validation Rule* property box that checks each entry against the acceptable range. Customize the error message that will display if incorrect data is entered in the field by typing that message in the *Validation Text* property box.

A validation rule and validation text can also be applied to a field in Datasheet view by selecting the field and then clicking the Table Tools Fields tab. Click the Validation button in the Field Validation group and then click *Field Validation Rule* at the drop-down list. Type the conditional statement in the Expression Builder dialog box and then click OK. Create validation text by clicking the Validation button and then clicking *Field Validation Message*. Type the message in the Enter Validation Message dialog box and then click OK.

Tutorial

Creating a Lookup Field

Using the Lookup Wizard

Like the Input Mask Wizard, the Lookup Wizard can be used to control data entered in a field. Use the Lookup Wizard to confine data entered into a field to a specific list of items. For example, in Activity 1c, the Lookup Wizard will be used to restrict the new *EmpCategory* field to one of three choices: *Salaried, Hourly,* and *Temporary*. When entering data, clicking in the field displays an option box arrow. Click the option box arrow and then click an option at the drop-down list.

Quick Steps

Use Lookup Wizard
1. Open table in Design view.
2. Type text in *Field Name* column.
3. Press Tab key.
4. Click option box arrow.
5. Click *Lookup Wizard*.
6. Complete wizard steps.

Figure 4.8 First Lookup Wizard Dialog Box

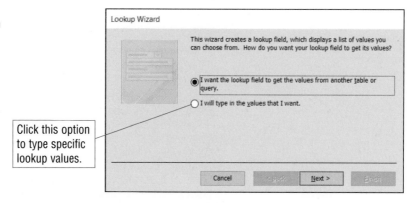

Click this option to type specific lookup values.

Figure 4.9 Second Lookup Wizard Dialog Box

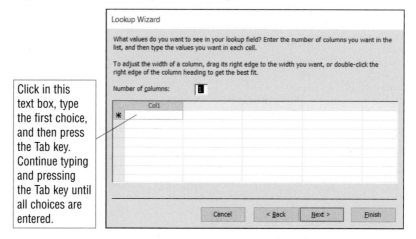

Click in this text box, type the first choice, and then press the Tab key. Continue typing and pressing the Tab key until all choices are entered.

Use the Lookup Wizard when assigning a data type to a field. Click in the field in the *Data Type* column, click the option box arrow, and then click *Lookup Wizard* at the drop-down list. This displays the first Lookup Wizard dialog box, as shown in Figure 4.8. At this dialog box, indicate that specific field choices will be entered by clicking the *I will type in the values that I want* option and then click the Next button. At the second Lookup Wizard dialog box, shown in Figure 4.9, click in the blank text box below *Col1* and then type the first choice. Press the Tab key and then type the second choice. Continue in this manner until all choices have been entered and then click the Next button. At the third Lookup Wizard dialog box, make sure the proper name is in the *What label would you like for your lookup column?* text box and then click the Finish button.

 Tutorial

Managing Fields in Design View

Inserting, Moving, and Deleting Fields in Design View

As shown in Chapter 1, field management tasks such as inserting, moving, and deleting fields can be completed in Datasheet view. These tasks can also be completed in Design view.

To insert a new field in a table in Design view, position the insertion point in a field in the row that will be immediately *below* the new field and then click the Insert Rows button in the Tools group on the Table Tools Design tab. Another option is to position the insertion point in text in the row that will be immediately *below* the new field, click the right mouse button, and then click *Insert Rows* at the shortcut menu. A row in the Design view creates a field in the table.

 Insert Rows

A field in a table can be moved to a different location in Datasheet view or Design view. To move a field in Design view, click the field selector bar at the left side of the row to be moved. With the row selected, position the mouse pointer in the field selector bar at the left side of the selected row, click and hold down the left mouse button, drag the mouse pointer with a gray square attached until a thick black line displays in the desired position, and then release the mouse button.

Delete a field in a table and all data entered in that field is also deleted. When a field is deleted, the deletion cannot be undone with the Undo button. Delete a field only if all data associated with it should be removed from the table. To delete a field in Design view, click in the field selector bar at the left side of the row to be deleted and then click the Delete Rows button in the Tools group. At the confirmation message, click Yes. A row can also be deleted by positioning the mouse pointer in the row to be deleted, clicking the right mouse button, and then clicking *Delete Rows* at the shortcut menu.

Inserting a *Total* Row

A *Total* row can be added to a table in Datasheet view and then used to perform functions such as finding the sum, average, maximum, minimum, count, standard deviation, or variance results in a numeric column. To insert a *Total* row, click the Totals button in the Records group on the Home tab. Access adds a row to the bottom of the table with the label *Total*. Click in the *Total* row, click the option box arrow that displays, and then click a specific function at the drop-down list.

Activity 1c **Validating Field Entries; Using the Lookup Wizard; and Inserting, Moving, and Deleting a Field**

Part 3 of 9

1. With **4-SunProperties** open, open the Employees table.
2. Insert a new field in the Employees table and apply a validation rule by completing the following steps:
   a. Click the View button to switch to Design view.
   b. Click in the empty field immediately below the *DentalIns* field in the *Field Name* column and then type LifeIns.
   c. Press the Tab key.
   d. Click the option box arrow in the field in the *Data Type* column and then click *Currency* at the drop-down list.
   e. Click in the *Validation Rule* property box, type <=100000, and then press the Enter key.
   f. With the insertion point positioned in the *Validation Text* property box, type Enter a value that is equal to or less than $100,000.

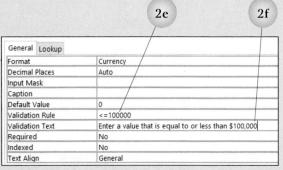

   g. Click in the *LifeIns* field in the *Description* column and then type Enter optional life insurance amount.
   h. Click the Save button on the Quick Access Toolbar. Since the validation rule was created *after* data was entered into the table, Access displays a warning message indicating that some data may not be valid. At this message, click No.
   i. Click the View button to switch to Datasheet view.

3. Click in the first empty field in the *LifeIns* column, type 200000, and then press the Down Arrow key.
4. Access displays the error message prompting you to enter an amount that is equal to or less than $100,000. At this error message, click OK.
5. Edit the amount in the field to be *100000* and then press the Down Arrow key.
6. Type the following entries in the remaining fields in the *LifeIns* column:

Record 2	25000	Record 7	100000
Record 3	0	Record 8	50000
Record 4	50000	Record 9	25000
Record 5	50000	Record 10	0
Record 6	0	Record 11	100000

7. Insert the *EmpCategory* field in the Employees table and use the Lookup Wizard to specify field choices by completing the following steps:

a. Click the View button to change to Design view.

b. Click in the *EmpFirstName* field in the *Field Name* column.

c. Click the Insert Rows button in the Tools group.

d. With the insertion point positioned in the new empty field in the *Field Name* column, type EmpCategory.

e. Press the Tab key. (This moves the insertion point to the *Data Type* column.)

f. Click the option box arrow in the field in the *Data Type* column and then click *Lookup Wizard* at the drop-down list.

g. At the first Lookup Wizard dialog box, click the *I will type in the values that I want* option and then click the Next button.

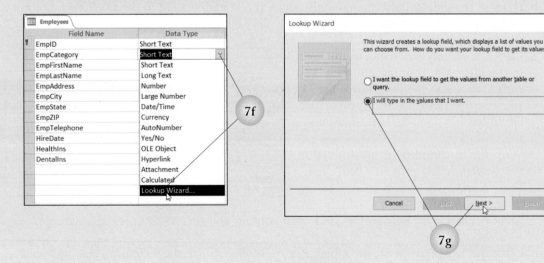

h. At the second Lookup Wizard dialog box, click in the blank text box below *Col1*, type Salaried, and then press the Tab key.

i. Type Hourly and then press the Tab key.

j. Type Temporary.

k. Click the Next button.

l. At the third Lookup Wizard dialog box, click the Finish button. ***Note: After completing the Lookup Wizard, the data type for* EmpCategory *will change to Short Text*.**

m. Press the Tab key and then type Click option box arrow and then click employee category in the *Description* column.

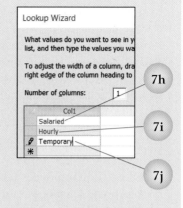

8. Click the Save button on the Quick Access Toolbar.

9. Click the View button to switch to Datasheet view.

10. Insert information in the *EmpCategory* column by completing the following steps:

a. Click in the first empty field in the new *EmpCategory* column.

b. Click the option box arrow in the field and then click *Hourly* at the drop-down list.

c. Click in the next empty field in the *EmpCategory* column, click the option box arrow, and then click *Salaried* at the drop-down list.

d. Continue entering information in the *EmpCategory* column by completing similar steps as in 10a-10b. Choose the following in the specified record:

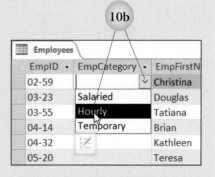

> Record 3: *Hourly*
> Record 4: *Salaried*
> Record 5: *Temporary*
> Record 6: *Hourly*
> Record 7: *Salaried*
> Record 8: *Temporary*
> Record 9: *Hourly*
> Record 10: *Salaried*
> Record 11: *Salaried*

11. Print the Employees table. (The table will print on two pages.)

12. After looking at the printed table, you decide to move the *EmpCategory* field. Move the *EmpCategory* field in Design view by completing the following steps:

a. With the Employees table open, click the View button to switch to Design view.

b. Click in the field selector bar at the left side of the *EmpCategory* field to select the row.

c. Position the mouse pointer in the *EmpCategory* field selector bar, click and hold down the left mouse button, drag down until a thick black line displays below the *EmpTelephone* field, and then release the mouse button.

Field Name	Data Type
EmpID	Short Text
EmpCategory	Short Text
EmpFirstName	Short Text
EmpLastName	Short Text
EmpAddress	Short Text
EmpCity	Short Text
EmpState	Short Text
EmpZIP	Short Text
EmpTelephone	Short Text
HireDate	Date/Time
HealthIns	Yes/No

13. Delete the *DentalIns* field by completing the following steps:
   a. Click in the field selector bar at the left side of the *DentalIns* field. (This selects the row.)
   b. Click the Delete Rows button in the Tools group.
   c. At the message asking if you want to permanently delete the field and all of the data in the field, click Yes.
14. Click the Save button on the Quick Access Toolbar.
15. Click the View button to switch to Datasheet view.
16. Print the Employees table. (The table will print on two pages.)
17. Close the Employees table.
18. Open the Payments table and then insert a new field in Design view by completing the following steps:
   a. Click the View button to switch to Design view.
   b. Click in the empty field immediately below the *PymntAmount* field in the *Field Name* column and then type LateFee.
   c. Press the Tab key.
   d. Click the option box arrow in the field in the *Data Type* column and then click *Currency* at the drop-down list.
   e. Press the Tab key and then type Enter a late fee amount if applicable in the Description column.
   f. Click the Save button on the Quick Access Toolbar.
   g. Click the View button to switch to Datasheet view.
19. Apply a validation rule and message for the LateFee field by completing the following steps:
   a. Click the *LateFee* column heading to select the column.
   b. Click the Table Tools Fields tab, click the Validation button in the Field Validation group, and then click *Field Validation Rule* at the drop-down list.
   c. Type <=50 in the Expression Builder dialog box and then click OK.
   d. Click Yes at the message that displays.
   e. Click the Validation button and then click *Field Validation Message*.
   f. Type Late fee must be $50 or less and then click OK.
20. Insert late fees for the last three records by completing the following steps:
   a. Click in the *LateFee* field for record 15, type 25, and then press the Down Arrow key.
   b. With the *LateFee* field for record 16 active, type 25 and then press the Down Arrow key.
   c. With the *LateFee* field for record 17 active, type 50 and then press the Up Arrow key.

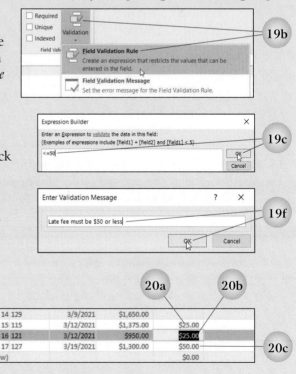

21. Insert a *Total* row by completing the following steps:
   a. In Datasheet view, click the Totals button in the Records group on the Home tab.
   b. Click in the empty field in the *PymntAmount* column in the *Total* row.
   c. Click the option box arrow in the field and then click *Sum* at the drop-down list.
   d. Click in the empty field in the *LateFee* column in the *Total* row.
   e. Click the option box arrow in the field and then click *Sum* at the drop-down list.
   f. Click in any other field.
22. Save, print, and then close the Payments table.

PymntIE	RenterID	PymntDate	PymntAmount	LateFee	Click to Add
1	130	3/1/2021	$1,800.00		
2	111	3/1/2021	$1,900.00		
3	136	3/1/2021	$1,250.00		
4	110	3/1/2021	$1,300.00		
5	135	3/2/2021	$1,900.00		
6	123	3/2/2021	$1,000.00		
7	117	3/2/2021	$1,100.00		
8	134	3/3/2021	$1,400.00		
9	131	3/3/2021	$1,200.00		
10	118	3/3/2021	$900.00		
11	125	3/5/2021	$1,650.00		
12	119	3/5/2021	$1,500.00		
13	133	3/8/2021	$1,650.00		
14	129	3/9/2021	$1,650.00		
15	115	3/12/2021	$1,375.00	$25.00	
16	121	3/12/2021	$950.00	$25.00	
17	127	3/19/2021	$1,300.00	$50.00	
* (New)				$0.00	
Total					

None
Sum
Average

Check Your Work

Sorting Records

Tutorial

Sorting Records in a Table

 Ascending

 Descending

The Sort & Filter group on the Home tab contains two buttons for sorting data in records. Click the Ascending button to sort data in the active field and text is sorted in alphabetical order from A to Z, numbers are sorted from lowest to highest, and dates are sorted from earliest to latest. Click the Descending button to sort data in the active field and text is sorted in alphabetical order from Z to A, numbers from highest to lowest, and dates from latest to earliest.

Printing Specific Records

Tutorial

Printing Specific Records in a Table

Quick Steps

Sort Records
1. Open table in Datasheet view.
2. Click field in column.
3. Click Ascending or Descending button.

Print Selected Records
1. Open table and select records.
2. Click File tab.
3. Click *Print* option.
4. Click next *Print* option.
5. Click *Selected Record(s)*.
6. Click OK.

Specific records in a table can be printed by selecting the records and then displaying the Print dialog box. Display this dialog box by clicking the File tab, clicking the *Print* option, and then clicking the next *Print* option in the center panel. At the Print dialog box, click the *Selected Record(s)* option in the *Print Range* section and then click OK.

To select specific records, open the table in Datasheet view, click the record selector for the first record, click and hold the left mouse button, drag to select the specific records, and then release the mouse button. The record selector is the light gray square at the left side of the record. When the mouse pointer is positioned on the record selector, the pointer turns into a right-pointing black arrow.

Formatting Table Data

A table in Datasheet view can be formatted with options available in the Text Formatting group on the Home tab as shown in Figure 4.10 and described in Table 4.2. (Some of the buttons in the Text Formatting group are dimmed and unavailable. These buttons are only available for fields formatted as rich text.) Alignment buttons such as Align Left, Center, or Align Right, apply formatting to data in the currently active column. Click one of the other options or buttons in the Text Formatting group and the formatting is applied to data in all columns and rows.

Figure 4.10 Text Formatting Group on the Home Tab

Table 4.2 Text Formatting Buttons and Option Boxes

Button/Option Box	Name	Description
Calibri (Detail)	*Font*	Change the text font.
11	*Font Size*	Change the text size.
B	Bold	Bold the text.
I	Italic	Italicize the text.
U	Underline	Underline the text.
A	Font Color	Change the text color.
(fill)	Background Color	Apply a background color to all fields.
(align left)	Align Left	Align all text in the currently active column at the left side of the fields.
(center)	Center	Center all text in the currently active column in the center of the fields.
(align right)	Align Right	Align all text in the currently active column at the right side of the fields.
(gridlines)	Gridlines	Specify whether to display vertical and/or horizontal gridlines.
(alt row)	Alternate Row Color	Apply a specified color to alternating rows in the table.

When creating a table, a data type is specified for a field, such as the Short Text, Date/Time, or Currency data type. To format text in a field rather than all of the fields in a column or the entire table, choose the Long Text data type and then specify rich text formatting. For example, in Activity 1d, specific credit scores will be formatted in the *RenCreditScore* column. To be able to format specific scores, the data type will be changed to Long Text with rich text formatting. Use the Long Text data type only for fields containing text—not for fields containing currency amounts, numbers, or dates.

By default, the Long Text data type uses plain text formatting. To change to rich text, click in the *Text Format* property box in the *Field Properties* section (displays with the text *Plain Text*), click the option box arrow at the right side of the property box, and then click *Rich Text* at the drop-down list.

Activity 1d Sorting, Printing, and Formatting Records and Fields in Tables   Part 4 of 9

1. With **4-SunProperties** open, open the Renters table.
2. With the table in Datasheet view, sort records in ascending alphabetical order by last name by completing the following steps:
   a. Click in any last name in the *RenLastName* column in the table.
   b. Click the Ascending button in the Sort & Filter group on the Home tab.
   c. Print the Renters table in landscape orientation.
3. Sort records in descending order (highest to lowest) by credit score number by completing the following steps:
   a. Click in any number in the *RenCreditScore* column.
   b. Click the Descending button in the Sort & Filter group.
   c. Print the Renters table in landscape orientation.
4. Close the Renters table without saving the changes.
5. Open the Properties table.
6. Sort and then print selected records with a specific apartment property type by completing the following steps:
   a. Click in any entry in the *CatID* column.
   b. Click the Ascending button in the Sort & Filter group.
   c. Position the mouse pointer on the record selector of the first record with a category ID of *A*, click and hold down the left mouse button, drag to select the four records with a category ID of *A*, and then release the mouse button.
   d. Click the File tab and then click the *Print* option.
   e. Click the next *Print* option in the center panel.

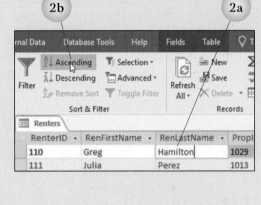

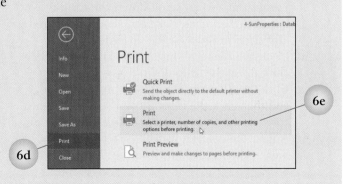

f. At the Print dialog box, click the *Selected Record(s)* option in the *Print Range* section.

g. Click OK.

7. With the Properties table open, apply the following text formatting:

a. Click in any field in the *CatID* column and then click the Center button in the Text Formatting group on the Home tab.

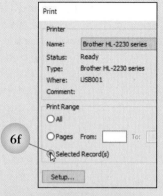

6f

7a

b. Click in any field in the *PropID* column and then click the Center button in the Text Formatting group.

c. Click the Bold button in the Text Formatting group. (This applies bold to all text in the table.)

d. Click the Font Color button arrow and then click the *Dark Blue* color option (fourth column, first row in the *Standard Colors* section).

e. Adjust the column widths.

f. Save, print, and then close the Properties table.

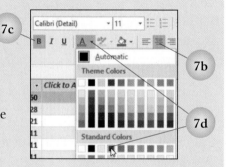

7c

7b

7d

8. Open the Payments table and apply the following text formatting:

a. With the first field active in the *PymntID* column, click the Center button.

b. Click in any field in the *RenterID* column and then click the Center button.

c. Click the *Font* option box arrow in the Text Formatting group, scroll down the drop-down list, and then click *Candara*. (Fonts are listed in alphabetical order in the drop-down list.)

d. Click the *Font Size* option box arrow and then click *12* at the drop-down list.

e. Click the Alternate Row Color button arrow and then click the *Green 2* color option (seventh column, third row in the *Standard Colors* section).

f. Adjust the column widths.

g. Save, print, and then close the Payments table.

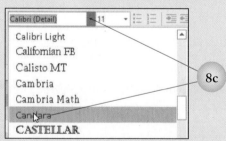

8c

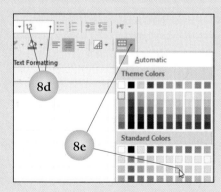

8d

8e

9. Open the Renters table and then apply the following formatting to columns in the table:
   a. With the first field active in the *RenterID* column, click the Center button.
   b. Click in any field in the *PropID* column and then click the Center button.
   c. Click in any field in the *EmpID* column and then click the Center button.
   d. Click in any field in the *RenCreditScore* column and then click the Center button.
10. Change the data type for the *RenCreditScore* field to Long Text with rich text formatting and apply formatting by completing the following steps:
   a. Click the View button to switch to Design view.
   b. Click in the *RenCreditScore* field in the *Data Type* column, click the option box arrow in the field, and then click *Long Text* at the drop-down list.
   c. Click in the *Text Format* property box in the *Field Properties* section (displays with the words *Plain Text*), click the option box arrow in the property box, and then click *Rich Text* at the drop-down list.

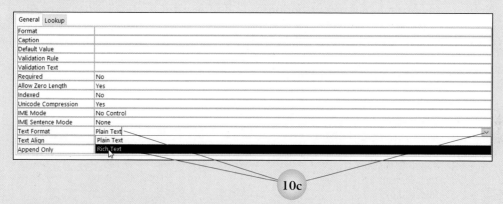

10c

   d. At the message stating that the field will be converted to rich text, click the Yes button.
   e. Click the Save button on the Quick Access Toolbar.
   f. Click the View button to switch to Datasheet view.
   g. Double-click the field value *538* in the *RenCreditScore* column in the row for Dana Rozinski. (Double-clicking in the field selects the field value *538*.)
   h. With *538* selected, click the Font Color button in the Text Formatting group. (This changes the color of the number to standard red. If the font color does not change to red, click the Font Color button arrow and then click the *Red* option [second column, bottom row in the *Standard Colors* section].)
   i. Change the font to standard red for any credit scores below 600.
   j. Save and print the Renters table in landscape orientation and then close the table.

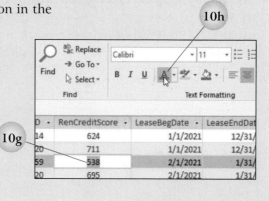

10h

10g

Check Your Work

Tutorial

Completing a
Spelling Check

💡 **Hint** Begin a
spelling check with the
keyboard shortcut F7.

Spelling

Quick Steps

Complete Spelling Check
1. Open table in Datasheet view.
2. Click Spelling button.
3. Change or ignore spelling.
4. Click OK.

Completing a Spelling Check

The spelling feature in Access finds misspelled words and suggests replacement words. It also finds duplicate words and irregular capitalizations. When checking the spelling of an object in a database, such as a table, the words in a table are compared with the words in the spelling dictionary. If a match is found, the word is passed over. If no match is found, then the word is selected and possible replacements are suggested.

To complete a spelling check, open a table in Datasheet view and then click the Spelling button in the Records group on the Home tab. If no match is found for a word in the table, the Spelling dialog box displays with replacement options. Figure 4.11 displays the Spelling dialog box with the word *Citruis* selected and possible replacements display in the *Suggestions* list box. Use options in the Spelling dialog box, to ignore the word (for example, if a proper name is selected), change to one of the replacement options, or add the word to the dictionary or AutoCorrect feature. A spelling check also can be completed on other objects in a database, such as a query, form, and report. (Forms and reports are covered in later chapters.)

Figure 4.11 Spelling Dialog Box

The spelling check selects this word in the table and suggests possible replacements in this list box.

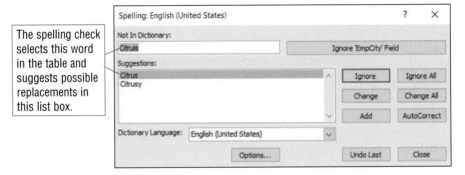

Activity 1e Checking Spelling in a Table   Part 5 of 9

1. With **4-SunProperties** open, open the Employees table.
2. Add the following record to the Employees table. (Type the misspelled words as shown below. You will correct the spelling in a later step.)

EmpID	02-72
EmpFirstName	Roben
EmpLastName	Wildre
EmpAddress	9945 Valley Avenue
EmpCity	Citruis Heights
EmpState	(CA automatically inserted)
EmpZIP	95610
EmpTelephone	9165556522
EmpCategory	(choose *Salaried*)
HireDate	01may20
HealthIns	No (Remove check mark)
LifeIns	50000

3. Save the Employees table.

4. Click in the first field in the *EmpID* column.
5. Click the Spelling button in the Records group on the Home tab.
6. The name *Kasadev* is selected. This is a proper name, so click the Ignore button to leave the name as written.
7. The name *Roben* is selected. Although this is a proper name, it is spelled incorrectly. Click *Robin* (the proper spelling) in the *Suggestions* list box and then click the Change button.
8. The name *Wildre* is selected. Although this is a proper name, it is spelled incorrectly. The proper spelling *(Wilder)* is selected in the *Suggestions* list box, so click the Change button.
9. The word *Citruis* is selected. The proper spelling *(Citrus)* is selected in the *Suggestions* list box, so click the Change button.
10. At the message stating that the spelling check is complete, click OK.
11. Print the Employees table and then close the table.

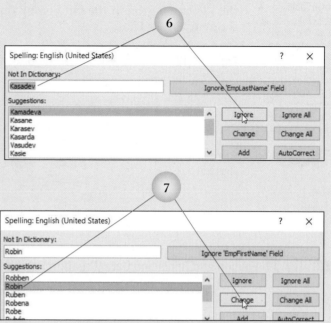

Check Your Work

Finding and Replacing Data

Tutorial

Finding Data

Tutorial

Finding and Replacing Data

Find

To find a specific entry in a field in a table, consider using options at the Find and Replace dialog box with the Find tab selected, as shown in Figure 4.12. Display this dialog box by clicking the Find button in the Find group on the Home tab or with the keyboard shortcut Ctrl + F. At the Find and Replace dialog box, enter the data to be found in the *Find What* text box. By default, Access looks only in the specific column where the insertion point is positioned. Click the Find Next button to find the next occurrence of the data or click the Cancel button to close the Find and Replace dialog box.

The *Look In* option defaults to the column where the insertion point is positioned. This can be changed to search the entire table by clicking the *Look In* option box arrow and then clicking the table name at the drop-down list. The *Match* option has a default setting of *Whole Field*. This can be changed to *Any Part of Field* or *Start of Field*. The *Search* option has a default setting of *All*, which means that Access will search all of the data in a specific column. This can be changed to *Up* or *Down*. To find data that contains specific uppercase and lowercase letters, insert a check mark in the *Match Case* check box and Access will return results that match the case formatting of the text entered in the *Find What* text box.

Quick Steps

Find Data in Table
1. Click Find button.
2. Type data in *Find What* text box.
3. Click Find Next button.

Quick Steps

Find and Replace Data in Table
1. Click Replace button.
2. Type find data in *Find What* text box.
3. Type replacement data in *Replace With* text box.
4. Click Find Next button.
5. Click Replace button or Find Next button.

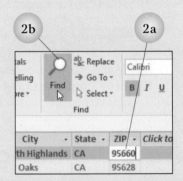

Replace

Use the Find and Replace dialog box with the Replace tab selected to search for specific data and replace it with other data. Display this dialog box by clicking the Replace button in the Find group on the Home tab or with the keyboard shortcut Ctrl + H.

Figure 4.12 Find and Replace Dialog Box with Find Tab Selected

Enter the data to be found in this text box.

Activity 1f Finding and Replacing Data, Creating Relationships, and Performing Queries

Part 6 of 9

1. With **4-SunProperties** open, open the Properties table.
2. Find records containing the zip code *95610* by completing the following steps:
   a. Click in the first field in the *ZIP* column.
   b. Click the Find button in the Find group on the Home tab.
   c. At the Find and Replace dialog box with the Find tab selected, type *95610* in the *Find What* text box.
   d. Click the Find Next button. (Access finds and selects the first occurrence of *95610*. If the Find and Replace dialog box covers the data, position the mouse pointer on the dialog box title bar, click and hold down the left mouse button, and then drag the dialog box to a different location on the screen.)
   e. Continue clicking the Find Next button until a message displays stating that Access has finished searching the records. At this message, click OK.
   f. Click the Cancel button to close the Find and Replace dialog box.
3. Suppose a new zip code has been added to the city of North Highlands and you need to change to this new zip code for some of the North Highlands properties. Complete the following steps to find *95660* and replace it with *95668*:
   a. Click in the first field in the *ZIP* column.
   b. Click the Replace button in the Find group.

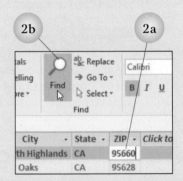
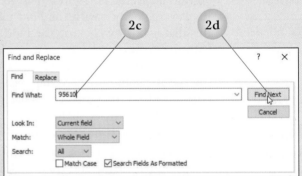

c. At the Find and Replace dialog box with the Replace tab selected, delete the existing text in the *Find What* text box, and then type 95660 in the *Find What* text box.

d. Press the Tab key. (This moves the insertion point to the *Replace With* text box.)

e. Type 95668 in the *Replace With* text box.

f. Click the Find Next button.

g. When Access selects the first occurrence of *95660*, click the Replace button.

h. When Access selects the second occurrence of *95660*, click the Find Next button.

i. When Access selects the third occurrence of *95660*, click the Replace button.

j. When Access selects the fourth occurrence of *95660*, click the Find Next button.

k. When Access selects the fifth occurrence of *95660*, click the Find Next button.

l. When Access selects the sixth occurrence of *95660*, click the Replace button.

m. Access goes back and selects the first occurrence of *95660* (record 1003) in the table. Click the Cancel button to close the Find and Replace dialog box.

4. Print and then close the Properties table.

5. Open the Relationships window and then create the following relationships (enforce referential integrity and cascade fields and records):

a. Create a one-to-many relationship with the *CatID* field in the Categories table field list box the "one" and the *CatID* field in the Properties table field list box the "many."

b. Create a one-to-many relationship with the *EmpID* field in the Employees table field list box the "one" and the *EmpID* field in the Renters table field list box the "many."

c. Create a one-to-many relationship with the *PropID* field in the Properties table field list box the "one" and the *PropID* field in the Renters table field list box the "many."

d. Create a one-to-many relationship with the *RenterID* field in the Renters table field list box the "one" and the *RenterID* field in the Payments table field list box the "many."

e. Save the relationships and then print the relationship report in landscape orientation.

f. Close the relationship report without saving it and then close the Relationships window.

6. Design a query that returns employees with health insurance benefits with the following specifications:

a. Insert the Employees table in the query window.

b. Insert the *EmpID* field in the first field in the *Field* row.

c. Insert the *EmpFirstName* field in the second field in the *Field* row.

d. Insert the *EmpLastName* field in the third field in the *Field* row.

e. Insert the *HealthIns* field in the fourth field in the *Field* row.

f. Click in the check box in the *EmpID* field in the *Show* row to remove the check mark. (This hides the EmpID numbers in the query results.)

g. Extract those employees with health benefits. (Type 1 in the *HealthIns* field in the *Criteria* row.)

h. Run the query.

i. Save the query, typing Emp_wHealthIns as the query name.

j. Print and then close the query.

7. Design a query that returns all properties in the city of Citrus Heights with the following specifications:
   a. Insert the Properties table and the Categories table field list box in the query window.
   b. Insert the *PropID* field from the Properties table field list box in the first field in the *Field* row.
   c. Insert the *Category* field from the Categories table in the second field in the *Field* row.
   d. Insert the *Address*, *City*, *State*, and *ZIP* fields from the Properties table field list box to the third, fourth, fifth, and sixth fields in the *Field* row, respectively.
   e. Extract those properties in the city of Citrus Heights.
   f. Run the query.
   g. Save the query, typing Prop_CitrusHeights as the query name.
   h. Print and then close the query.
8. Design a query that returns rent payments made between 3/1/2021 and 3/5/2021 with the following specifications:
   a. Insert the Payments table and the Renters table in the query window.
   b. Insert the *PymntID*, *PymntDate*, and *PymntAmount* fields from the Payments table field list box in the first, second, and third fields in the *Field* row fields, respectively.
   c. Insert the *RenFirstName* and *RenLastName* fields from the Renters table field list box in the fourth and fifth fields in the *Field* row, respectively.
   d. Extract those payments made between 3/1/2021 and 3/5/2021.
   e. Run the query.
   f. Save the query, typing Pymnts_March1to5 as the query name.
   g. Print and then close the query.
9. Design a query that returns properties in Citrus Heights or Orangevale that rent for less than $1,501 a month as well as the type of property with the following specifications:
   a. Insert the Categories table and the Properties table in the query window.
   b. Insert the *Category* field from the Categories table field list box.
   c. Insert the *PropID*, *MoRent*, *Address*, *City*, *State*, and *ZIP* fields from the Properties table field list box.
   d. Extract those properties in Citrus Heights or Orangevale that rent for less than $1,501.
   e. Run the query.
   f. Save the query, typing Prop_CitrusOrange<$1501 as the query name.
   g. Print the query in landscape orientation and then close the query.
10. Design a query that returns properties in Citrus Heights assigned to employee identification number *07-20* with the following specifications:
    a. Insert the Employees table and Properties table in the query window.
    b. Insert the *EmpID*, *EmpFirstName*, and *EmpLastName* fields from the Employees table field list box.
    c. Insert the *Address*, *City*, *State*, and *ZIP* fields from the Properties table field list box.
    d. Extract those properties in Citrus Heights assigned to employee identification number 07-20.
    e. Run the query.
    f. Save the query, typing Emp_07-20Citrus as the query name.
    g. Print and then close the query.

Using the Help and Tell Me Features

Microsoft Access includes a Help feature that contains information about Access features and commands. This on-screen reference manual is similar to Windows Help and the Help features in Word, PowerPoint, and Excel. The Tell Me feature provides information and guidance on how to complete a function.

Getting Help at the Help Task Pane

Click the Help button in the Help group on the Help tab or press the F1 function key to display the Help task pane, as shown in Figure 4.13. In this task pane, type a topic, feature, or question in the search text box and then press the Enter key. Articles related to the search text display in the Help task pane. Open an article by clicking the article's hyperlink.

The Help task pane contains buttons above the search text box. Use the Back and three dots buttons to navigate within the task pane. Click the Move and Size button (displays as a down arrow at the top right of the task pane) to show sizing and moving options for the Help task pane.

Getting Help on a Button

Position the mouse pointer on a button and a ScreenTip displays with information about the button. Some button ScreenTips display with a Help icon and the text *Tell me more*. Click this hyperlinked text or press the F1 function key and the Help task pane opens with information about the button feature.

Figure 4.13 Help Task Pane

1. With **4-SunProperties** open, click
   the Help tab and then click the Help
   button in the Help group.
2. At the Help task pane, type input
   mask in the search text box and then
   press the Enter key.
3. When the list of articles displays,
   click the <u>Control data entry formats with</u>
   <u>input masks</u> hyperlink. (If this article is
   not available, choose a similar article.)
4. Read the information on creating an input
   mask.
5. Close the Help task pane by clicking the
   Close button in the upper right corner of
   the task pane.
6. Click the Create tab.
7. Hover the mouse over the Table button and then
   click the <u>Tell me more</u> hyperlink at the bottom of the
   ScreenTip.
8. At the Help task pane, read the information on
   tables and then close the task pane.

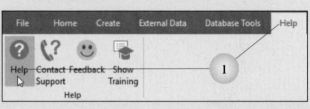
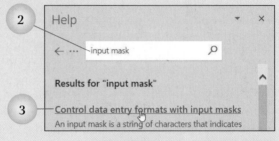
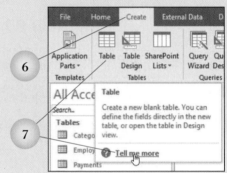
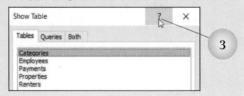

Getting Help in a Dialog Box or Backstage Area

Some dialog boxes and backstage areas provide a Help button that, when clicked,
displays the Microsoft Office support website with information about the dialog
box or backstage area. After reading the information, close the browser window
and return to Access.

1. With **4-SunProperties** open, click the Database Tools tab.
2. Click the Relationships button. (Make sure the Show Table dialog box displays. If it does
   not, click the Show Table button in the Relationships group.)
3. Click the Help button in the upper right corner of the Show Table dialog box.

4. Look at the information at the Microsoft Office support website and then close the window.
5. Close the Show Table dialog box and then close the Relationships window.
6. Click the File tab and then click the *Open* option.
7. At the Open backstage area, click the Microsoft Access Help button in the upper right corner.
8. Look at the information at the Microsoft Office support website and then close the window.
9. Press the Esc key to return to the database.

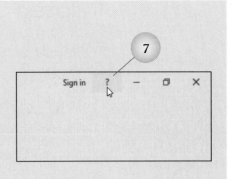

Using the Tell Me Feature

Hint Alt + Q is the keybord shortcut to make the *Tell Me* text box active.

Access includes a Tell Me feature that provides information as well as guidance on how to complete a function. To use Tell Me, click in the *Tell Me* text box on the ribbon to the right of the View tab and then type the function. Type text in the *Tell Me* text box and a drop-down list displays with options that are refined as the text is typed. The drop-down list displays options for completing the function or for displaying information on the function in the Help task pane.

Activity 1i Using the Tell Me Feature
Part 9 of 9

1. With **4-SunProperties** open, open the query named *Prop_CitrusHeights*.
2. Use the Tell Me feature to display the Find and Replace dialog box with the Replace tab selected and apply vertical gridlines by completing the following steps:
   a. Click in the *Tell Me* text box.
   b. Type replace.
   c. Click the *Replace* option at the drop-down list. (This displays the Find and Replace dialog box with the Replace tab selected.)
   d. Click the Cancel button to close the Find and Replace dialog box.
   e. Click in the *Tell Me* text box.
   f. Type gridlines.
   g. Click the arrow at the right side of the *Gridlines* option at the drop-down list.
   h. Click the *Gridlines: Vertical* option at the side menu.
3. Save, print, and then close the Prop_CitrusHeights query.
4. Close the **4-SunProperties** database.

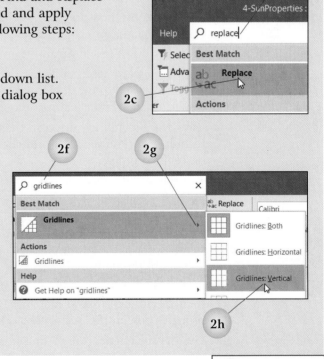

Check Your Work

Chapter Summary

- A table can be created in Datasheet view or Design view. Click the View button on the Table Tools Fields tab or the Home tab to switch between Datasheet view and Design view.

- Define each field in a table in the rows in the top section of Design view. Access automatically assigns the first field the name *ID* and assigns the AutoNumber data type.

- In Design view, specify a field name, data type, and description for each field.

- Assign a data type in Design view by clicking in a specific field in the *Data Type* column, clicking the option box arrow at the right side of the field, and then clicking the data type at the drop-down list.

- Create a default value for a field in Design view with the *Default Value* property box in the *Field Properties* section.

- Use the Input Mask Wizard to set a pattern for how data is entered in a field. Use the *Format* property box to control what displays in a field.

- Use the *Validation Rule* property box in the *Field Properties* section in Design view to enter a statement containing a conditional test. Customize the error message if the data entered violates the validation rule by typing that message in the *Validation Text* property box. In Datasheet view, use the Validation button in the Field Validation group on the Table Tools Fields tab to apply a validation rule and message.

- Use the Lookup Wizard to confine data entered in a field to a specific list of items.

- Insert a field in Design view by clicking in the row immediately below where the new field is to be inserted and then clicking the Insert Rows button.

- Move a field in Design view by clicking in the field selector bar of the field to be moved and then dragging to the new position.

- Delete a field in Design view by clicking in the field selector bar at the left side of the field to be deleted and then clicking the Delete Rows button.

- Insert a *Total* row in a table in Datasheet view by clicking the Totals button in the Records group on the Home tab, clicking the option box arrow in the *Total* row field, and then clicking a function at the drop-down list.

- Click the Ascending button in the Sort & Filter group on the Home tab to sort records in ascending order and click the Descending button to sort records in descending order.

- To print specific records in a table, select the records, display the Print dialog box, make sure *Selected Record(s)* is selected, and then click OK.

- Apply formatting to a table in Datasheet view with options and buttons in the Text Formatting group on the Home tab. Depending on the option or button selected, formatting is applied to all of the data in a table or data in a specific column in the table.

- To format text in a specific field, change the data type to Long Text and then specify rich text formatting. Do this in Design view with the *Text Format* property box in the *Field Properties* section.

- Use the spelling feature to find misspelled words in a table and consider possible replacement words. Begin checking the spelling in a table by clicking the Spelling button in the Records group on the Home tab.

- Use options at the Find and Replace dialog box with the Find tab selected to search for specific field entries in a table. Use options at the Find and Replace dialog box with the Replace tab selected to search for specific data and replace it with other data.
- Click the Help tab and then click the Help button or press the F1 function key to display the Help task pane. At the Help task pane, type a topic in the search text box and then press the Enter key.
- The ScreenTip for some buttons displays with a Help icon and the text *Tell me more*. Click this hyperlinked text or press the F1 function key and the Help task pane opens with information about the button.
- Some dialog boxes and backstage areas contain a Help button that links to the Microsoft Office support website to provide more information about available functions and features.
- The Tell Me feature, located on the ribbon, provides information and guidance on how to complete a function.

Commands Review

FEATURE	RIBBON TAB, GROUP	BUTTON	KEYBOARD SHORTCUT
align text left	Home, Text Formatting		
align text right	Home, Text Formatting		
alternate row color	Home, Text Formatting		
background color	Home, Text Formatting		
bold formatting	Home, Text Formatting	B	
center text	Home, Text Formatting		
delete field	Table Tools Design, Tools		
Design view	Home, Views OR Table Tools Fields, Views		
Find and Replace dialog box with Find tab selected	Home, Find		Ctrl + F
Find and Replace dialog box with Replace tab selected	Home, Find		Ctrl + H
font	Home, Text Formatting	Calibri (Detail)	
font color	Home, Text Formatting	A	
font size	Home, Text Formatting	11	
gridlines	Home, Text Formatting		
Help task pane	Help, Help		F1

FEATURE	RIBBON TAB, GROUP	BUTTON	KEYBOARD SHORTCUT
insert field	Table Tools Design, Tools		
italic formatting	Home, Text Formatting	*I*	
sort records ascending	Home, Sort & Filter		
sort records descending	Home, Sort & Filter		
spelling check	Home, Records		F7
Tell Me feature			Alt + Q
Total row	Home, Records	Σ	
underline formatting	Home, Text Formatting	U	

Microsoft®

Access® Level 1

Unit 2

Creating Forms and Reports

Microsoft®
Access®

CHAPTER 5

Creating Forms

Performance Objectives

Upon successful completion of Chapter 5, you will be able to:

1 Create a form using the Form button

2 Change views in a form

3 Print a form

4 Navigate in a form

5 Delete a form

6 Add records to and delete records from a form

7 Sort records in a form

8 Create a form with a related table

9 Manage control objects in a form

10 Format a form

11 Apply conditional formatting to data in a form

12 Add an existing field to a form

13 Insert a calculation in a form

14 Create a split form and multiple items form

15 Create a form using the Form Wizard

In this chapter, you will learn how to create forms from database tables, improving the data display and making data entry easier. Access offers several methods for presenting data on the screen for easier data entry. You will create a form using the Form button, create a split form and multiple items form, and use the Form Wizard to create a form. You will also learn how to customize control objects, insert control objects and fields, and apply formatting to a form.

 Data Files

Before beginning chapter work, copy the AL1C5 folder to your storage medium and then make AL1C5 the active folder.

The online course includes additional training and assessment resources.

Creating a Form

Hint A form allows you to focus on a single record at a time.

Access offers a variety of options for presenting data in a clear and attractive format. For instance, data can be viewed, added, or edited in a table in Datasheet view. When data is entered in a table in Datasheet view, multiple records are visible at the same time. If a record contains several fields, not all of the fields in the record may be visible at the same time. Create a form, however, and all of the fields for a record are generally visible on the screen.

Hint Save a form before making changes or applying formatting to it.

A form is an object used to enter and edit data in a table or query. It is a user-friendly interface for viewing, adding, editing, and deleting records. A form is also useful in helping to prevent incorrect data from being entered and it can be used to control access to specific data. Several methods are available for creating forms. This chapter covers creating forms using the Form, Split Form, and Multiple Items buttons, as well as the Form Wizard.

 Tutorial

Creating a Form Using the Form Button

 Form

Creating a Form Using the Form Button

The simplest method for creating a form is to click a table in the Navigation pane, click the Create tab, and then click the Form button in the Forms groups. Figure 5.1 shows the form that will be created in Activity 1a. Access creates the form using all fields in the table in a vertical layout and opens the form in Layout view with the Form Layout Tools Design tab active.

Quick Steps
Create Form with Form Button
1. Click table in Navigation pane.
2. Click Create tab.
3. Click Form button.

 Form View

 Layout View

 Design View

Changing Views

Click the Form button to create a form and the form opens in Layout view. This is one of three views for working with forms. Use the Form view to enter and manage records. Use the Layout view to view the data and modify the appearance and contents of the form. Use the Design view to view the form's structure and modify it. Change views with the View button in the Views group on the Form Layout Tools Design tab or with buttons in the view area at the right side of the Status bar. An existing form can be opened in Layout view by right-clicking the form name in the Navigation pane and then clicking *Layout View* at the shortcut menu.

Printing a Form

Print all of the records in a form by clicking the File tab, clicking the *Print* option, and then clicking the *Quick Print* option. To print a specific record in a form, click the File tab, click the *Print* option, and then click the *Print* option at the Print backstage area. At the Print dialog box, click the *Selected Record(s)* option and then click OK. Print a range of records by clicking the *Pages* option in the *Print Range*

Figure 5.1 Form Created with the Sales Table

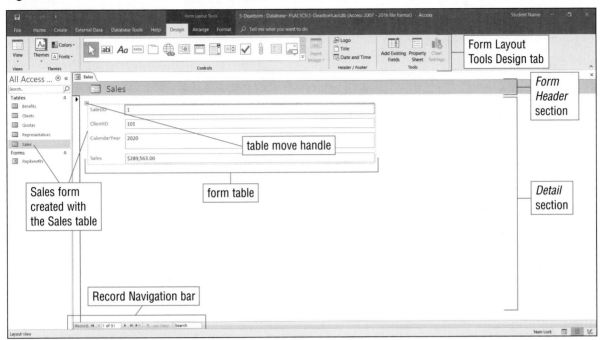

Quick Steps

Print Specific Record
1. Click File tab.
2. Click *Print* option.
3. Click next *Print* option.
4. Click *Selected Record(s)*.
5. Click OK.

section of the Print dialog box and then entering the beginning record number in the *From* text box and the ending record number in the *To* text box.

Before printing a form, display the form in Print Preview. If column shading displays on the second page without any other data, decrease the width of the column. To do this, click the Columns button in the Page Layout group on the Print Preview tab. At the Page Setup dialog box with the Columns tab selected, decrease the measurement in the *Width* measurement box, and then click OK.

Deleting a Form

If a form is no longer needed in a database, delete the form. Delete a form by clicking the form name in the Navigation pane, clicking the Delete button in the Records group on the Home tab, and then clicking the Yes button at the confirmation message. Another method is to right-click the form name in the Navigation pane, click *Delete* at the shortcut menu, and then click Yes at the message. If a form is being deleted from a computer's hard drive, the confirmation message will not display. This is because Access automatically sends the deleted form to the Recycle Bin, where it can be restored if necessary.

> Tutorial
>
> Navigating in Objects

Navigating in a Form

When a form displays in Form view or Layout view, navigation buttons display along the bottom of the form in the Record Navigation bar, as identified in Figure 5.1. Use these navigation buttons to display the first, previous, next, or last record in the form or add a new record. Navigate to a specific record by clicking in the *Current Record* box, selecting the current number, typing the number of the record to view, and then pressing the Enter key. The keyboard can also be used to navigate in a form. Press the Page Down key to move forward or the Page Up key to move back a single record. Press Ctrl + Home to go to the first record or press Ctrl + End to display the last record.

1. Display the Open dialog box with your AL1C5 folder active.
2. Open **5-Dearborn** and enable the content.
3. Create a form with the Sales table by completing the following steps:
   a. Click *Sales* in the Tables group in the Navigation pane.
   b. Click the Create tab.
   c. Click the Form button in the Forms group.

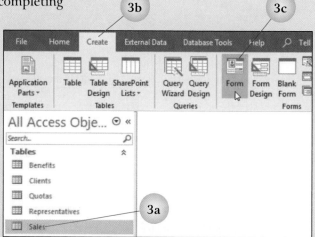

4. Switch to Form view by clicking the View button in the Views group on the Form Layout Tools Design tab.
5. Navigate in the form by completing the following steps:
   a. Click the Next record button in the Record Navigation bar to display the next record.
   b. Click in the *Current Record* box, select its contents, type 15, and then press the Enter key.
   c. Click the First record button in the Record Navigation bar to display the first record.

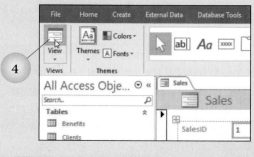

6. Save the form by completing the following steps:
   a. Click the Save button on the Quick Access Toolbar.
   b. At the Save As dialog box, with *Sales* inserted in the *Form Name* text box, click OK.
7. Print the current record in the form by completing the following steps:
   a. Click the File tab and then click the *Print* option.
   b. At the Print backstage area, click the *Print* option.
   c. At the Print dialog box, click the *Selected Record(s)* option in the *Print Range* section and then click OK.

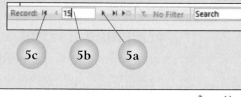

8. Close the Sales form.
9. Delete the RepBenefits form by right-clicking *RepBenefits* in the Navigation pane, clicking *Delete* at the shortcut menu, and then clicking Yes at the confirmation message.

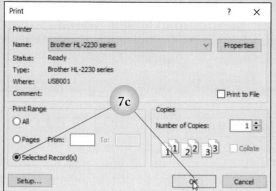

Check Your Work

Tutorial

Adding and
Deleting Records in
a Form

 New (blank)
record

 Delete

Quick Steps

Add Record
Click New (blank)
record button on
Record Navigation bar.
OR
1. Click Home tab.
2. Click New button.

Delete Record
1. Click Home tab.
2. Click Delete button
   arrow.
3. Click *Delete Record*.
4. Click Yes.

Adding and Deleting Records

Add a new record to the form by clicking the New (blank) record button (contains a right-pointing arrow and a yellow asterisk) on the Record Navigation bar along the bottom of the form. A new record can also be added to a form by clicking the Home tab and then clicking the New button in the Records group.

To delete a record, display the record, click the Home tab, click the Delete button arrow in the Records group, and then click *Delete Record* at the drop-down list. At the confirmation message, click the Yes button. Add records to or delete records from the table from which the form was created and the form will reflect the additions or deletions. Also, if additions or deletions are made to the form, the changes are reflected in the table from which the form was created.

Sorting Records

Sort data in a form by clicking in the field containing data on which to sort and then clicking the Ascending button or Descending button in the Sort & Filter group on the Home tab. Click the Ascending button to sort text in alphabetic order from A to Z, numbers from lowest to highest, and dates from earliest to latest. Click the Descending button to sort text in alphabetic order from Z to A, numbers from highest to lowest, and dates from latest to earliest.

Activity 1b Adding, Deleting, and Sorting Records in a Form

Part 2 of 7

1. With **5-Dearborn** open, open the Sales table (not the form) and add a new record by completing the following steps:
   a. Click the New (blank) record button in the Record Navigation bar.

1a

   b. At the new blank record, type the following information in the specified fields. (Move to the next field by pressing the Tab key or the Enter key; move to the previous field by pressing Shift + Tab.)

SalesID	(This is an AutoNumber field, so press the Tab key.)
ClientID	127
CalendarYear	2021
Sales	176420

2. Close the Sales table.
3. Open the Sales form.
4. Click the Last record button on the Record Navigation bar and notice that the new record you added to the table also has been added to the form.

5. Delete the second record (*SalesID 3*) in the form by completing the following steps:
   a. Click the First record button in the Record Navigation bar.
   b. Click the Next record button in the Record Navigation bar.
   c. With Record 2 active, click the Delete button arrow in the Records group on the Home tab and then click *Delete Record* at the drop-down list.

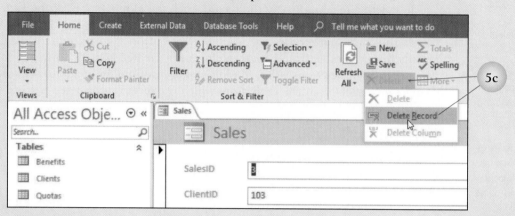

d. At the confirmation message, click the Yes button.
6. Click the New (blank) record button in the Record Navigation bar and then type the following information in the specified fields:

SalesID	(Press the Tab key.)
ClientID	103
CalendarYear	2021
Sales	110775

7. Sort the records in the form by completing the following steps:
   a. Click in the field containing *103* and then click the Ascending button in the Sort & Filter group on the Home tab.
   b. Click in the field containing *$289,563.00* and then click the Descending button in the Sort & Filter group.
   c. Click in the field containing *36* and then click the Ascending button.
8. Close the Sales form.

Creating a Form with a Related Table

Creating a Form with a Related Table

When the form was created with the Sales table, the form contained only the Sales table fields. If a form is created with a table that has a one-to-many relationship established, Access adds a datasheet to the form that is based on the related table.

For example, in Activity 1c, the form shown in Figure 5.2 will be created from the Representatives table and, since it is related to the Clients table by a one-to-many relationship, Access inserts a datasheet at the bottom of the form containing all of the records in the Clients table. Notice the datasheet at the bottom of the form.

If only a single one-to-many relationship has been created in a database, the datasheet for the related table displays in the form. If multiple one-to-many relationships have been created in a table, Access will not display any datasheets in a form created with that table.

Figure 5.2 Representatives Form with Clients Datasheet

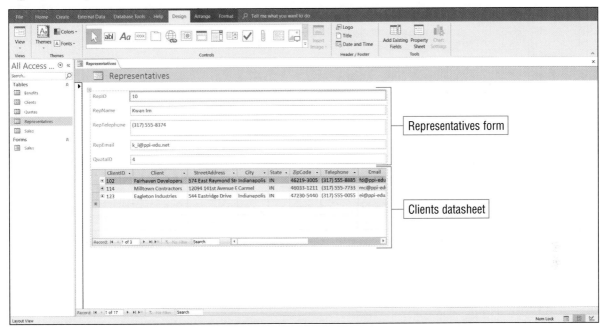

Activity 1c Creating a Form with a Related Table

1. With **5-Dearborn** open, create a form with the Representatives table by completing the following steps:
   a. Click *Representatives* in the Tables group in the Navigation pane.
   b. Click the Create tab.
   c. Click the Form button in the Forms group.
2. Insert a new record in the Clients table for representative *12* (Catherine Singleton) by completing the following steps:
   a. Click two times on the Next record button in the Record Navigation bar at the bottom of the form window (not the Record Navigation bar in the Clients datasheet) to display the record for Catherine Singleton.
   b. Click in the cell immediately below *127* in the *ClientID* field in the Clients datasheet.

c. Type the following information in the specified fields:

ClientID	129	State	IN
Client	Dan-Built Construction	ZipCode	460339050
StreetAddress	903 James Street	Telephone	3175551122
City	Carmel	Email	dc@ppi-edu.net

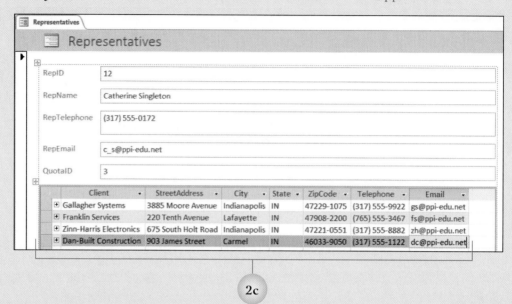

2c

3. Click the Save button on the Quick Access Toolbar and then, at the Save As dialog box with *Representatives* in the *Form Name* text box, click OK.
4. Print the current record in the form by completing the following steps:
   a. Click the File tab and then click the *Print* option.
   b. At the Print backstage area, click the *Print* option.
   c. At the Print dialog box, click the *Selected Record(s)* option in the *Print Range* section and then click OK.
5. Close the Representatives form.

 Check Your Work

 Tutorial

Managing Control Objects in a Form

Hint Almost all changes can be made to a form in Layout view.

Managing Control Objects

A form, like a table in a Word document, is made up of cells that are arranged in rows and columns. Each cell in a form can contain one control object, which is an object that displays a title or description, accepts data, or performs an action. For example, a cell can contain a label control object that displays a field name from the table used to create the form, a text box control that displays and accepts data, or a logo control that displays a logo image. Control objects are contained in the *Form Header* and *Detail* sections of the form. (Refer to Figure 5.1 on page 137.) The control objects in the *Detail* section are contained within a form table.

Manage control objects with buttons on the Form Layout Tools ribbon with the Design tab, Arrange tab, or Format tab selected. When a form is opened in Layout view, the Form Layout Tools Design tab is active.

Inserting Data in a Control Object

 Logo

 Title

Date and Time

Use buttons in the Header/Footer group on the Form Layout Tools Design tab to insert a logo, form title, or date and time. Click the Logo button and the Insert Picture dialog box displays. Browse to the folder containing the image and then double-click the image file. Click the Title button and the current title is selected. Type the new title and then press the Enter key. Click the Date and Time button in the Header/Footer group and the Date and Time dialog box displays. At this dialog box, choose a date and time format and then click OK. The date and time are inserted at the right side of the *Form Header* section.

Sizing Control Objects

When Access creates a form from a table, the cells in the first column in the *Detail* section of the form contain the label control objects with the field names from the table. The second column of cells contains the text box control objects with the field values entered in the table. The control objects in the *Form Header* section and the columns in the *Detail* section can be sized by dragging the border of a selected control object or with the *Width* property box in the Property Sheet task pane with the Format tab selected.

To size a control object by dragging, select the object (displays with an orange border) and then position the mouse pointer on the left or right border of the object until the pointer displays as a left-and-right pointing arrow. Click and hold down the left mouse button, drag left or right to change the width of the column, and then release the mouse button. Complete similar steps to change the height of a control object. When dragging a border, a line and character count displays at the left side of the Status bar. Use the line and character count numbers to move the border to a precise location. When dragging the border of a label control or text box control object, the entire column width is sized.

In addition to dragging a control object border, the column width can be adjusted with the *Width* property box in the Property Sheet task pane with the Format tab selected and the height can be adjusted with the *Height* property box. Display this task pane by clicking the Property Sheet button in the Tools group. In the Property Sheet task pane with the Format tab selected, select the current measurement in the *Width* or *Height* property box, type the new measurement, and then press the Enter key. Close the Property Sheet task pane by clicking the Close button in the upper right corner of the task pane.

Deleting a Control Object

To delete a control object from the form, click the object and then press the Delete key. Or, right-click the object and then click *Delete* at the shortcut menu. To delete a form row, right-click an object in the row to be deleted and then click *Delete Row* at the shortcut menu. To delete a column, right-click one of the objects in the column to be deleted and then click *Delete Column* at the shortcut menu.

Inserting Control Objects

Tutorial

Inserting Control Objects

Select

The Controls group on the Form Layout Tools Design tab contains a number of control objects that can be inserted in a form. By default, the Select button is active. With this button active, use the mouse pointer to select control objects. A new label control and text box control object can be inserted in a form by

abl Text Box

clicking the Text Box button in the Controls group and then clicking in the desired position in the form. Click the label control object, select the default text, and then type the label text.

Text can be entered in a label control object in Layout view, but not in a text box control object. In Form view, data can be entered in a text box control object but text in a label control object cannot be edited. The Controls group contains a number of additional buttons for inserting control objects in a form, such as a hyperlink, combo box, or image.

Activity 1d Creating a Form and Customizing the Design of a Form

Part 4 of 7

1. With **5-Dearborn** open, create a form with the Clients table and delete the accompanying datasheet by completing the following steps:
   a. Click *Clients* in the Tables group in the Navigation pane.
   b. Click the Create tab.
   c. Click the Form button.
   d. Click in the *SalesID* field in the datasheet that displays below the form.
   e. Click the table move handle in the upper left corner of the datasheet (see image at right).
   f. Press the Delete key.

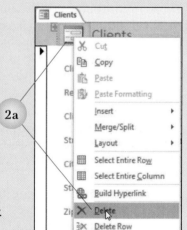

2. Insert a logo image in the *Form Header* section by completing the following steps:
   a. Right-click the logo object that displays in the *Form Header* section (to the left of the title *Clients*) and then click *Delete* at the shortcut menu.
   b. Click the Logo button in the Header/Footer group on the Form Layout Tools Design tab.
   c. At the Insert Picture dialog box, navigate to your AL1C5 folder and then double-click the file named **DearbornLogo**.
3. Change the title by completing the following steps:
   a. Click the Title button in the Header/Footer group. (This selects *Clients* in the *Form Header* section.)
   b. Type Dearborn Clients Form and then press the Enter key.

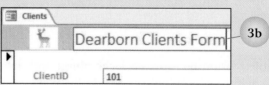

4. Insert the date and time in the *Form Header* section by completing the following steps:
   a. Click the Date and Time button in the Header/Footer group.
   b. At the Date and Time dialog box, click OK.
5. Size the control object containing the title by completing the following steps:
   a. Click any field outside the title and then click the title to select the control object.
   b. Position the mouse pointer on the right border of the selected object until the pointer displays as a black left-and-right-pointing arrow.

c. Click and hold down the left mouse button, drag to the left until the right border is immediately right of the title, and then release the mouse button.

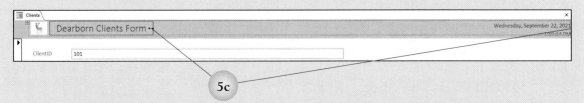

6. Size and move the control objects containing the date and time by completing the following steps:
   a. Click the date to select the control object.
   b. Press and hold down the Shift key, click the time, and then release the Shift key. (Both control objects should be selected.)
   c. Position the mouse pointer on the left border of the selected objects until the pointer displays as a black left-and-right-pointing arrow.
   d. Click and hold down the left mouse button, drag to the right until the border is immediately left of the date, and then release the mouse button.
   e. Position the mouse pointer in the selected objects until the pointer displays with a four-headed arrow attached.
   f. Click and hold down the left mouse button, drag the outline of the date and time objects to the left until the outline displays near the title, and then release the mouse button.
7. Decrease the size of the second column of cells containing control objects in the *Detail* section by completing the following steps:
   a. Click the text box control object containing the client number *101*. (This selects and inserts an orange border around the object.)
   b. Position the mouse pointer on the right border of the selected object until the pointer displays as a black left-and-right-pointing arrow.
   c. Click and hold down the left mouse button, drag to the left until *Lines: 1 Characters: 30* displays at the left side of the Status bar, and then release the mouse button.

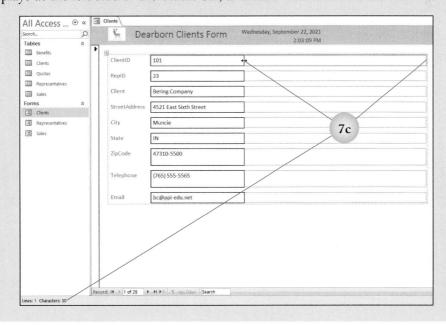

8. Insert a label control object by completing the following steps:
   a. Click the Label button in the Controls group.
   b. Click immediately right of the text box control object containing the telephone number *(765) 555-5565*. (This inserts the label to the right of the *Telephone* text box control object.)
   c. With the insertion point positioned inside the label, type Type the telephone number without symbols or spaces and then press the Enter key.
9. Change the width and height of the new label control object by completing the following steps:
   a. Click the Property Sheet button in the Tools group.
   b. In the Property Sheet task pane that displays, if necessary, click the Format tab.

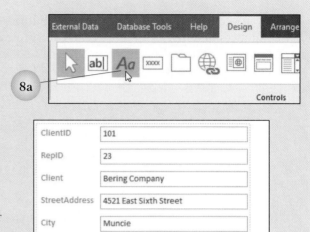

   c. Select the current measurement in the fixed *Width* property box, type 2, and then press the Enter key.
   d. With the current measurement selected in the *Height* property box, type 0.4.
   e. Close the Property Sheet task pane by clicking the Close button in the upper right corner of the task pane.

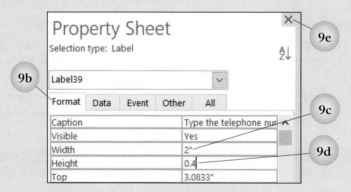

10. Delete the control object containing the time by clicking the time to select the object and then pressing the Delete key.
11. Click the Save button on the Quick Access Toolbar.
12. At the Save As dialog box with *Clients* in the *Form Name* text box, click OK.

Check Your Work

Moving a Form Table

Hint You can move a control object by dragging it to a new location.

The control objects in the *Detail* section in a form in Layout view are contained in cells within the form table. Click in a control object and the table move handle is visible. The table move handle is a small square with a four-headed arrow inside that displays in the upper left corner of the table. (Refer to Figure 5.1 on page 137.) To move the table, position the mouse pointer on the table move handle, click and hold down the left mouse button, drag the table to the new position, and then release the mouse button.

Arranging a Control Object

 Tutorial

Arranging Control Objects in a Form

The Form Layout Tools Arrange tab contains options for selecting, inserting, deleting, arranging, merging, and splitting cells. When a label control object was inserted to the right of the *Telephone* text box control object in Activity 1d, empty cells were inserted in the form above and below the new label control object. Select a control object or cell by clicking in the object or cell. Select adjacent control objects or cells by pressing and holding down the Shift key while clicking in each of the objects or cells. To select nonadjacent control objects or cells, press and hold down the Ctrl key while clicking in each of the objects or cells.

 Select Row

 Select Column

Select a row of cells by clicking the Select Row button in the Rows & Columns group or by right-clicking in a cell and then clicking *Select Entire Row* at the shortcut menu. To select a column of cells, click the Select Column button in the Rows & Columns group or right-click an object or cell and then click *Select Entire Column* at the shortcut menu. A column of cells can also be selected by positioning the mouse pointer at the top of the column until the pointer displays as a small, black, down arrow and then clicking the left mouse button.

 Insert Above

 Insert Below

The Rows & Columns group contains buttons for inserting a row or column of blank cells. To insert a new row, select a cell in a row and then click the Insert Above button to insert a row of blank cells above the current row or click the Insert Below button to insert a row of blank cells below the current row. Complete similar steps to insert a new column of blank cells to the left or right of the current column.

 Merge

 Split Vertically

 Split Horizontally

Merge adjacent selected cells by clicking the Merge button in the Merge/Split group on the Form Layout Tools Arrange tab. A cell can contain only one control object. So, merging two cells, each containing a control object, is not possible. A cell containing a control object can be merged with an empty cell or cells. Split a cell by clicking in the cell to make it active and then clicking the Split Vertically button or Split Horizontally button in the Merge/Split group. When a cell is split, an empty cell is created to the right of the cell or below the cell.

 Control Margins

Control Padding

A row of cells can be moved up or down by selecting the row and then clicking the Move Up button in the Move group or the Move Down button. Use the Control Margins button in the Position group to increase or decrease margins within cells. The Position group also contains a Control Padding button for increasing or decreasing spacing between cells.

The Table group at the left side of the Form Layout Tools Arrange tab contains buttons for applying gridlines to cells and changing the layout of the cells to a stacked or columnar layout.

1. With the Clients form in **5-Dearborn** open in Layout view, select and merge cells by completing the following steps:
   a. Click to the right of the text box control object containing the text *101*. (This selects the empty cell.)
   b. Press and hold down the Shift key, click to the right of the text box control object containing the text *Muncie*, and then release the Shift key. (This selects five adjacent cells.)
   c. Click the Form Layout Tools Arrange tab.
   d. Click the Merge button in the Merge/Split group.

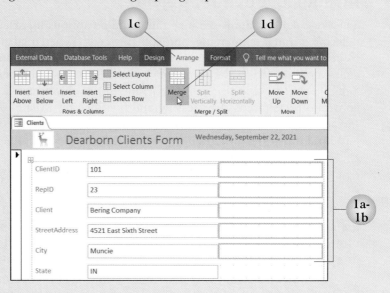

2. With the cells merged, insert an image control object and then insert an image by completing the following steps:
   a. Click the Form Layout Tools Design tab.
   b. Click the Image button in the Controls group.
   c. Move the mouse pointer (which displays as crosshairs with an image icon next to the crosshairs) to the location of the merged cell until the cell displays with pink fill color and then click the left mouse button.

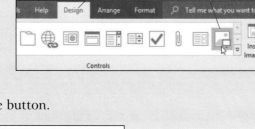

   d. At the Insert Picture dialog box, navigate to your AL1C5 folder and then double-click *Dearborn*.

3. Move down a row of cells by completing the following steps:
   a. Click the Form Layout Tools Arrange tab.
   b. Click in the control object containing the text *Telephone*.
   c. Click the Select Row button in the Rows & Columns group.
   d. Click the Move Down button in the Move group.
4. Decrease the margins within cells, increase the spacing (padding) between cells in the form, and apply gridlines by completing the following steps:
   a. If necessary, click the Form Layout Tools Arrange tab.
   b. Click the Select Layout button in the Rows & Columns group. (This selects all cells in the form table.)
   c. Click the Control Margins button in the Position group and then click *Narrow* at the drop-down list.

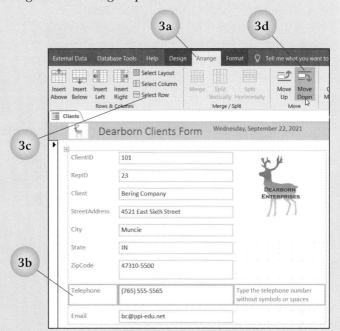

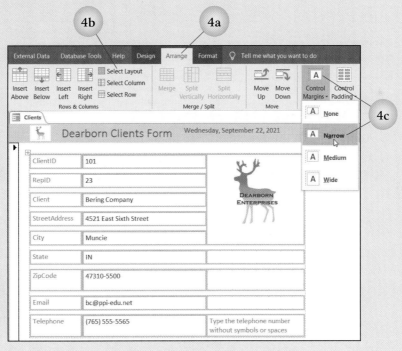

d. Click the Control Padding button in the Position group and then click *Medium* at the drop-down list.

e. Click the Gridlines button in the Table group and then click *Top* at the drop-down list.

f. Click the Gridlines button in the Table group, point to *Color*, and then click the *Orange, Accent 2, Darker 50%* option (sixth column, bottom row in the *Theme Colors* section).

5. Move the form table by completing the following steps:

a. Position the mouse pointer on the table move handle (which displays as a small square with a four-headed arrow inside in the upper left corner of the table).

b. Click and hold down the left mouse button, drag the form table up and to the left so it is positioned close to the top left border of the *Detail* section, and then release the mouse button.

6. Click in the control object containing the field name *ClientID*.

7. Save the Clients form.

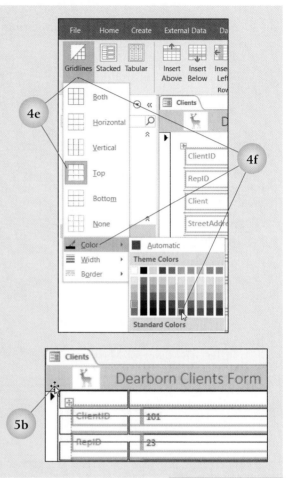

 Check Your Work

 Tutorial

Formatting a Form

Formatting a Form

Apply formatting to enhance the appearance of a form. Format a form by applying a theme, theme colors, and theme fonts; applying formatting with options on the Form Layout Tools Format tab; and applying conditional formatting that meets a specific criterion.

Hint Themes available in Access are the same as the themes available in Word, Excel, and PowerPoint.

Applying Themes

Access provides a number of themes for formatting objects in a database. A theme is a set of formatting choices that include a color theme (a set of colors) and a font theme (a set of heading and body text fonts). To apply a theme to a form, click the Themes button in the Themes group on the Form Layout Tools Design tab and then click a theme at the drop-down gallery. Position the mouse pointer over a theme and the live preview feature will display the form with the theme formatting applied. When a theme is applied, any new objects created in the database will be formatted with that theme.

 Themes

Further customize the formatting of a form with the Colors button and the Fonts button in the Themes group on the Form Layout Tools Design tab. To customize the theme colors, click the Colors button in the Themes group and then click an option at the drop-down list. Change the theme fonts by clicking the Fonts button in the Themes group and then clicking an option at the drop-down list.

 Colors

 Fonts

Formatting with the Form Layout Tools Format Tab

Click the Form Layout Tools Format tab and buttons and options display for applying formatting to a form or specific cells in a form. To apply formatting to a specific cell, click the cell in the form or click the Object button arrow in the Selection group and then click the control object at the drop-down list. To format all cells in the form, click the Select All button in the Selection group. This selects all cells in the form, including cells in the *Form Header* section. To select all of the cells in the *Detail* section (and not the *Form Header* section), click in a cell in the *Detail* section and then click the table move handle.

 Object

 Select All

Use options and buttons in the Font, Number, Background, and Control Formatting groups to apply formatting to a cell or selected cells in a form. Use options and buttons in the Font group to change the font, change the font size, apply text effects (such as bold and underline), and change the alignment of data in cells. If the form contains data with a Number or Currency data type, use buttons in the Number group to apply specific formatting to numbers. Insert a background image in the form using the Background Image button and apply formatting to cells with buttons in the Control Formatting group. Depending on what is selected in the form, some of the buttons may not be active.

 Background Image

Activity 1f Formatting a Form

Part 6 of 7

1. With the Clients form in **5-Dearborn** open and in Layout view, apply a theme by completing the following steps:
   a. Click the Form Layout Tools Design tab.
   b. Click the Themes button and then click the *Facet* option (second column, first row in the *Office* section).
2. Change the theme fonts by clicking the Fonts button in the Themes group and then clicking *Gill Sans MT* at the drop-down gallery. (You will need to scroll down the list to display *Gill Sans MT*.)

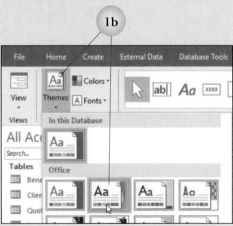

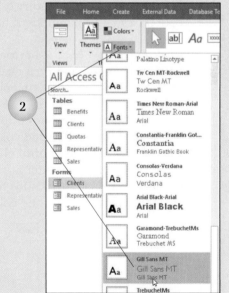

3. Change the theme colors by clicking the Colors button in the Themes group and then clicking *Orange* at the drop-down gallery.

4. Change the font and font size of text in the form table by completing the following steps:
   a. Click in any cell containing a control object in the form table.
   b. Select all cells in the form table by clicking the table move handle in the upper left corner of the *Detail* section.
   c. Click the Form Layout Tools Format tab.
   d. Click the *Font* option box arrow, scroll down the drop-down list, and then click *Tahoma*. (Fonts are alphabetized in the drop-down list.)
   e. Click the *Font Size* option box arrow and then click *10* at the drop-down list.

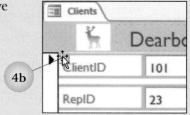

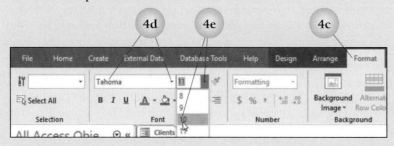

5. Apply formatting and change the alignment of the first column of cells by completing the following steps:
   a. Click in the control object containing the field name *ClientID*, press and hold down the Shift key, click in the bottom control object containing the field name *Telephone*, and then release the Shift key.
   b. Click the Bold button in the Font group.
   c. Click the Shape Fill button in the Control Formatting group and then click the *Brown, Accent 3, Lighter 60%* option (seventh column, third row in the *Theme Colors* section).
   d. Click the Shape Outline button in the Control Formatting group and then click the *Brown, Accent 3, Darker 50%* option (seventh column, bottom row in the *Theme Colors* section).
   e. Click the Align Right button in the Font group.
6. Apply shape fill to the second column of cells by completing the following steps:
   a. Click in the text box control object containing the text *101*.
   b. Position the mouse pointer at the top border of the selected cell until the pointer displays as a small, black, down arrow and then click the left mouse button. (Make sure all of the cells in the second column are selected.)

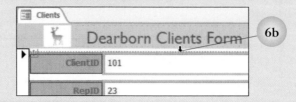

   c. Click the Shape Fill button in the Control Formatting group and then click the *Brown, Accent 3, Lighter 80%* option (seventh column, second row in the *Theme Colors* section).

7. Remove the gridlines by completing the following steps:
   a. Click the Form Layout Tools Arrange tab.
   b. Click the Select Layout button in the Rows & Columns group.
   c. Click the Gridlines button in the Table group and then click *None* at the drop-down list.
8. Click the Save button on the Quick Access Toolbar to save the Clients form.
9. Insert a background image by completing the following steps:
   a. Click the Form Layout Tools Format tab.
   b. Click the Background Image button in the Background group and then click *Browse* at the drop-down list.
   c. Navigate to your AL1C5 folder and then double-click **Mountain**.
   d. View the form and background image in Print Preview. (To display Print Preview, click the File tab, click the *Print* option, and then click the *Print Preview* option.)
   e. After viewing the form in Print Preview, return to the form by clicking the Close Print Preview button.
10. Click the Undo button on the Quick Access Toolbar to remove the background image. (If this does not remove the image, close the form without saving it and then reopen the form.)
11. Save the Clients form.

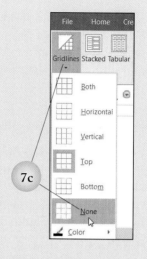

7c

⬤ Check Your Work >

⬤ Tutorial >

Applying Conditional Formatting to a Form

Conditional Formatting

Quick Steps
Apply Conditional Formatting
1. Click Form Layout Tools Format tab.
2. Click Conditional Formatting button.
3. Click New Rule button.
4. Specify formatting.
5. Click OK.
6. Click OK.

Applying Conditional Formatting

Use the Conditional Formatting button in the Control Formatting group on the Form Layout Tools Format tab to apply formatting to data that meets a specific criterion. For example, conditional formatting can be applied to display sales amounts higher than a certain number in a different color, or to display certain state names in a specific color. Conditional formatting can also be applied that inserts data bars that visually compare data among records. The data bars provide a visual representation of the comparison. For example, in Activity 1g, data bars will be inserted in the *Sales* field that provide a visual representation of how the sales amount in one record compares to the sales amounts in other records.

To apply conditional formatting, click the Conditional Formatting button in the Control Formatting group and the Conditional Formatting Rules Manager dialog box displays. At this dialog box, click the New Rule button and the New Formatting Rule dialog box displays, as shown in Figure 5.3. In the *Select a rule type* list box, choose the *Check values in the current record or use an expression* option if the conditional formatting is applied to a field in the record that matches a specific condition. Click the *Compare to other records* option to insert data bars in a field in all records that compare the data among the records.

Apply conditional formatting to a field by specifying the field and field condition with options in the *Edit the rule description* section of the dialog box. Specify the type of formatting to be applied to data in a field that meets the specific criterion. For example, in Activity 1g, conditional formatting will be applied that changes the shape fill to a light green for all *City* fields containing the text *Indianapolis*. When all changes have been made at the dialog box, click OK to close the dialog box and then click OK to close the Conditional Formatting Rules Manager dialog box.

To insert data bars in a field, click the Conditional Formatting button, click the New Rule button at the Conditional Formatting Rules Manager dialog box, and then click the *Compare to other records* option in the *Select a rule type* list box. This changes the options in the dialog box, as shown in Figure 5.4. Make specific changes in the *Edit the rule description* section.

Figure 5.3 New Formatting Rule Dialog Box with the *Check values in the current record or use an expression* Option Selected

Specify the field value in this option box.

Use options in this section to specify conditional formatting.

Specify the field condition in this option box.

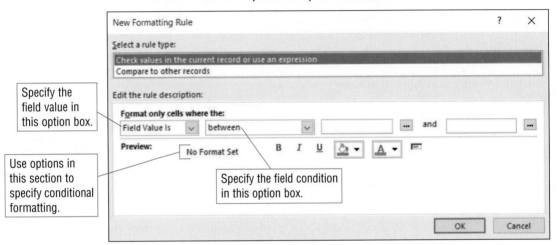

Figure 5.4 New Formatting Rule Dialog Box with the *Compare to other records* Option Selected

Click this option to insert data bars that visually compare data among records.

Use options in this section to specify the formatting of data bars.

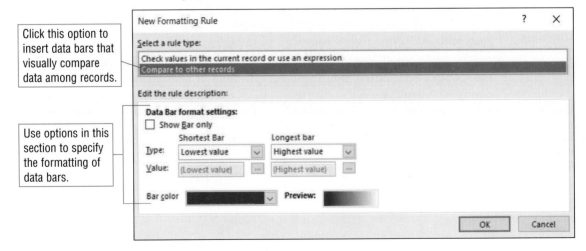

1. With the Clients form in **5-Dearborn** open and in Layout view, apply conditional formatting so that the *City* field displays all Indianapolis entries with a light green shape fill by completing the following steps:
   a. Click in the text box control object containing the text *Muncie*.
   b. Click the Form Layout Tools Format tab.
   c. Click the Conditional Formatting button in the Control Formatting group.
   d. At the Conditional Formatting Rules Manager dialog box, click the New Rule button.

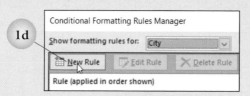

   e. At the New Formatting Rule dialog box, click the option box arrow for the option box containing the word *between* and then click *equal to* at the drop-down list.
   f. Click in the text box to the right of the *equal to* option box and then type Indianapolis.
   g. Click the Background color button arrow and then click the *Green 3* option (seventh column, fourth row).
   h. Click OK to close the New Formatting Rule dialog box.
   i. Click OK to close the Conditional Formatting Rules Manager dialog box.

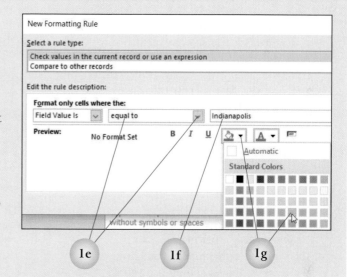

2. Click the Home tab and then click the View button to switch to Form view.
3. Click the Next record button to display the next record in the form. Continue clicking the Next record button to view records and notice that *Indianapolis* entries display with a light green shape fill.
4. Click the First record button in the Record Navigation bar.
5. Click the Save button on the Quick Access Toolbar.
6. Print page 2 of the form by completing the following steps:
   a. Click the File tab and then click the *Print* option.
   b. At the Print backstage area, click the *Print* option.
   c. At the Print dialog box, click the *Pages* option in the *Print Range* section, type 2 in the *From* text box, press the Tab key, and then type 2 in the *To* text box.
   d. Click OK.
7. Close the Clients form.
8. Open the Sales form and switch to Layout View by clicking the View button in the Views group on the Home tab.

9. With the text box control object containing the sales ID number *1* selected, drag the right border to the left until *Lines: 1 Characters: 21* displays at the left side of the Status bar.

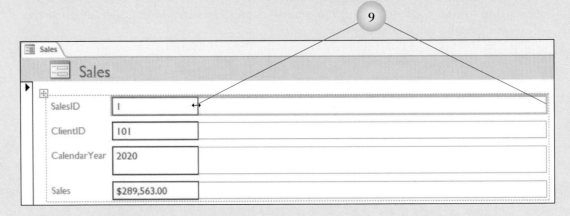

10. Change the alignment of text by completing the following steps:
    a. Right-click the selected text box control object (the object containing *1*) and then click *Select Entire Column* at the shortcut menu.
    b. Click the Form Layout Tools Format tab.
    c. Click the Align Right button in the Font group.
11. Apply data bars to the *Sales* field by completing the following steps:
    a. Click in the text box control object containing the amount *$289,563.00*.
    b. Make sure the Form Layout Tools Format tab is active.
    c. Click the Conditional Formatting button.
    d. At the Conditional Formatting Rules Manager dialog box, click the New Rule button.
    e. At the New Formatting Rule dialog box, click the *Compare to other records* option in the *Select a rule type* list box.
    f. Click the *Bar color* option box arrow and then click the *Green 4* option (seventh column, fifth row).
    g. Click OK to close the New Formatting Rule dialog box and then click OK to close the Conditional Formatting Rules Manager dialog box.

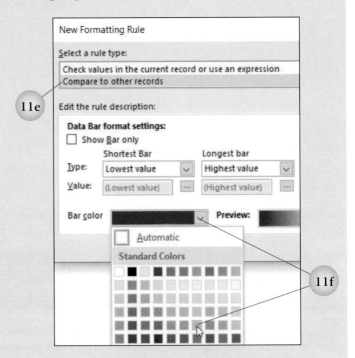

12. Click the Next record button in the Record Navigation bar to display the next record. Continue clicking the Next record button and notice the data bars in the *Sales* field.
13. Click the First record button in the Record Navigation bar.
14. Click the Save button on the Quick Access Toolbar.

15. Print page 1 of the form by completing the following steps:
   a. Click the File tab and then click the *Print* option.
   b. At the Print backstage area, click the *Print* option.
   c. At the Print dialog box, click the *Selected Record(s)* option in the *Print Range* section and then click OK.
16. Close the Sales form.
17. Close **5-Dearborn**.

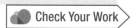 Check Your Work

---

Activity 2 **Add Fields, Create a Split Form and Multiple Items Form, and Use the Form Wizard**   **6 Parts**

You will open the Skyline database, create a form and add related fields and a calculation to the form, create a split and multiple items form, and create a form using the Form Wizard.

 Tutorial

Adding an Existing Field to a Form

 Add Existing Fields

♡ **Hint** Alt + F8 is the keyboard shortcut to display the Field List task pane.

♡ **Hint** Use the Field List task pane to add fields from a table or query to a form.

Quick Steps

Add Existing Field to Form
1. Click Add Existing Fields button on Form Layout Tools Design tab.
2. Drag field from Field List task pane to specific location in form.

Adding an Existing Field

A field can be inserted into an existing form by opening the form in Layout view and then clicking the Add Existing Fields button in the Tools group on the Form Layout Tools Design tab. Clicking the Add Existing Field button displays the Field List task pane at the right side of the screen. This task pane, shown in Figure 5.5, displays the fields available in the current view, fields available in related tables, and fields available in other tables.

In the *Fields available for this view* section, Access displays all fields in any tables used to create the form. So far, forms in activities have been created using all fields in one table. In the *Fields available in related tables* section, Access displays tables that are related to the table(s) used to create the form. To display the fields in the related table, click the expand button (plus symbol in a square) that displays to the left of the table name in the Field List task pane and the list expands to display all of the field names.

To add a field to the form, double-click the field in the Field List task pane. This inserts the field below the active control object in the form. Another method for inserting a field is to drag the field from the Field List task pane into the form. To do this, position the mouse pointer on the field in the Field List task pane, click and hold down the left mouse button, drag into the form, and then release the mouse button. A pink insert indicator bar displays when dragging the field into the existing fields in the form. Drag over an empty cell and the cell displays with pink fill. When the pink insert indicator bar is in the desired position or the cell is selected, release the mouse button.

Multiple fields can be inserted in a form from the Field List task pane. To do this, press and hold down the Ctrl key while clicking specific fields and then drag the fields into the form. Dragging a field from a table in the *Fields available in other tables* section displays the Specify Relationship dialog box. To move a field from the Field List task pane to the form, the field must be in a table that is related to the table(s) used to create the form.

Figure 5.5 Field List Task Pane

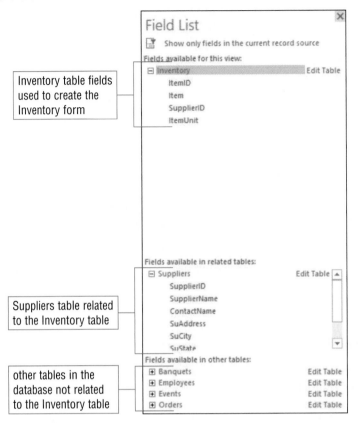

Inventory table fields used to create the Inventory form

Suppliers table related to the Inventory table

other tables in the database not related to the Inventory table

Activity 2a Adding Existing Fields to a Form

Part 1 of 6

1. Open **5-Skyline** from your AL1C5 folder and enable the content.
2. Create a form with the Inventory table by clicking *Inventory* in the Tables group in the Navigation pane, clicking the Create tab, and then clicking the Form button.
3. With the text box control object containing the text *001* selected, drag the right border to the left until the selected object is approximately one-half the original width.

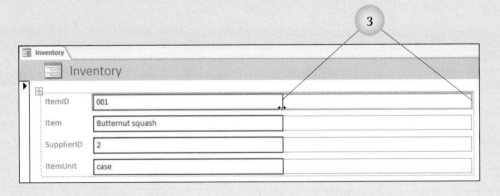

4. With the text box control object still selected, click the Form Layout Tools Arrange tab and then click the Split Horizontally button in the Merge/Split group. (This splits the text box control object into one object and one empty cell.)

5. You decide that you want to add the supplier name to the form so the name displays when entering data in the form. Add the *SupplierName* field by completing the following steps:
   a. Click the Form Layout Tools Design tab.
   b. Click the Add Existing Fields button in the Tools group.
   c. Click the Show all tables hyperlink in the Field List task pane.
   d. Click the expand button immediately left of the Suppliers table name in the *Fields available in related tables* section of the Field List task pane.

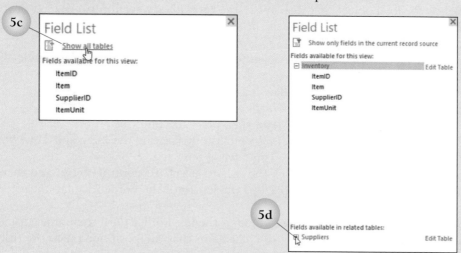

   e. Position the mouse pointer on the *SupplierName* field, click and hold down the left mouse button, drag into the form until the pink insert indicator bar displays immediately right of the text box control containing the *2* (the text box control at the right side of the *SupplierID* label control), and then release the mouse button. Access inserts the field as a Lookup field (a down arrow displays at the right side of the field).

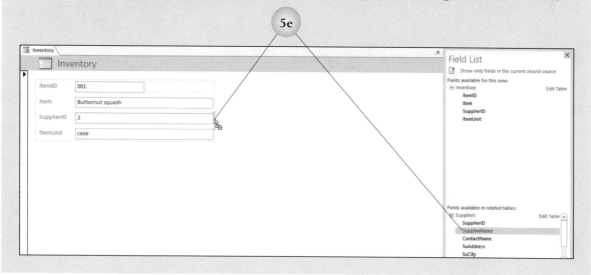

f. Change the *SupplierName* field from a Lookup field to a text box by clicking the Options button that displays below the field and then clicking *Change to Text Box* at the drop-down list. (This removes the down arrow at the right side of the field.)

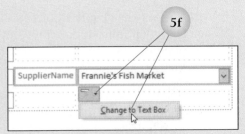

g. Close the Field List task pane by clicking the Close button in the upper right corner of the task pane.

6. Insert a logo image in the *Form Header* section by completing the following steps:
   a. Right-click the logo object that displays in the *Form Header* section (to the left of the title *Inventory*) and then click *Delete* at the shortcut menu.
   b. Click the Logo button.
   c. At the Insert Picture dialog box, navigate to your AL1C5 folder and then double-click the file named **Cityscape**.

7. Change the title by completing the following steps:
   a. Click the Title button. (This selects *Inventory* in the *Form Header* section.)
   b. Type Skyline Inventory Input Form and then press the Enter key.

8. Insert the date and time in the *Form Header* section by clicking the Date and Time button and then clicking OK at the Date and Time dialog box.

9. Click in any field outside the title, click the title to select the control object, and then drag the right border of the title control object to the left until the border displays near the title.

10. Select the date and time control objects, drag in the left border until the border displays near the date and time, and then drag the objects so they are positioned near the title.

11. Scroll through the records in the form.

12. Click the First record button in the Record Navigation bar.

13. Click the Save button on the Quick Access Toolbar and save the form with the name *Inventory*.

14. Print the current record.

15. Close the Inventory form.

Inserting a Calculation in a Form

Inserting a Calculation in a Form

A calculation can be inserted in a form in a text box control object. To insert a text box control object as well as a label control object, click the Text Box button in the gallery in the Controls group on the Form Layout Tools Design tab, and then click in the location in the form where the two objects are to display. Click in the label control box and then type a label for the calculated field.

Insert a calculation by clicking in the text box control object and then clicking the Property Sheet button in the Tools group on the Form Layout Tools Design tab. This displays the Property Sheet task pane at the right side of the screen. The Property Sheet task pane contains options for setting the form's properties. To insert a calculation in the text box, click the Data tab in the Property Sheet task pane, click in the *Control Source* property box, and then type the calculation.

Quick Steps

Insert Calculation in Form
1. Click Text Box button on Form Layout Tools Design tab.
2. Click in form to insert label and text box control objects.
3. Click in text box control object.
4. Click Property Sheet button in Tools group.
5. Click Data tab.
6. Click in *Control Source* property box.
7. Type calculation.
8. Click Close button.

Type a calculation in the *Control Source* property box using mathematical operators such as the plus symbol (+) for addition, hyphen (-) for subtraction, the asterisk (*) for multiplication, and the forward slash symbol (/) for division. Type field names in the calculation inside square brackets. A field name must be typed in the calculation as it appears in the source object. For example, in Activity 2b the calculation =*[AmountTotal]-[AmountPaid]* will be inserted in a text box control object to determine the amount due on banquet reservations. Notice that the calculation begins with the equals sign, the field names are typed inside brackets, and the hyphen is used to indicate subtraction. This calculation will subtract the amount paid for a banquet event from the banquet event total amount.

If a calculation result is currency, apply currency formatting to the text box control object. Apply currency formatting by clicking the Form Layout Tools Format tab and then clicking the Apply Currency Format button in the Number group.

Activity 2b Inserting a Calculation in a Form

Part 2 of 6

1. With **5-Skyline** open, create a form with the AmountDue query using the Form button.
2. With the form open in Layout view, insert a text box control by completing the following steps:
   a. Click the Text Box button in the gallery in the Controls group on the Form Layout Tools Design tab.
   b. Position the crosshairs below the *AmountPaid* field (the pink insert indicator bar displays below the field) and then click the left mouse button. (This inserts a label control object and a text box control object in the form.)

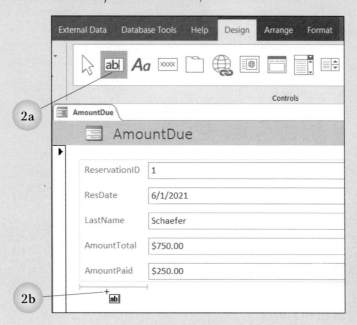

3. Name the new field by clicking in the label control object, double-clicking the text in the label control object, and then typing AmountDue.
4. Click in the text box control object and then insert a calculation by completing the following steps:
   a. Click the Property Sheet button in the Tools group on the Form Layout Tools Design tab.
   b. At the Property Sheet task pane, click the Data tab.
   c. Click in the *Control Source* property box.
   d. Type =[AmountTotal]-[AmountPaid] and then press the Enter key.
   e. Close the Property Sheet task pane by clicking the Close button in the upper right corner of the task pane.

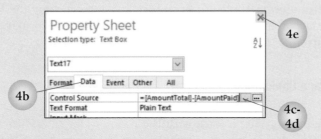

5. With the text box control object still selected, apply currency formatting by completing the following steps:
   a. Click the Form Layout Tools Format tab.
   b. Click the Apply Currency Format button in the Number group.

6. Scroll through the records in the form by clicking the Next record button and then click the First record button.
7. Save the form with the name *BanquetAmountsDue*.
8. Print the first page of the form.
9. Close the BanquetAmountsDue form.

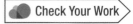 **Check Your Work**

 Tutorial

Creating a Split Form and Multiple Items Form

 More Forms

Quick Steps

Create Split Form
1. Click table.
2. Click Create tab.
3. Click More Forms button.
4. Click *Split Form*.

Creating a Split Form

Another method for creating a form is to use the *Split Form* option at the More Forms button drop-down list in the Forms group on the Create tab. Use this option to create a form and Access splits the screen in the work area and provides two views of the form. The top half of the work area displays the form in Layout view and the bottom half of the work area displays the form in Datasheet view. The two views are connected and are synchronous, which means that displaying or modifying a specific field in the Layout view portion will cause the same action to occur in the field in the Datasheet view portion. Figure 5.6 displays the split form that will be created in Activity 2c.

Figure 5.6 Split Form

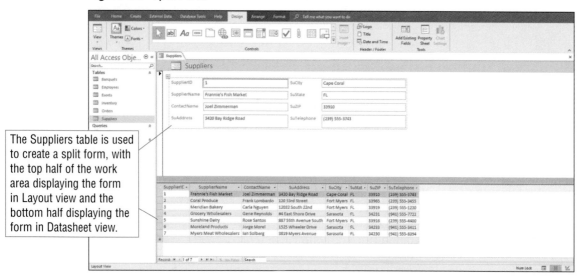

The Suppliers table is used to create a split form, with the top half of the work area displaying the form in Layout view and the bottom half displaying the form in Datasheet view.

Activity 2c Creating a Split Form

1. With **5-Skyline** open, create a split form with the Suppliers table by completing the following steps:
   a. Click *Suppliers* in the Tables group in the Navigation pane.
   b. Click the Create tab.
   c. Click the More Forms button in the Forms group and then click *Split Form* at the drop-down list.

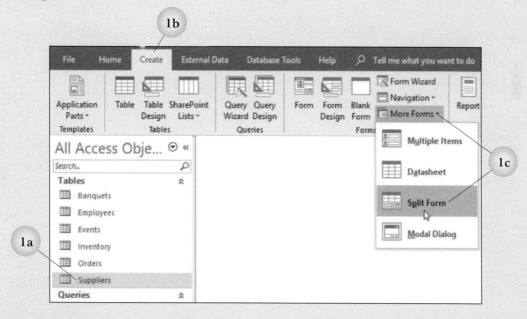

   d. Click several times on the Next record button in the Record Navigation bar. (As you display records, notice that the current record in the Form view in the top portion of the window is the same record selected in Datasheet view in the lower portion of the window.)
   e. Click the First record button.

2. Apply a theme by clicking the Themes button in the Themes group on the Form Layout Tools Design tab and then clicking *Integral* at the drop-down gallery (fourth column, first row in the *Office* section).

3. Insert a logo image in the *Form Header* section by completing the following steps:
   a. Right-click the logo object that displays in the *Form Header* section (to the left of the title *Suppliers*) and then click *Delete* at the shortcut menu.
   b. Click the Logo button.
   c. At the Insert Picture dialog box, navigate to your AL1C5 folder and then double-click **Cityscape**.

4. Change the title by completing the following steps:
   a. Click the Title button.
   b. Type Skyline Suppliers Input Form and then press the Enter key.
   c. Click in any field outside the title, click the title again to select the control object, and then drag the right border to the left until the border displays near the title.

5. Click the text box control object containing the supplier identification number *1* and then drag the right border of the text box control object to the left until *Lines: 1 Characters: 25* displays at the left side of the Status bar.

6. Click the text box control object containing the city *Cape Coral* and drag the right border of the text box control object to the left until *Lines: 1 Characters: 25* displays at the left side of the Status bar.

7. Insert a new record in the Suppliers form by completing the following steps:
   a. Click the View button to switch to Form view.
   b. Click the New (blank) record button in the Record Navigation bar.
   c. Click in the *SupplierID* field in the Form view portion of the window and then type the following information in the specified fields:

SupplierID	8
SupplierName	Jackson Produce
ContactName	Marshall Jackson
SuAddress	5790 Cypress Avenue
SuCity	Fort Myers
SuState	FL
SuZIP	33917
SuTelephone	2395555002

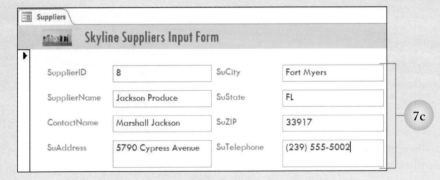

8. Click the Save button on the Quick Access Toolbar and save the form with the name *Suppliers*.
9. Print the current form by completing the following steps:
    a. Click the File tab and then click the *Print* option.
    b. At the Print backstage area, click the *Print* option.
    c. At the Print dialog box, click the Setup button.
    d. At the Page Setup dialog box, click the *Print Form Only* option in the *Split Form* section of the dialog box and then click OK.
    e. At the Print dialog box, click the *Selected Record(s)* option and then click OK.
10. Close the Suppliers form.

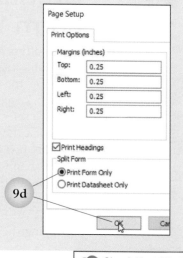

9d

Check Your Work

Creating a Multiple Items Form

Quick Steps

Create Multiple Items Form
1. Click table.
2. Click Create tab.
3. Click More Forms button.
4. Click *Multiple Items*.

When a form is created with the Form button, a single record displays. Use the *Multiple Items* option at the More Forms button drop-down list to create a form that displays multiple records. The advantage to creating a multiple items form over displaying the table in Datasheet view is that the form can be customized using buttons on the Form Layout Tools ribbon with the Design, Arrange, or Format tab selected.

Activity 2d Creating a Multiple Items Form

Part 4 of 6

1. With **5-Skyline** open, create a multiple items form by completing the following steps:
    a. Click *Orders* in the Tables group in the Navigation pane.
    b. Click the Create tab.
    c. Click the More Forms button and then click *Multiple Items* at the drop-down list.
2. Delete the existing logo and then insert the **Cityscape** image as the logo.
3. Type Skyline Orders as the title.
4. Click in any field outside the title, click the title again to select the control object, and then drag the right border to the left until the border displays near the title.
5. Save the form with the name *Orders*.
6. Print the first page of the form by completing the following steps:
    a. Click the File tab and then click the *Print* option.
    b. At the Print backstage area, click the *Print* option.
    c. At the Print dialog box, click the *Pages* option in the *Print Range* section.
    d. Type 1 in the *From* text box, press the Tab key, and then type 1 in the *To* text box.
    e. Click OK.
7. Close the Orders form.

Check Your Work

Form Wizard

Quick Steps

**Create Form Using
Form Wizard**
1. Click Create tab.
2. Click Form Wizard
   button.
3. Choose options at
   each Form Wizard
   dialog box.

💡*Hint* With the
Form Wizard, you can
be more selective
about which fields you
insert in a form.

Creating a Form Using the Form Wizard

Access offers a Form Wizard that provides steps for creating a form. To create a form using the Form Wizard, click the Create tab and then click the Form Wizard button in the Forms group. At the first Form Wizard dialog box, shown in Figure 5.7, specify the table or query and then the fields to be included in the form. To select the table or query, click the *Table/Queries* option box arrow and then click the table or query at the drop-down list. Select a field in the *Available Fields* list box and then click the One Field button (the button containing the greater-than [>] symbol). This inserts the field in the *Selected Fields* list box. Continue in this manner until all of the fields have been inserted in the *Selected Fields* list box. To insert all of the fields into the *Selected Fields* list box at one time, click the All Fields button (the button containing two greater-than symbols). After specifying the fields, click the Next button.

At the second Form Wizard dialog box, specify the layout for the records. Choose from these layout type options: *Columnar, Tabular, Datasheet,* and *Justified.* Click the Next button and the third and final Form Wizard dialog box displays. It offers a title for the form and also provides the option *Open the form to view or enter information.* Make any necessary changes in this dialog box and then click the Finish button.

Figure 5.7 First Form Wizard Dialog Box

Click the *Table/Queries* option box arrow and then click a table or query at the drop-down list.

Add a field to the *Selected Fields* list box by clicking the field in the *Available Fields* list box and then clicking the One Field button.

Insert all of the fields into the *Selected Fields* list box at one time by clicking the All Fields button.

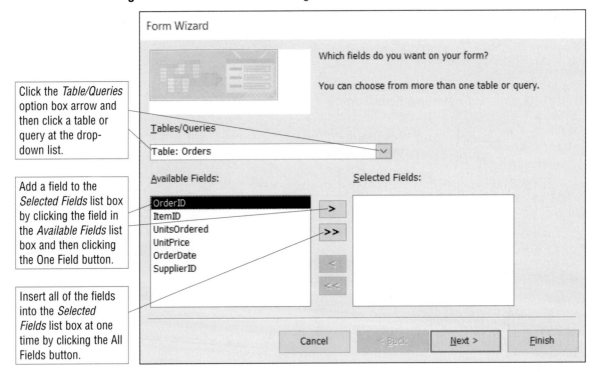

1. With **5-Skyline** open, create a form with the Form Wizard by completing the following steps:
   a. Click the Create tab.
   b. Click the Form Wizard button in the Forms group.
   c. At the first Form Wizard dialog box, click the *Tables/Queries* option box arrow and then click *Table: Employees* at the drop-down list.
   d. Specify that you want all of the fields included in the form by clicking the All Fields button (the button containing the two greater-than symbols).

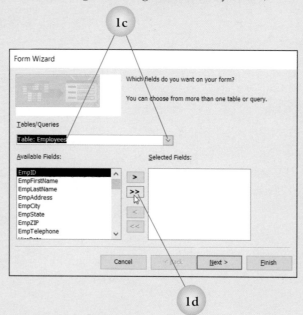

e. Click the Next button.
f. At the second Form Wizard dialog box, click the *Justified* option and then click the Next button.
g. At the third and final Form Wizard dialog box, click the Finish button.
2. Format the field headings by completing the following steps:
   a. Click the View button to switch to Layout view.
   b. Click the *EmpID* label control object. (This selects the object.)

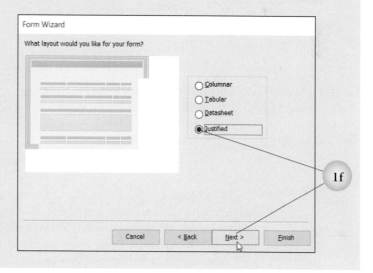

c. Press and hold down the Ctrl key and then click each of the following label control objects: *EmpFirstName, EmpLastName, EmpAddress, EmpCity, EmpState, EmpZIP, EmpTelephone, HireDate,* and *HealthIns.*

d. With all of the label control objects selected, release the Ctrl key.

e. Click the Form Layout Tools Format tab.

f. Click the Shape Fill button and then click the *Aqua Blue 2* option (ninth column, third row in the *Standard Colors* section).

g. Click the Form Layout Tools Design tab and then click the View button to switch to Form view.

3. In Form view, click the New (blank) record button and then add the following records:

EmpID	13
EmpFirstName	Carol
EmpLastName	Thompson
EmpAddress	6554 Willow Drive, Apt. B
EmpCity	Fort Myers
EmpState	FL
EmpZIP	33915
EmpTelephone	2395553719
HireDate	10/1/2021
HealthIns	(Click the check box to insert a check mark.)

EmpID	14
EmpFirstName	Eric
EmpLastName	Hahn
EmpAddress	331 South 152nd Street
EmpCity	Cape Coral
EmpState	FL
EmpZIP	33906
EmpTelephone	2395558107
HireDate	10/1/2021
HealthIns	(Leave blank.)

4. Click the Save button on the Quick Access Toolbar.

5. Print the record for Eric Hahn and then print the record for Carol Thompson.

6. Close the Employees form.

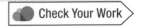 Check Your Work

In Activity 2e, the Form Wizard was used to create a form with all of the fields in one table. If tables are related, a form can be created using fields from related tables. At the first Form Wizard dialog box, choose fields from the selected table and then choose fields from a related table. To change to a related table, click the *Tables/Queries* option box arrow and then click the name of the related table.

1. With **5-Skyline** open, create a form with related tables by completing the following steps:
   a. Click the Create tab.
   b. Click the Form Wizard button.
   c. At the first Form Wizard dialog box, click the *Tables/Queries* option box arrow and then click *Table: Banquets*.
   d. Click *ResDate* in the *Available Fields* list box and then click the One Field button. (This inserts *ResDate* in the *Selected Fields* list box.)
   e. Click *AmountTotal* in the *Available Fields* list box and then click the One Field button.
   f. With *AmountPaid* selected in the *Available Fields* list box, click the One Field button.
   g. Click the *Tables/Queries* option box arrow and then click *Table: Events* at the drop-down list.
   h. Click *Event* in the *Available Fields* list box and then click the One Field button.
   i. Click the *Tables/Queries* option box arrow and then click *Table: Employees* at the drop-down list.
   j. Click *EmpLastName* in the *Available Fields* list box and then click the One Field button.
   k. Click the Next button.
   l. At the second Form Wizard dialog box, click the Next button.
   m. At the third Form Wizard dialog box, click the Next button.
   n. At the fourth Form Wizard dialog box, select the text in the *What title do you want for your form?* text box, type Upcoming Banquets, and then click the Finish button.

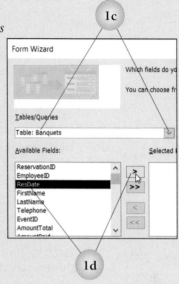

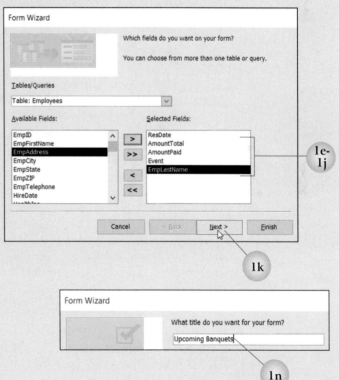

2. When the first record displays, print the record.
3. Save and then close the form.
4. Close **5-Skyline**.

> Check Your Work

Chapter Summary

- Creating a form generally improves the ease of entering data into a table. Some methods for creating a form include using the Form, Split Form, and Multiple Items buttons or the Form Wizard.

- A form is an object used to enter and edit data in a table or query and to help prevent incorrect data from being entered in a database.

- The simplest method for creating a form is to click a table in the Navigation pane, click the Create button, and then click the Form button in the Forms group.

- Create a form and it displays in Layout view. Use this view to display data and modify the appearance and contents of the form. Other form views include Form view and Design view. Use Form view to enter and manage records and use Design view to view and modify the structure of the form.

- Open an existing form in Layout view by right-clicking the form in the Navigation pane and then clicking *Layout View* at the shortcut menu.

- Print a form with options at the Print dialog box. To print an individual record, display the Print dialog box, click the *Selected Record(s)* option, and then click OK.

- Delete a form with the Delete button in the Records group on the Home tab or by right-clicking the form in the Navigation pane and then clicking *Delete* at the shortcut menu. A message may display asking to confirm the deletion.

- Navigate in a form with buttons in the Record Navigation bar.

- Add a new record to a form by clicking the New (blank) record button in the Record Navigation bar or by clicking the Home tab and then clicking the New button in the Records group.

- Delete a record from a form by displaying the record, clicking the Home tab, clicking the Delete button arrow, and then clicking *Delete Record* at the drop-down list.

- If a form is created with a table that has a one-to-many relationship established, Access adds a datasheet at the bottom of the form.

- A form is made up of cells arranged in rows and columns and each cell can contain one control object. Customize control objects with buttons on the Form Layout Tools ribbon with the Design tab, Arrange tab, or Format tab selected. These tabs are available when a form displays in Layout view.

- Apply a theme to a form with the Themes button in the Themes group on the Form Layout Tools Design tab. Use the Colors and Fonts buttons in the Themes group to further customize a theme.

- Use buttons in the Header/Footer group on the Form Layout Tools Design tab to insert a logo, form title, and the date and time.

- Control objects can be sized, deleted, and inserted in Layout view.

- Use buttons in the Rows & Columns group on the Form Layout Tools Arrange tab to select or insert rows or columns of cells.

- The Controls group on the Form Layout Tools Design tab contains control objects that can be inserted in a form.

- Merge cells in a form by selecting cells and then clicking the Merge button in the Merge/Split group on the Form Layout Tools Arrange tab. Split selected cells by clicking the Split Vertically or Split Horizontally button.

- Format cells in a form with buttons on the Form Layout Tools Format tab.

- Use the Conditional Formatting button in the Control Formatting group on the Form Layout Tools Format tab to apply formatting to data that matches a specific criterion.
- Click the Add Existing Fields button in the Tools group on the Form Layout Tools Design tab to display the Field List task pane. Add fields to the form by double-clicking a field or dragging the field from the task pane to the form.
- Insert a calculation in a form by inserting a text box control object, displaying the Property Sheet task pane with the Data tab selected, and then typing the calculation in the *Control Source* property box. If a calculation result is currency, apply currency formatting with the Apply Currency Format button on the Form Layout Tools Format tab.
- Create a split form by clicking the More Forms button on the Create tab and then clicking *Split Form* in the drop-down list. Access displays the form in Layout view in the top portion of the work area and in Datasheet view in the bottom portion of the work area. The two views are connected and synchronous.
- Create a Multiple Items form by clicking the More Forms button on the Create tab and then clicking *Multiple Items* in the drop-down list.
- The Form Wizard provides steps for creating a form such as specifying the fields to be included in the form, a layout for the records, and a name for the form.
- A form can be created with the Form Wizard that contains fields from tables connected by a one-to-many relationship.

Commands Review

FEATURE	RIBBON TAB, GROUP	BUTTON, OPTION	KEYBOARD SHORTCUT
Conditional Formatting Rules Manager dialog box	Form Layout Tools Format, Control Formatting		
Field List task pane	Form Layout Tools Design, Tools		
form	Create, Forms		
Form Wizard	Create, Forms		
multiple items form	Create, Forms	, *Multiple Items*	
Property Sheet task pane	Form Layout Tools Design, Tools		Alt + Enter
split form	Create, Forms	, *Split Form*	

Microsoft®

Access®

Creating Reports
and Mailing Labels

Performance Objectives

Upon successful completion of Chapter 6, you will be able to:

1   Create a report using the Report button

2   Modify the record source

3   Select, edit, size, move, and delete control objects

4   Sort records

5   Find data

6   Display and customize a report in Print Preview

7   Delete a report

8   Format a report

9   Apply conditional formatting to data in a report

10   Group and sort records in a report

11   Insert a calculation in a report

12   Create a report using the Report Wizard

13   Create mailing labels using the Label Wizard

In this chapter, you will learn how to prepare reports from data in a table or query using the Report button in the Reports group on the Create tab and using the Report Wizard. You will also learn how to manage control objects, format, and insert a calculation in a report and create mailing labels using the Label Wizard.

 Data Files

Before beginning chapter work, copy the AL1C6 folder to your storage medium and then make AL1C6 the active folder.

The online course includes additional training and assessment resources.

Tutorial

Creating a Report

Quick Steps

Create Report
1. Click table or query in Navigation pane.
2. Click Create tab.
3. Click Report button.

Report

💡 **Hint** Create a report to control what data appears on the page when printed.

Creating a Report

Create a report in a database to control what data appears on the page when printed and how the data is formatted. Reports generally answer specific questions (queries). For example, a report could answer the question *What customers have submitted claims?* or *What products do we currently have on order?* The record source for a report can be a table or query. Create a report with the Report button in the Reports group or use the Report Wizard that provides steps for creating a report.

Creating a Report with the Report Button

To create a report with the Report button, click a table or query in the Navigation pane, click the Create tab, and then click the Report button in the Reports group. This displays the report in columnar style in Layout view with the Report Layout Tools Design tab active, as shown in Figure 6.1. Access creates the report using all of the fields in the table or query.

Figure 6.1 Report Created with Sales Table

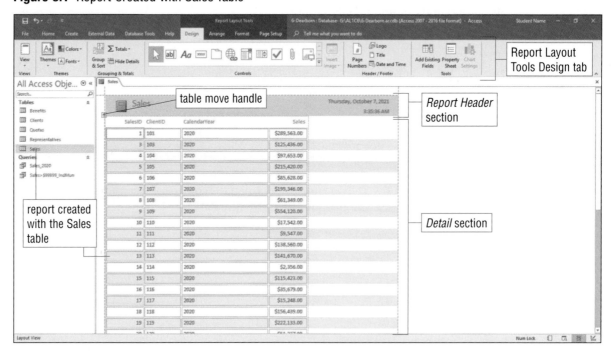

Modifying the Record Source

The record source for a report is the table or query used to create the report. If changes are made to the record source, such as adding or deleting records, those changes are reflected in the report. For example, in Activity 1a, a report will be created based on the Sales table. A record will be added to the Sales table (the record source for the report) and the added record will display in the Sales report.

Activity 1a Creating Reports with the Report Button   Part 1 of 5

1. Open **6-Dearborn** from your AL1C6 folder and enable the content.
2. Create a report based on the Sales table by completing the following steps:
   a. Click *Sales* in the Tables group in the Navigation pane.
   b. Click the Create tab.
   c. Click the Report button in the Reports group.

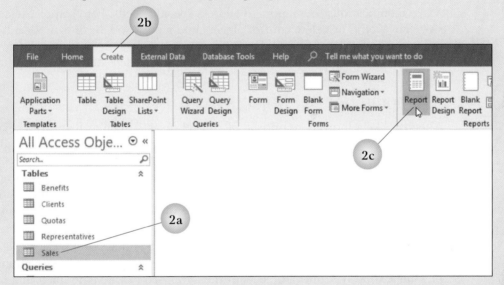

   d. Save the report by clicking the Save button on the Quick Access Toolbar and then clicking OK at the Save As dialog box. (This saves the report with the default name *Sales*.)
   e. Close the Sales report.
3. Add a record to the Sales table by completing the following steps:
   a. Double-click *Sales* in the Tables group in the Navigation pane. (Make sure you open the Sales table and not the Sales report.)
   b. Click the New button in the Records group on the Home tab.
   c. Press the Tab key to accept the default number in the *SalesID* field.
   d. Type 127 in the *ClientID* field and then press the Tab key.
   e. Type 2021 in the *CalendarYear* field and then press the Tab key.
   f. Type 176420 in the *Sales* field.
   g. Close the Sales table.
4. Open the Sales report and then scroll down to the bottom. Notice that the new record you added to the Sales table displays in the report.
5. Close the Sales report.

6. Use the Sales_2020 query to create a report by completing the following steps:
   a. Click *Sales_2020* in the Queries group in the Navigation pane.
   b. Click the Create tab.
   c. Click the Report button in the Reports group.
7. Access automatically inserted a total amount for the *Sales* column of the report. Delete this amount by scrolling down to the bottom of the report, clicking the total amount at the bottom of the *Sales* column, and then pressing the Delete key. (This deletes the total amount but not the underline above the amount.)
8. Save the report by clicking the Save button on the Quick Access Toolbar, typing 2020Sales in the *Report Name* text box in the Save As dialog box, and then clicking OK.

Modifying a Report

Make modifications to a report to address specific needs. For example, select, size, move, edit, or delete control objects in a report; sort records in ascending or descending order, and find data in a report. Use options in Print Preview to customize a report and, if the report is no longer needed, delete the report.

Managing Control
Objects in a Report

Managing Control Objects

A report, like a form, is comprised of control objects, such as logos, titles, labels, and text boxes. Select a control object in a report by clicking the object and the object displays with an orange border. Click in a cell in the report and Access selects all of the objects in the column except the column heading. Select adjacent control objects by pressing and holding down the Shift key and then clicking objects, or select nonadjacent control objects by pressing and holding down the Ctrl key and then clicking objects.

Like a form, a report contains a *Header* section and a *Detail* section. Select all of the control objects in the report in both the *Header* and *Detail* sections by pressing Ctrl + A. Control objects in the *Detail* section are contained in a report table. To select the control objects in the report table, click in any cell in the report and then click the table move handle. The table move handle is a small square with a four-headed arrow inside that displays in the upper left corner of the table as shown in Figure 6.1 (on page 174). Move the table and all of the control objects within the table by dragging the table move handle using the mouse.

Change the size of control objects by dragging the border of a selected control object or with the *Width* and *Height* property boxes in the Property Sheet task pane with the Data tab selected. To change the size of a control object by dragging, select the object (displays with an orange border) and then, using the mouse, drag a left or right border to increase or decrease the width or drag a top or bottom border to increase or decrease the height of the control object. When dragging the border of a control object, a line and character count displays at the left side of the Status bar. Use the line and character count numbers to adjust the width and/or height of the object by a precise line and character count number.

The width and height of a control object or column of control objects can be adjusted with the *Width* and *Height* property boxes in the Property Sheet task pane. Display this task pane by clicking the Property Sheet button in the Tools group. In the Property Sheet task pane, click the Data tab, select the current measurement in the *Width* or *Height* property box, type the new measurement, and then press the Enter key.

A selected control object can be moved by positioning the mouse pointer in the object until the pointer displays with a four-headed arrow attached. Click and hold down the left mouse button, drag to the new location, and then release the mouse button. To move a column of selected control objects, position the mouse pointer in the column heading until the pointer displays with a four-headed arrow attached and then click and drag to the new location. While dragging a control object(s), a pink insert indicator bar displays indicating where the control object(s) will be positioned when the mouse button is released.

Some control objects in a report, such as a column heading or title, are label control objects. Edit a label control by double-clicking in the label control object and then making the specific changes. For example, to rename a label control, double-click in the label control and then type the new name.

Sorting Records

Ascending

Descending

Sort data in a report by clicking in the field containing the data to be sorted and then clicking the Ascending button or the Descending button in the Sort & Filter group on the Home tab. Click the Ascending button to sort text in alphabetic order from A to Z, numbers from lowest to highest, and dates from earliest to latest. Click the Descending button to sort text in alphabetic order from Z to A, numbers from highest to lowest, and dates from latest to earliest.

Quick Steps

Sort Records
1. Click in field containing data.
2. Click Ascending button or Descending button.

 Find

Finding Data in a Report

Find specific data in a report with options at the Find dialog box. Display this dialog box by clicking the Find button in the Find group on the Home tab. At the Find dialog box, enter the search data in the *Find What* text box.

The *Match* option at the Find dialog box is set at *Whole Field* by default. At this setting, the data entered must match the entire entry in a field. To search for partial data in a field, change the *Match* option to *Any Part of Field* or *Start of Field*. If the text entered in the *Find What* text box needs to match the case in a field entry, click the *Match Case* check box to insert a check mark.

Access searches the entire report by default. This can be changed to *Up* to tell Access to search from the currently active field to the beginning of the report or *Down* to search from the currently active field to the end of the report. Click the Find Next button to find data that matches the data in the *Find What* text box.

 Tutorial

Customizing a Report in Print Preview

Print Preview

Displaying and Customizing a Report in Print Preview

When a report is created, the report displays in the work area in Layout view. In addition to Layout view, three other views are available: Report, Print Preview, and Design. Use Print Preview to display the report as it will appear when printed. To change to Print Preview, click the Print Preview button in the view area at the right side of the Status bar. Another method for displaying the report in Print Preview is to click the View button arrow in the Views group on the Home tab or Report Layout Tools Design tab and then click *Print Preview* at the drop-down list.

In Print Preview, send the report to the printer by clicking the Print button on the Print Preview tab. Use options in the Page Size group to change the page size and margins. To print only the report data and not the column headings, report title, shading, and gridlines, insert a check mark in the *Print Data Only* check box. Use options in the Page Layout group to specify the page orientation, specify the number and size of columns, and display the Page Setup dialog box. Click the Page Setup button and the Page Setup dialog box displays with options for customizing margins, orientation, size, and columns. The Zoom group contains options and buttons for specifying a zoom percentage and for displaying one, two, or multiple pages of the report.

Deleting a Report

Quick Steps

Delete Report
1. Click report name in Navigation pane.
2. Click Delete button on Home tab.
3. Click Yes.
OR
1. Right-click report name in Navigation pane.
2. Click *Delete*.
3. Click Yes.

If a report is no longer needed in a database, delete the report. Delete a report by clicking the report name in the Navigation pane, clicking the Delete button in the Records group on the Home tab, and then clicking the Yes button at the confirmation message. Another method is to right-click the report in the Navigation pane, click *Delete* at the shortcut menu, and then click Yes at the message. If a report is being deleted from the computer's hard drive, the confirmation message will not display. This is because Access automatically sends the deleted report to the Recycle Bin, where it can be retrieved at a later time, if necessary.

Activity 1b Adjusting Control Objects, Renaming Labels, Finding and Sorting
Data, Displaying a Report in Print Preview, and Deleting a Report   Part 2 of 5

1. With the 2020Sales report open, reverse the order of the *RepName* and *Client* columns by completing the following steps:
   a. Make sure the report displays in Layout view.
   b. Click the *RepName* column heading.
   c. Press and hold down the Shift key, click in the first control object below the *RepName* column heading (the control object containing *Linda Foster*), and then release the Shift key.
   d. Position the mouse pointer inside the *RepName* column heading until the pointer displays with a four-headed arrow attached.
   e. Click and hold down the left mouse button, drag to the left until the vertical pink insert indicator bar displays to the left of the *Client* column, and then release the mouse button.

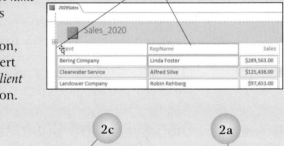

2. Sort the data in the *Sales* column in descending order by completing the following steps:
   a. Click in any field in the *Sales* column.
   b. Click the Home tab.
   c. Click the Descending button in the Sort & Filter group.

3. Rename the *RepName* label control as *Representative* by double-clicking in the label control object containing the text *RepName*, selecting *RepName*, and then typing Representative.

4. Double-click in the *Sales* label control and then rename it *Sales 2020*.
5. Move the report table by completing the following steps:
   a. Click in a cell in the report.
   b. Position the mouse pointer on the table move handle (the small square with a four-headed arrow inside that displays in the upper left corner of the table).
   c. Click and hold down the left mouse button, drag the report table to the right until it is centered between the left and right sides of the *Detail* section, and then release the mouse button. (When you drag with the mouse, you will see only outlines of some of the control objects.)

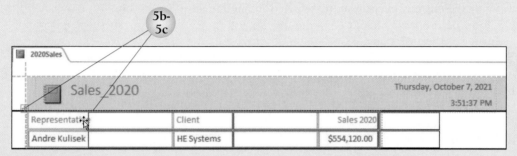

6. Display the report in Print Preview by clicking the Print Preview button in the view area at the right side of the Status bar.

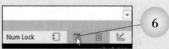

7. Click the One Page button in the Zoom group to display the entire page. (Even though the button displays with a darker gray background and appears to be active, you need to click it to display the entire page.)
8. Click the Zoom button arrow in the Zoom group and then click *50%* at the drop-down list.

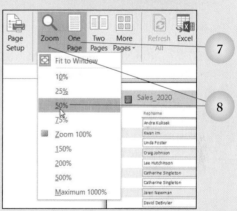

9. Click the One Page button in the Zoom group.
10. Print the report by clicking the Print button on the Print Preview tab and then clicking OK at the Print dialog box.
11. Close Print Preview by clicking the Close Print Preview button at the right side of the Print Preview tab.
12. Save and then close the 2020Sales report.
13. Create a report with the Representatives table by completing the following steps:
    a. Click *Representatives* in the Tables group in the Navigation pane.
    b. Click the Create tab.
    c. Click the Report button in the Reports group.
14. Adjust the width of the second column by completing the following steps:
    a. Click in the *RepName* column heading.
    b. Drag the right border of the selected column heading to the left until *Lines: 1 Characters: 18* displays at the left side of the Status bar, and then release the mouse button.
15. Complete steps similar to those in Step 14 to decrease the width of the third column (*RepTelephone*) to *Lines: 1 Characters: 15* and the fourth column (*RepEmail*) to *Lines: 1 Characters: 16*.

16. Adjust the width of the *QuotaID* column and the width and height of the title control object by completing the following steps:

   a. Click in the *QuotaID* column heading.

   b. Click the Property Sheet button in the Tools group and, if necessary, click the Format tab.

   c. Select the current measurement in the *Width* property box, type 0.9, and then press the Enter key.

   d. Click *Representatives* in the title control object.

   e. Select the current measurement in the *Width* property box, type 2, and then press the Enter key.

   f. With the current measurement selected in the *Height* property box, type 0.6.

   g. Close the Property Sheet task pane by clicking the Close button in the upper right corner of the task pane.

17. Search for fields containing a quota of *2* by completing the following steps:

   a. Click in the *RepID* column heading.

   b. Click the Home tab and then click the Find button in the Find group.

   c. At the Find dialog box, type 2 in the *Find What* text box.

   d. Make sure the *Match* option is set to *Whole Field*. (If not, click the *Match* option box arrow and then click *Whole Field* at the drop-down list.)

   e. Click the Find Next button.

   f. Continue clicking the Find Next button until a message displays stating that Access has finished searching the records. Click OK at the message.

   g. Click the Cancel button to close the Find dialog box.

18. Suppose you want to find information on a representative and you remember the first name but not the last name. Search for a field containing the first name *Lydia* by completing the following steps:

   a. Click in the *RepID* column heading.

   b. Click the Find button in the Find group.

   c. At the Find dialog box, type Lydia in the *Find What* text box.

   d. Click the *Match* option box arrow and then click *Any Part of Field* at the drop-down list.

   e. Click the Find Next button. (Access will find and select the representative name *Lydia Alvarado*.)

   f. Click the Cancel button to close the Find dialog box.

19. Click the control object at the bottom of the *RepID* column containing the number *17* and then press the Delete key. (This does not delete the underline above the amount.)

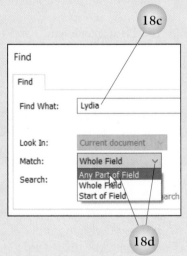

20. Switch to Print Preview by clicking the View button arrow in the Views group on the Home tab and then clicking *Print Preview* at the drop-down list.
21. Click the Margins button in the Page Size group and then click *Normal* at the drop-down list.

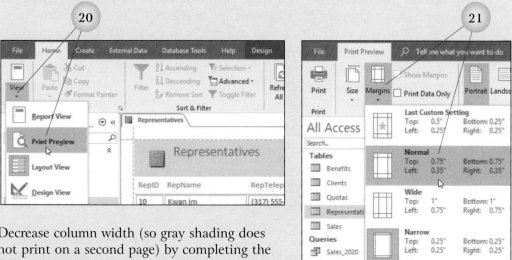

22. Decrease column width (so gray shading does not print on a second page) by completing the following steps:
    a. Click the Columns button in the Page Layout group on the Print Preview tab.
    b. Select the current measurement in the *Width* measurement box in the *Column Size* section of the dialog box and then type 7.5.
    c. Click OK to close the dialog box.

23. Print the report by clicking the Print button at the left side of the Print Preview tab and then clicking OK at the Print dialog box.
24. Close Print Preview by clicking the Close Print Preview button.
25. Save the report with the name *Representatives*.
26. Close the Representatives report.
27. Delete the Sales report by right-clicking *Sales* in the Reports group in the Navigation pane, clicking *Delete* at the shortcut menu, and then clicking Yes at the confirmation message.

Formatting a Report

Customize a report in much the same manner as customizing a form. When a report is created, the report displays in Layout view and the Report Layout Tools Design tab is active. Customize control objects in the *Detail* section and the *Header* section with buttons on the Report Layout Tools ribbon with the Design tab, Arrange tab, Format tab, or Page Setup tab selected.

Totals

The Report Layout Tools Design tab contains many of the same options as the Form Layout Tools Design tab. Use options on this tab to apply a theme, insert controls, insert header or footer data, and add existing fields. Use the Totals button in the Grouping & Totals group to perform calculations, such as finding the sum, average, maximum, or minimum of the numbers in a column. To use the Totals button, click in the column heading of the column containing the data to be totaled, click the Totals button, and then click a function at the drop-down list.

Hint The themes available in Access are the same as the themes available in Word, Excel, and PowerPoint.

The Report Layout Tools Arrange tab contains options for inserting and selecting rows, splitting cells horizontally and vertically, moving data up or down, controlling margins, and changing the padding between cells. The options on the Report Layout Tools Arrange tab are the same as the options on the Form Layout Tools Arrange tab.

Select and format data in a report with options on the Report Layout Tools Format tab. The options on this tab are the same as the options on the Form Layout Tools Format tab. Use options to apply formatting to a report or specific objects in a report. To apply formatting to a specific object, click the object in the report or click the Object button arrow in the Selection group on the Report Layout Tools Format tab and then click the object at the drop-down list. To format all objects in the report, click the Select All button in the Selection group. This selects all objects in the report, including objects in the *Header* section. To select all of the objects in the report table, click the table move handle.

Use options and buttons in the Font, Number, Background, and Control Formatting groups to apply formatting to a cell or selected cells in a report. Use options and buttons in the Font group to change the font, apply a different font size, apply text effects (such as bold and underline), and change the alignment of data in objects. Insert a background image in the report using the Background button and apply formatting to cells with buttons in the Control Formatting group. Depending on what is selected in the report, some of the buttons may not be active.

Background

The buttons on the Report Layout Tools Page Setup tab are also available in Print Preview. For example, the Print Preview tab contains buttons for changing the page size and page layout of the report and displaying the Page Setup dialog box.

Applying Conditional Formatting to a Report

Apply conditional formatting to a report in the same manner as applying conditional formatting to a form (covered in Chapter 5). Click the Conditional Formatting button in the Control Formatting group on the Report Layout Tools Format tab and the Conditional Formatting Rules Manager dialog box displays. Click the New Rule button and then use options in the New Formatting Rule dialog box to specify the conditional formatting.

Conditional
Formatting

1. With **6-Dearborn** open, open the 2020Sales report.
2. Display the report in Layout view.
3. Click the Themes button in the Themes group on the Report Layout Tools Design tab and then click *Ion* at the drop-down gallery.
4. Click the Title button in the Header/Footer group (which selects the current title), type 2020 Sales, and then press the Enter key.
5. Insert a row of empty cells in the report by completing the following steps:
   a. Click in the *Representative* cell.
   b. Click the Report Layout Tools Arrange tab.
   c. Click the Insert Above button in the Rows & Columns group.

6. Merge the cells in the new row by completing the following steps:
   a. Click in the empty cell immediately above the *Representative* cell.
   b. Press and hold down the Shift key, click immediately above the *Sales 2020* cell, and then release the Shift key. (This selects three cells.)
   c. Click the Merge button in the Merge/Split group.
   d. Type Dearborn 2020 Sales in the new cell.
7. Split a cell by completing the following steps:
   a. Click in the *2020 Sales* title in the *Header* section.

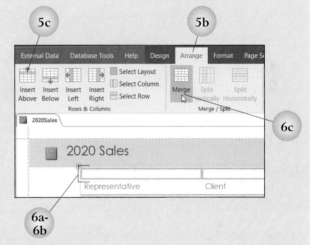

   b. Split the cell containing the title by clicking the Split Horizontally button in the Merge/Split group.
   c. Click in the empty cell immediately right of the cell containing the title *2020 Sales* and then press the Delete key. (Deleting the empty cell causes the date and time to move to the left in the *Header* section.)
8. Change the report table margins and padding by completing the following steps:
   a. Click in any cell in the *Detail* section and then click the table move handle in the upper left corner of the *Dearborn 2020 Sales* cell. (This selects all of the control objects in the report table in the *Detail* section.)
   b. Click the Control Margins button in the Position group and then click *Narrow* at the drop-down list.
   c. Click the Control Padding button in the Position group and then click *Medium* at the drop-down list.

9. Click in the *Dearborn 2020 Sales* cell and then drag down the bottom border so all of the text in the cell is visible.
10. Change the font for all of the data in the report by completing the following steps:
    a. Press Ctrl + A to select all control objects in the report. (An orange border displays around selected objects.)
    b. Click the Report Layout Tools Format tab.
    c. Click the *Font* option box arrow in the Font group and then click *Cambria* at the drop-down list. (You may need to scroll down the list to find *Cambria*.)

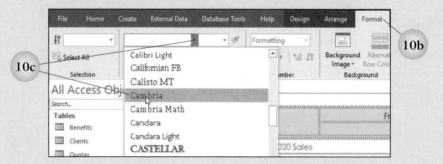

11. Apply bold formatting and change the alignment of the column headings by completing the following steps:
    a. Click *Dearborn 2020 Sales* to select the control object.
    b. Press and hold down the Shift key, click *Sales 2020*, and then release the Shift key. (This selects four cells.)
    c. Click the Bold button in the Font group.
    d. Click the Center button in the Font group.

12. Format and apply conditional formatting to the amounts by completing the following steps:
    a. Click the first field value below the *Sales 2020* column heading. (This selects all of the amounts in the column.)
    b. Click the Decrease Decimals button in the Number group two times.

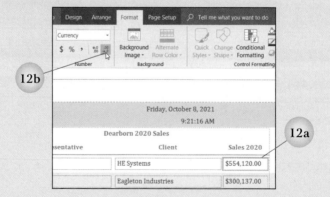

    c. Click the Conditional Formatting button in the Control Formatting group.
    d. At the Conditional Formatting Rules Manager dialog box, click the New Rule button.

e. At the New Formatting Rule dialog box, click the option box arrow for the second option box in the *Edit the rule description* section and then click *greater than* at the drop-down list.

f. Click in the text box immediately right of the option box containing *greater than* and then type 199999.

g. Click the Background color button arrow and then click the *Green 2* color option (seventh column, third row).

h. Click OK.

i. At the Conditional Formatting Rules Manager dialog box, click the New Rule button.

j. At the New Formatting Rule dialog box, click the option box arrow for the second option box in the *Edit the rule description* section and then click *less than* at the drop-down list.

k. Click in the text box immediately right of the option box containing *less than* and then type 200000.

l. Click the Background color button arrow and then click the *Maroon 2* color option (sixth column, third row in the *Standard Colors* section).

m. Click OK to close the New Formatting Rule dialog box.

n. Click OK to close the Conditional Formatting Rules Manager dialog box.

13. Sum the totals in the *Sales 2020* column by completing the following steps:

a. Click in the *Sales 2020* column heading.

b. Click the Report Layout Tools Design tab.

c. Click the Totals button in the Grouping & Totals group and then click *Sum* at the drop-down list.

14. Click in the *Sales 2020* sum amount (at the bottom of the *Sales 2020* column) and then drag down the bottom border so the entire amount is visible in the cell.

15. Change the top margin by completing the following steps:

a. Click in the *Representative* column heading and then click the Report Layout Tools Page Setup tab.

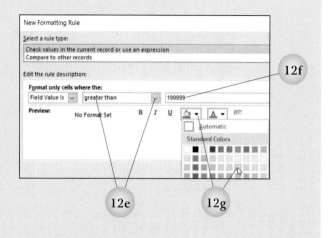

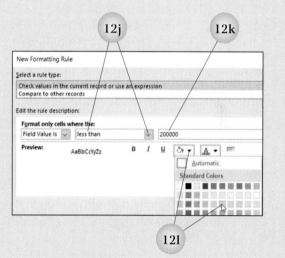

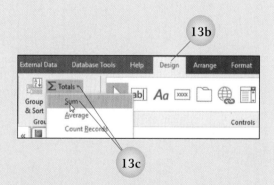

b. Click the Page Setup button in the Page Layout group.

c. At the Page Setup dialog box with the Print Options tab selected, select the current measurement in the *Top* measurement box and then type 0.5.

d. Click OK to close the Page Setup dialog box.

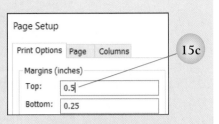

16. Change the page size by clicking the Size button in the Page Size group and then clicking *Legal* at the drop-down list.

17. Display the report in Print Preview by clicking the File tab, clicking the *Print* option, and then clicking the *Print Preview* option.

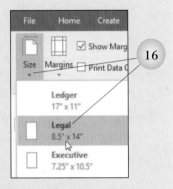

18. Click the One Page button in the Zoom group and notice that the entire report will print on one legal-sized page.

19. Click the Close Print Preview button to return to the report.

20. Change the page size by clicking the Report Layout Tools Page Setup tab, clicking the Size button in the Page Size group, and then clicking *Letter* at the drop-down list.

21. Insert and then remove a background image by completing the following steps:

a. Click the Report Layout Tools Format tab.

b. Click the Background Image button in the Background group and then click *Browse* at the drop-down list.

c. At the Insert Picture dialog box, navigate to your AL1C6 folder and then double-click ***Mountain***.

d. Scroll through the report and notice how the image displays in the report.

e. Click the Undo button on the Quick Access Toolbar to remove the background image. (You may need to click the Undo button more than once.)

22. Print the report by clicking the File tab, clicking the *Print* option, and then clicking the *Quick Print* option.

23. Save and then close the report.

Check Your Work

Grouping and Sorting Records in a Report

Grouping and Sorting Records in a Report

  Group & Sort

  Add a group

Quick Steps

Group and Sort Records
1. Open report in Layout view.
2. Click Group & Sort button.
3. Click Add a group button.
4. Click group field.

A report presents database information in a printed form and generally contains data that answers a specific question. To make the data in a report easy to understand, divide the data into groups. For example, data can be divided in a report by regions, sales, dates, or any other division that helps clarify the data for the reader. Access contains a group and sort feature for dividing data into groups and sorting the data.

Click the Group & Sort button in the Grouping & Totals group on the Report Layout Tools Design tab and the Group, Sort, and Total pane displays at the bottom of the work area, as shown in Figure 6.2. Click the Add a group button in the Group, Sort, and Total pane and Access adds a new grouping level row to the pane, along with a list of available fields. Click the field by which data is to be grouped in the report and Access adds the grouping level in the report. With options in the grouping level row, change the group, specify the sort order, and expand the row to display additional options.

Figure 6.2 Group, Sort, and Total Pane

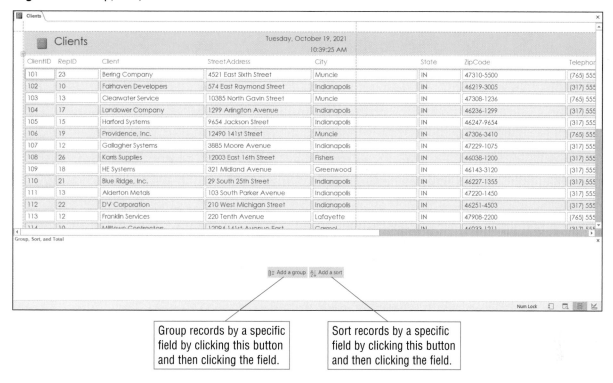

Group records by a specific field by clicking this button and then clicking the field.

Sort records by a specific field by clicking this button and then clicking the field.

Add a sort

💡 **Hint** Grouping allows you to separate groups of records visually.

When a grouping level is specified, Access automatically sorts that level in ascending order (from A to Z or lowest to highest). Additional data can be sorted within the report by clicking the Add a sort button in the Group, Sort, and Total pane. This inserts a sorting row in the pane below the grouping level row, along with a list of available fields. At this list, click the field by which to sort. For example, in Activity 1d, one of the reports will be grouped by city (which will display in ascending order) and then the client names will display in alphabetic order within the city.

To delete a grouping or sorting level in the Group, Sort, and Total pane, click the Delete button at the right side of the level row. After specifying the grouping and sorting levels, close the Group, Sort, and Total pane by clicking the Close button in the upper right corner of the pane or by clicking the Group & Sort button in the Grouping & Totals group.

Activity 1d Grouping and Sorting Records in a Report

Part 4 of 5

1. With **6-Dearborn** open, create a report with the Clients table using the Report button on the Create tab.
2. Click in each column heading individually and then decrease the size of each column so the right border is just right of the longest entry.
3. Change to landscape orientation by completing the following steps:
   a. Click the Report Layout Tools Page Setup tab.
   b. Click the Landscape button in the Page Layout group.
4. Group the report by representative ID and then sort by clients by completing the following steps:
   a. Click the Report Layout Tools Design tab.
   b. Click the Group & Sort button in the Grouping & Totals group.

c. Click the Add a group button in the Group, Sort, and Total pane.

d. Click the *RepID* field in the list box.
e. Scroll through the report and notice that the records are grouped by the *RepID* field. Also, notice that the client names within each *RepID* field group are not in alphabetic order.

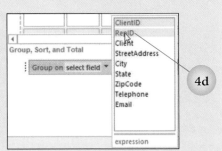

f. Click the Add a sort button in the Group, Sort, and Total pane.
g. Click the *Client* field in the list box.
h. Scroll through the report and notice that client names are now alphabetized within *RepID* field groups.

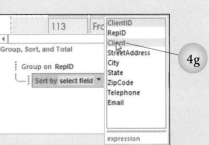

i. Close the Group, Sort, and Total pane by clicking the Group & Sort button in the Grouping & Totals group.
5. Save the report with the name *RepIDGroupedRpt*.
6. Change column width and print the first page of the report by completing the following steps:
   a. Click the File tab, click the *Print* option, and then click the *Print Preview* option.
   b. Click the Columns button on the Print Preview tab, select the current measurement in the *Width* measurement box, type 10, and then click OK. (The Email field will not fully display.)
   c. Click the Print button on the Print Preview tab.
   d. At the Print dialog box, click the *Pages* option in the *Print Range* section.
   e. Type 1 in the *From* text box, press the Tab key, and then type 1 in the *To* text box.
   f. Click OK.
7. Close the RepIDGroupedRpt report.
8. Create a report with the Sales>$99999_IndMun query using the Report button on the Create tab. Make sure the report displays in Layout view.
9. Group the report by city and then sort by clients by completing the following steps:
   a. Click the Group & Sort button in the Grouping & Totals group on the Report Layout Tools Design tab.
   b. Click the Add a group button in the Group, Sort, and Total pane.
   c. Click the *City* field in the list box.
   d. Click the Add a sort button in the Group, Sort, and Total pane.
   e. Click the *Client* field in the list box.
   f. Close the Group, Sort, and Total pane by clicking the Group & Sort button.
10. Print the first page of the report.
11. Save the report, name it *Sales>$99999_IndMun*, and then close the report.
12. Close **6-Dearborn**.
13. Open **6-WarrenLegal** from your AL1C6 folder and enable the content.

14. Design a query that extracts records from three tables with the following specifications:
    a. Add the Billing, Clients, and Rates tables to the query window.
    b. Insert the *LastName* field from the Clients table field list box in the first field in the *Field* row.
    c. Insert the *BillDate* field from the Billing table field list box in the second field in the *Field* row.
    d. Insert the *BilledHours* field from the Billing table field list box in the third field in the *Field* row.
    e. Insert the *Rate* field from the Rates table field list box in the fourth field in the *Field* row.
    f. Click in the fifth field in the *Field* row, type Total: [BilledHours]*[Rate], and then press the Enter key.

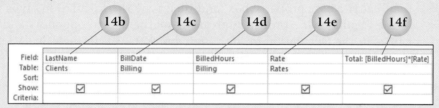

    g. Run the query.
    h. Save the query, typing Clients_Billing as the query name, and then close the query.
15. Create a report with the Clients_Billing query using the Report button on the Create tab.
16. Click in each column heading individually and then decrease the size of each column so the right border is near the longest entry.
17. Apply currency formatting to the numbers in the *Total* column by completing the following steps:
    a. Click the Report Layout Tools Format tab.
    b. Click in the first field below the *Total* column (the field containing the number *350*).
    c. Click the Apply Currency Format button in the Number group.
    d. If necessary, increase the size of the *Total* column so the entire amounts (including the dollar symbols ($)) are visible.

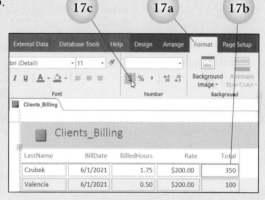

18. Group the report by last name and then sort by date by completing the following steps:
    a. Click the Report Layout Tools Design tab.
    b. Click the Group & Sort button.
    c. Click the Add a group button in the Group, Sort, and Total pane.
    d. Click the *LastName* field in the list box.
    e. Click the Add a sort button in the Group, Sort, and Total pane.
    f. Click the *BillDate* field in the list box.
    g. Close the Group, Sort, and Total pane by clicking the Close button in the upper right corner of the pane.
19. Scroll to the bottom of the report and delete the total amount in the *Rate* column and the line above the total. (Click the line and then press the Delete key.)
20. Save the report with the name *Clients_BillingRpt*.
21. Close the report.

Check Your Work

Inserting a Calculation in a Report

Like a form, a calculation can be inserted in a report in a text box control object. To insert a text box control object as well as a label control object, click the Text Box button in the gallery in the Controls group on the Report Layout Tools Design tab, and then click in a location in the report where the two objects are to display. Click in the label control box and then type a label for the calculated field.

Insert a calculation by clicking in the text box control object and then clicking the Property Sheet button in the Tools group on the Report Layout Tools Design tab. At the Property Sheet task pane, click the Data tab, click in the *Control Source* property box, and then type the calculation. Type a calculation in the *Control Source* property box using mathematical operators and type field names in the calculation inside square brackets. Begin the calculation with an equals sign (=). A field name must be typed in the calculation as it appears in the source object.

If a calculation result is currency, apply currency formatting to the text box control object. Apply currency formatting by clicking the Report Layout Tools Format tab and then clicking the Apply Currency Format button in the Number group.

Activity 1e Inserting a Calculation in a Report

Part 5 of 5

1. With **6-WarrenLegal** open, open the Clients_BillingRpt report in Layout view.
2. Insert label and text box control objects by completing the following steps:
   a. Click the Text Box button in the gallery in the Controls group on the Report Layout Tools Design tab.
   b. Position the crosshairs to the right of the *Total* field name and then click the left mouse button. (This inserts a label control object and text box object in the report.)

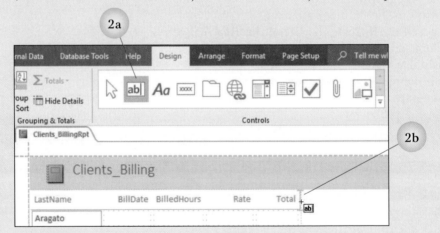

3. Name the new field by clicking in the label control object, double-clicking the text in the label control object, and then typing Total + 9% Tax.

4. Click in the text box control object and then insert a calculation by completing the following steps:

   a. Click the Property Sheet button in the Tools group on the Report Layout Tools Design tab.

   b. At the Property Sheet task pane, click the Data tab.

   c. Click in the *Control Source* property box.

   d. Type =[Total]*1.09 and then press the Enter key.

   e. Close the Property Sheet task pane by clicking the Close button in the upper right corner of the task pane.

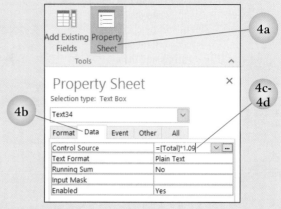

5. With the text box control object still selected, apply currency formatting by clicking the Report Layout Tools Format tab and then clicking the Apply Currency Format button.

6. Save and then print the report. (The report will print on three pages.)

7. Close the Clients_BillingRpt report and then close **6-WarrenLegal**.

 Check Your Work

---

Activity 2   Use Wizards to Create Reports and Labels   3 Parts

You will create reports using the Report Wizard and prepare mailing labels using the Label Wizard.

 Tutorial

Creating a Report Using the Report Wizard

 Report Wizard

Creating a Report Using the Report Wizard

Access offers a Report Wizard that provides the steps for creating a report. To create a report using the wizard, click the Create tab and then click the Report Wizard button in the Reports group. At the first Report Wizard dialog box, shown in Figure 6.3, choose a table or query with options from the *Tables/Queries* option box. Specify the fields to be included in the report by inserting them in the *Selected Fields* list box and then clicking the Next button.

At the second Report Wizard dialog box, shown in Figure 6.4, specify the grouping level of data in the report. To group data by a specific field, click the field in the list box at the left side of the dialog box and then click the One Field button. Use the button containing the left-pointing arrow to remove an option as a grouping level. Use the up and down arrows to change the priority of the field.

Hint Use the Report Wizard to select specific fields and specify how data is grouped and sorted.

Specify a field on which to sort and the sort order with options at the third Report Wizard dialog box, shown in Figure 6.5. To specify a field on which to sort, click the option box arrow for the option box preceded by the number *1* and then click the field name. The default sort order is ascending. This can be changed to descending by clicking the button at the right side of the option box. After identifying the field and sort order, click the Next button.

Figure 6.3 First Report Wizard Dialog Box

Choose a table or query to use to create the report by clicking this option box arrow and then clicking the table or query at the drop-down list.

Choose a field to be included in the report by clicking the field in this list box.

Click the One Field button to add a field to the *Selected Fields* list box.

Click the All Fields button to insert all fields in the *Available Fields* list box into the *Selected Fields* list box.

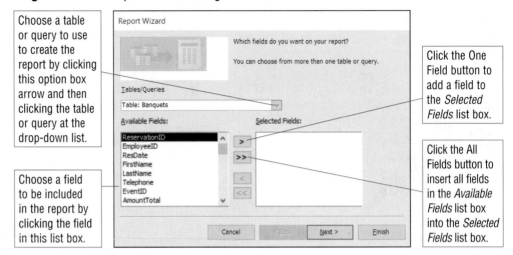

Figure 6.4 Second Report Wizard Dialog Box

Use these buttons to increase or decrease the field priority level.

Preview field priorities in this preview box.

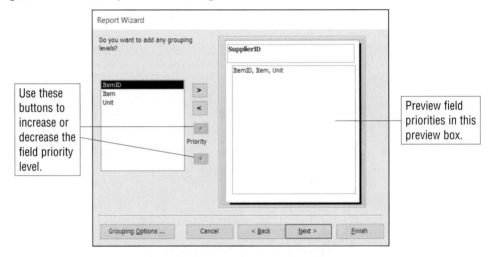

Quick Steps

Create Report Using Report Wizard

1. Click Create tab.
2. Click Report Wizard button.
3. Choose options at each Report Wizard dialog box.

Use options at the fourth Report Wizard dialog box, shown in Figure 6.6, to specify the layout and orientation of the report. The *Layout* section has the default setting of *Stepped*, which can be changed to *Block* or *Outline*. By default, the report will print in portrait orientation. Change to landscape orientation by clicking the *Landscape* option in the *Orientation* section of the dialog box. Access will adjust field widths in the report, so all of the fields fit on one page. To specify that field widths should not be adjusted, remove the check mark from the *Adjust the field width so all fields fit on a page* option.

At the fifth and final Report Wizard dialog box, type a name for the report and then click the Finish button.

Figure 6.5 Third Report Wizard Dialog Box

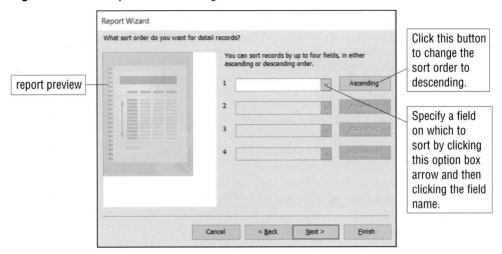

report preview

Click this button to change the sort order to descending.

Specify a field on which to sort by clicking this option box arrow and then clicking the field name.

Figure 6.6 Fourth Report Wizard Dialog Box

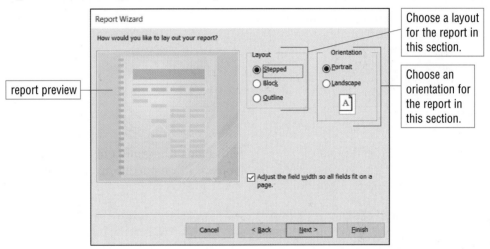

report preview

Choose a layout for the report in this section.

Choose an orientation for the report in this section.

Activity 2a Creating a Report Using the Report Wizard   Part 1 of 3

1. Open **6-Skyline** from your AL1C6 folder and enable the content.
2. Create a report using the Report Wizard by completing the following steps:
   a. Click the Create tab.
   b. Click the Report Wizard button in the Reports group.
   c. At the first Report Wizard dialog box, click the *Tables/Queries* option box arrow and then click *Table: Inventory* at the drop-down list.
   d. Click the All Fields button to insert all of the Inventory table fields in the *Selected Fields* list box.
   e. Click the Next button.

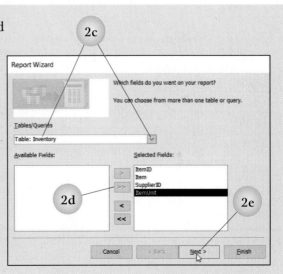

f. At the second Report Wizard dialog box, make sure *SupplierID* displays in blue at the top of the preview page at the right side of the dialog box and then click the Next button.

g. At the third Report Wizard dialog box, click the Next button. (You want to use the sorting defaults.)

h. At the fourth Report Wizard dialog box, click the *Block* option in the *Layout* section and then click the Next button.

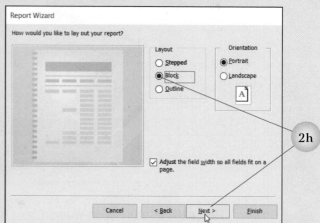

i. At the fifth Report Wizard dialog box, make sure *Inventory* displays in the *What title do you want for your report?* text box and then click the Finish button. (The report displays in Print Preview.)

3. With the report in Print Preview, click the Print button at the left side of the Print Preview tab and then click OK at the Print dialog box. (The report will print on two pages.)

4. Close Print Preview.

5. Switch to Report view by clicking the View button on the Report Design Tools Design tab.

6. Close the Inventory report.

Check Your Work

If a report is created with fields from only one table, options are specified in five Report Wizard dialog boxes. If a report is created with fields from more than one table, options are specified in six Report Wizard dialog boxes. After choosing the tables and fields at the first dialog box, the second dialog box asks how the data will be viewed. For example, if fields are selected from a Suppliers table and an Orders table, the second Report Wizard dialog box will ask if the data is to be viewed "by Suppliers" or "by Orders."

Activity 2b Creating a Report with Fields from Two Tables

Part 2 of 3

1. With **6-Skyline** open, create a report with the Report Wizard by completing the following steps:

a. Click the Create tab.

b. Click the Report Wizard button.

c. At the first Report Wizard dialog box, click the *Tables/Queries* option box arrow and then click *Table: Events* at the drop-down list.

d. Click the *Event* field in the *Available Fields* list box and then click the One Field button.

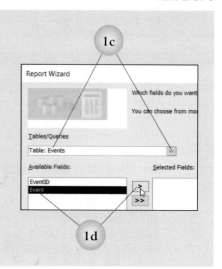

e. Click the *Tables/Queries* option box arrow and then click *Table: Banquets* at the drop-down list.

f. Insert the following fields in the *Selected Fields* list box:

> ResDate
> FirstName
> LastName
> AmountTotal
> AmountPaid

g. After inserting the fields, click the Next button.

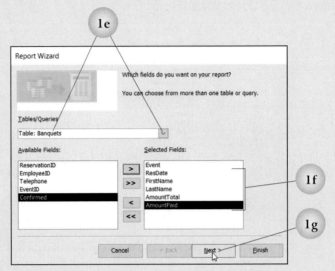

h. At the second Report Wizard dialog box, make sure *by Events* is selected and then click the Next button.

i. At the third Report Wizard dialog box, click the Next button. (The report preview shows that the report will be grouped by event.)

j. At the fourth Report Wizard dialog box, click the Next button. (You want to use the sorting defaults.)

k. At the fifth Report Wizard dialog box, click the *Block* option in the *Layout* section, click *Landscape* in the *Orientation* section, and then click the Next button.

l. At the sixth Report Wizard dialog box, select the current name in the *What title do you want for your report?* text box and then type BanquetEvents.

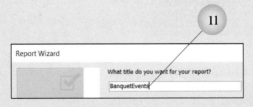

m. Click the Finish button.

2. Close Print Preview and then change to Layout view.

3. Print and then close the BanquetEvents report.

4. Close **6-Skyline**.

Labels

Quick Steps

**Create Mailing Labels
Using Label Wizard**
1. Click table.
2. Click Create tab.
3. Click Labels button.
4. Choose options at
   each Label Wizard
   dialog box.

Creating Mailing Labels

Access includes a mailing label wizard that provides the steps for creating mailing labels with fields in a table. To create mailing labels, click a table, click the Create tab, and then click the Labels button in the Reports group. At the first Label Wizard dialog box, shown in Figure 6.7, specify the label size, unit of measure, and label type and then click the Next button. At the second Label Wizard dialog box, shown in Figure 6.8, specify the font name, size, weight, and color and then click the Next button.

Figure 6.7 First Label Wizard Dialog Box

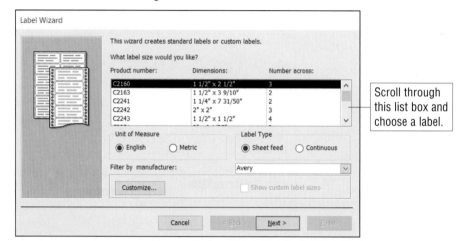

Figure 6.8 Second Label Wizard Dialog Box

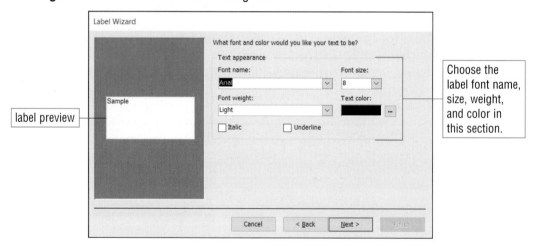

Specify the fields to be included in the mailing labels at the third Label Wizard dialog box, shown in Figure 6.9. To do this, click the field in the *Available fields* list box and then click the One Field button. This moves the field to the *Prototype label* box. Insert the fields in the *Prototype label* box as the text should display on the label. After inserting the fields in the *Prototype label* box, click the Next button.

At the fourth Label Wizard dialog box, shown in Figure 6.10, specify a field from the database by which the labels will be sorted. To sort labels (for example, by last name, postal code, etc.), insert that field in the *Sort by* list box and then click the Next button.

At the last Label Wizard dialog box, type a name for the label report and then click the Finish button. After a few moments, the labels display on the screen in Print Preview. Print the labels and/or close Print Preview.

Figure 6.9 Third Label Wizard Dialog Box

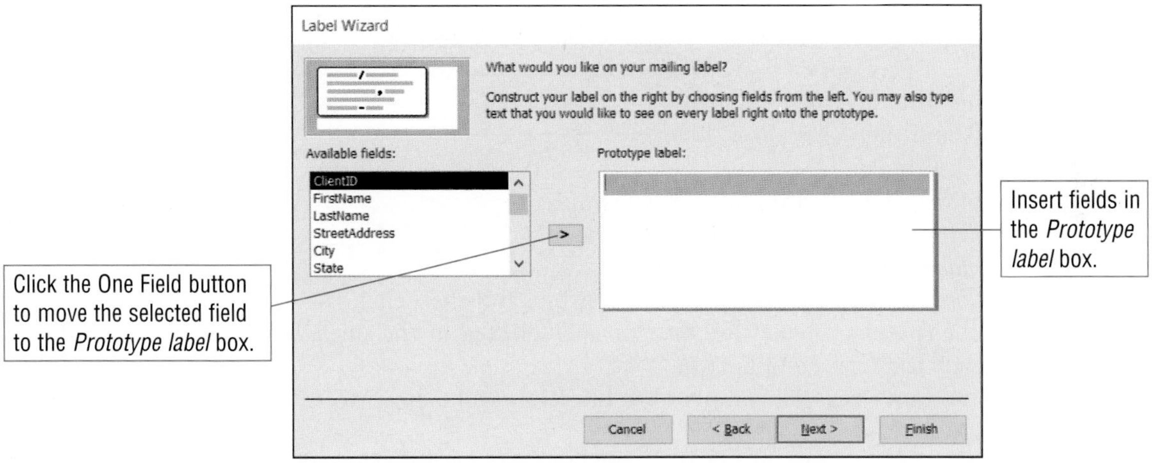

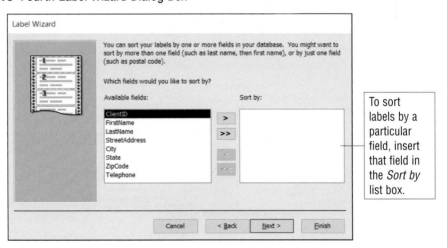

Figure 6.10 Fourth Label Wizard Dialog Box

1. Open **6-WarrenLegal** and enable the content.
2. Click *Clients* in the Tables group in the Navigation pane.
3. Click the Create tab and then click the Labels button in the Reports group.
4. At the first Label Wizard dialog box, make sure *English* is selected in the *Unit of Measure* section, *Avery* is selected in the *Filter by manufacturer* list box, *Sheet feed* is selected in the *Label Type* section, and *C2160* is selected in the *Product number* list box and then click the Next button.
5. At the second Label Wizard dialog box, if necessary, change the font size to 10 points and then click the Next button.
6. At the third Label Wizard dialog box, complete the following steps to insert the fields in the *Prototype label* box:
   a. Click *FirstName* in the *Available fields* list box and then click the One Field button.
   b. Press the spacebar, make sure *LastName* is selected in the *Available fields* list box, and then click the One Field button.
   c. Press the Enter key. (This moves the insertion point down to the next line in the *Prototype label* box.)
   d. With *StreetAddress* selected in the *Available fields* list box, click the One Field button.
   e. Press the Enter key.
   f. With *City* selected in the *Available fields* list box, click the One Field button.
   g. Type a comma (,) and then press the spacebar.
   h. With *State* selected in the *Available fields* list box, click the One Field button.
   i. Press the spacebar.
   j. With *ZipCode* selected in the *Available fields* list box, click the One Field button.

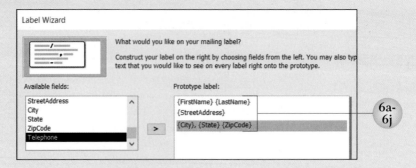

   k. Click the Next button.

7. At the fourth Label Wizard dialog box, sort by zip code. To do this, click *ZipCode* in the *Available fields* list box and then click the One Field button.
8. Click the Next button.
9. At the last Label Wizard dialog box, click the Finish button. (The Label Wizard automatically names the label report *Labels Clients*.)
10. Print the labels by clicking the Print button at the left side of the Print Preview tab and then click OK at the Print dialog box.
11. Close Print Preview.
12. Switch to Report view by clicking the View button on the Report Design Tools Design tab.
13. Close the labels report and then close **6-WarrenLegal**.

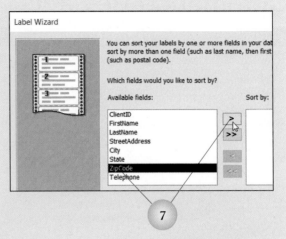

Chapter Summary

- Create a report with data in a table or query to control how data appears on the page and how the data is formatted when printed.
- Create a report with the Report button in the Reports group on the Create tab.
- Four views are available for viewing a report: Report view, Print Preview, Layout view, and Design view.
- Use options on the Print Preview tab to specify how a report prints.
- In Layout view, a report control object can be selected and then sized or moved.
- One method for changing column width in a report is to click a column heading and then drag the border to the desired width.
- Sort data in a record using the Ascending button or Descending button in the Sort & Filter group on the Home tab.
- Customize a report with options on the Report Layout Tools ribbon with the Design tab, Arrange tab, Format tab, or Page Setup tab selected.
- To make data in a report easier to understand, divide the data into groups using the Group, Sort, and Total pane. Display this pane by clicking the Group & Sort button in the Grouping & Totals group on the Report Layout Tools Design tab.
- Insert a calculation in a report by inserting a text box object, displaying the Property Sheet task pane with the Data tab selected, and then typing the calculation in the *Control Source* property box. If a calculation result is currency, apply currency formatting with the Apply Currency Format button on the Report Layout Tools Format tab.
- Use the Report Wizard to provide the steps for creating a report. Begin the wizard by clicking the Create tab and then clicking the Report Wizard button in the Reports group.
- Create mailing labels with fields in a table using the Label Wizard. Begin the wizard by clicking a table, clicking the Create tab, and then clicking the Labels button in the Reports group.

Commands Review

FEATURE	RIBBON TAB, GROUP	BUTTON	KEYBOARD SHORTCUT
Conditional Formatting Rules Manager dialog box	Report Layout Tools Format, Control Formatting		
Find dialog box	Home, Find		Ctrl + F
Group, Sort, and Total pane	Report Layout Tools Design, Grouping & Totals		
Labels Wizard	Create, Reports		
Property Sheet task pane	Report Layout Tools Design, Tools		Alt + Enter
report	Create, Reports		
Report Wizard	Create, Reports		
sort data in ascending order	Home, Sort & Filter		
sort data in descending order	Home, Sort & Filter		

Microsoft®

Access®

Modifying, Filtering, and Viewing Data

Performance Objectives

Upon successful completion of Chapter 7, you will be able to:

1 Filter records using the Filter button

2 Remove a filter

3 Filter on specific values, by selection, by shortcut menu, and using the *Filter By Form* option

4 View object dependencies

5 Compact and repair a database

6 Encrypt a database with a password

7 View and customize document properties

8 Save a database in an earlier version of Access

9 Save a database object as a PDF or XPS file

10 Back up a database

Data in a database object can be filtered to view specific records without having to change the design of the object. In this chapter, you will learn how to filter data by selection, by shortcut menu, and by form, as well as how to remove a filter. You will also learn how to view object dependencies, compact and repair a database, encrypt a database with a password, view and customize database properties, save a database in an earlier version of Access, save a database object as a PDF or XPS file, and back up a database.

 Data Files

Before beginning chapter work, copy the AL1C7 folder to your storage medium and then make AL1C7 the active folder.

The online course includes additional training and assessment resources.

Tutorial

Filtering Records

Q̃uick Steps

Filter Records
1. Open object.
2. Click in entry in field to filter.
3. Click Filter button.
4. Select sorting option at drop-down list.

 Filter

Filtering Records

A set of restrictions, called a *filter*, can be put on records in a table, query, form, or report to isolate temporarily specific records. A filter, like a query, displays specific records without having to change the design of the table, query, form, or report. Access provides a number of buttons and options for filtering data. Filter data using the Filter button in the Sort & Filter group on the Home tab, right-click specific data in a record and then specify a filter, and use the Selection and Advanced buttons in the Sort & Filter group.

Filtering Using the Filter Button

Use the Filter button in the Sort & Filter group on the Home tab to filter records in an object (a table, query, form, or report). To use this button, open the object, click in any entry in the field column to be filtered, and then click the Filter button. This displays a drop-down list with sorting options and a list of all of the field entries, as shown in Figure 7.1. In a table, display this drop-down list by clicking the filter arrow at the right side of a column heading. To filter on a specific

Figure 7.1 *City* Field Drop-down List

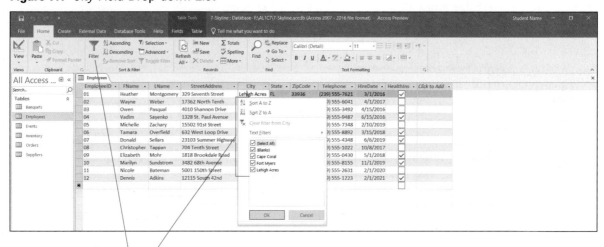

To filter on the *City* field, click in any entry in the field and then click the Filter button. This displays a drop-down list with sorting options and a list of all field entries.

criterion, click the *(Select All)* check box to remove all check marks from the list of field entries. Click the item to filter by in the list box and then click OK.

Open a table, query, or form and the Record Navigation bar contains the dimmed words *No Filter* preceded by a filter icon with a delete symbol (X). If records are filtered, *Filtered* displays in place of *No Filter*, the delete symbol is removed, and the text and filter icon display with an orange background. In a report, apply a filter to records and the word *Filtered* displays at the right side of the Status bar.

Removing a Filter

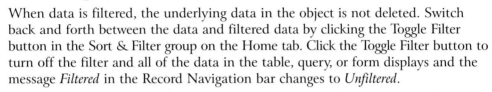

Toggle Filter

When data is filtered, the underlying data in the object is not deleted. Switch back and forth between the data and filtered data by clicking the Toggle Filter button in the Sort & Filter group on the Home tab. Click the Toggle Filter button to turn off the filter and all of the data in the table, query, or form displays and the message *Filtered* in the Record Navigation bar changes to *Unfiltered*.

Clicking the Toggle Filter button may redisplay all of the data in an object, but it does not remove the filter. To remove the filter, click in the field column containing the filter and then click the Filter button in the Sort & Filter group on the Home tab. At the drop-down list that displays, click *Clear filter from xxx* (where *xxx* is the name of the field). Remove all of the filters from an object by clicking the Advanced button in the Sort & Filter group and then clicking the *Clear All Filters* option. When all filters are removed (cleared) from an object, the *Unfiltered* message in the Record Navigation bar changes to *No Filter*.

Activity 1a Filtering Records in a Table and Form   Part 1 of 4

1. Open **7-Skyline** from your AL1C7 folder and enable the content.
2. Filter records in the Employees table by completing the following steps:
   a. Open the Employees table.
   b. Click in any entry in the *EmCity* field.
   c. Click the Filter button in the Sort & Filter group on the Home tab. (This displays a drop-down list of options for sorting and filtering the *EmCity* field.)

d. Click the *(Select All)* check box in the filter drop-down list box. (This removes all check marks from the list options.)

e. Click the *Fort Myers* check box in the list box. (This inserts a check mark in the check box.)

f. Click OK. (Access displays only those records with a city field entry of *Fort Myers* and also displays *Filtered* and the filter icon with an orange background in the Record Navigation bar.)

g. Print the filtered records by pressing Ctrl + P (the keyboard shortcut to display the print dialog box) and then clicking OK at the Print dialog box.

3. Toggle the display of filtered data by clicking the Toggle Filter button in the Sort & Filter group on the Home tab. (This redisplays all of the data in the table.)

4. Remove the filter by completing the following steps:

a. Click in any entry in the *EmCity* field.

b. Click the Filter button in the Sort & Filter group.

c. Click the *Clear filter from EmCity* option at the drop-down list. (Notice that the message on the Record Navigation bar changes to *No Filter* and dims the words.)

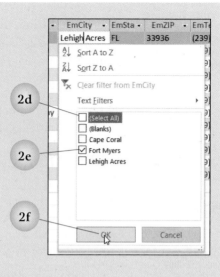

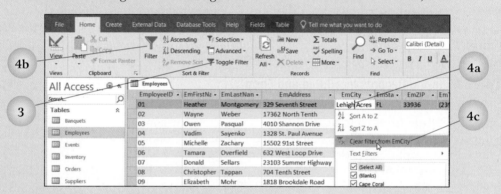

5. Save and then close the Employees table.

6. Create a form by completing the following steps:

a. Click *Orders* in the Tables group in the Navigation pane.

b. Click the Create tab and then click the Form button in the Forms group.

c. Click the Form View button in the view area at the right side of the Status bar.

d. Save the form with the name *Orders*.

7. Filter the records to show only those records with a supplier identification number of 2 by completing the following steps:

a. Click in the *SupplierID* field containing the text *2*.

b. Click the Filter button in the Sort & Filter group.

c. At the filter drop-down list, click *(Select All)* to remove all of the check marks from the list options.

d. Click the *2* option to insert a check mark.

e. Click OK.

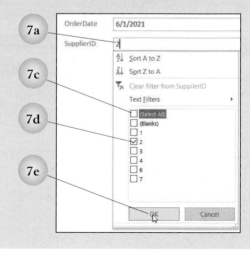

f. Navigate through the records and notice that only the records with a supplier identification number of 2 display.

8. Close the Orders form.

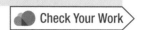

Filtering on Specific Values

Quick Steps
Filter on Specific Value
1. Click in entry in field.
2. Click Filter button on Home tab.
3. Point to filter option.
4. Click specific value.

When filtering on a specific field, a list of unique values for that field can be displayed. Click the Filter button for a field containing text and the drop-down list for the specific field will contain a *Text Filters* option. Click this option and a values list displays next to the drop-down list. The options in the values list vary depending on the type of data in the field. Click the Filter button for a field containing number values and the option in the drop-down list displays as *Number Filters*. If filtering dates, the Filter button's drop-down list displays as *Date Filters*. Use the options in the values list to refine a filter for a specific field. For example, the values list can be used to show only money amounts within a specific range or order dates from a certain time period. The values list can also be used to find fields that are "equal to" or "not equal to" data in the current field.

Activity 1b Filtering Records in a Query and Report

Part 2 of 4

1. With **7-Skyline** open, create a query in Design view with the following specifications:
   a. Add the Banquets and Events tables to the query window.
   b. Insert the *ResDate* field from the Banquets table field list box in the first field in the *Field* row.
   c. Insert the *ResFirstName* field from the Banquets table field list box in the second field in the *Field* row.
   d. Insert the *ResLastName* field from the Banquets table field list box in the third field in the *Field* row.
   e. Insert the *ResTelephone* field from the Banquets table field list box in the fourth field in the *Field* row.
   f. Insert the *Event* field from the Events table field list box in the fifth field in the *Field* row.
   g. Insert the *EmployeeID* field from the Banquets table field list box in the sixth field in the *Field* row.
   h. Run the query.
   i. Save the query, typing BanquetReservations as the query name.

2. Filter records of reservations on or before June 15, 2021, in the query by completing the following steps:
   a. With the BanquetReservations query open, make sure the first entry is selected in the *ResDate* field.
   b. Click the Filter button in the Sort & Filter group on the Home tab.
   c. Point to the *Date Filters* option in the drop-down list box.
   d. Click *Before* in the values list.

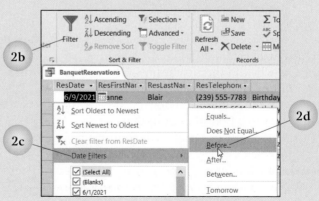

e. At the Custom Filter dialog box, type *6/15/2021* and then click OK.

f. Print the filtered query by pressing Ctrl + P and then clicking OK at the Print dialog box.

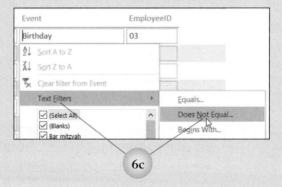

3. Remove the filter by clicking the filter icon at the right side of the *ResDate* column heading and then clicking *Clear filter from ResDate* at the drop-down list.

4. Save and then close the BanquetReservations query.

5. Create and format a report by completing the following steps:

a. Click *BanquetReservations* in the Queries group in the Navigation pane.

b. Click the Create tab and then click the Report button in the Reports group.

c. Delete the total amount and line at the bottom of the *ResDate* column.

d. Delete the page number control object at the bottom of the report.

e. With the report in Layout view, decrease the column widths so the right side column border displays near the longest entry in each column.

f. Move the date and time control objects so they align with the last column in the report.

g. Click the Report View button in the view area at the right of the Status bar.

h. Save the report with the name *BanquetReport*.

6. Filter the records to show all records of events except *Other* events by completing the following steps:

a. Click in the first entry in the *Event* field.

b. Click the Filter button.

c. Point to the *Text Filters* option in the drop-down list box and then click *Does Not Equal* at the values list.

d. At the Custom Filter dialog box, type *Other* and then click OK.

7. Further refine the filter by completing the following steps:

a. Click in the first entry in the *EmployeeID* field.

b. Click the Filter button.

c. At the filter drop-down list, click the *(Select All)* check box to remove all of the check marks from the list options.

d. Click the *03* check box to insert a check mark.

e. Click OK.

8. Print the filtered report by pressing Ctrl + P and then clicking OK at the Print dialog box.

9. Save and then close the BanquetReport report.

Filtering by Selection

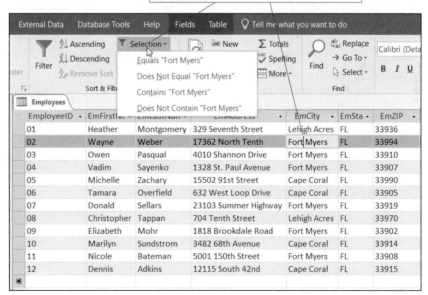
Ö̈uick Steps

Filter by Selection
1. Click in entry in field.
2. Click Selection button.
3. Click filtering option.

Filter by Shortcut Menu
1. Right-click in entry in field.
2. Click filtering option at shortcut menu.

Click in a field in an object and then click the Selection button in the Sort & Filter group on the Home tab and a drop-down list displays with options for filtering on the data in the field. For example, click in a field containing the city name *Fort Myers* and then click the Selection button and a drop-down list displays as shown in Figure 7.2. Click one of the options at the drop-down list to filter records.

Specific data can be selected in an object and then filtered by the selected data. For example, in Activity 1c, the word *peppers* will be selected in the entry *Green peppers* and then records will be filtered containing the word *peppers*.

Filtering by Shortcut Menu

Right-click in a field entry and a shortcut menu displays with options to sort the text, display a values list, or filter on a specific value. For example, right-click in the field entry *Birthday* in the *Event* field and the shortcut menu displays, as shown in Figure 7.3. Click a sort option to sort text in the field in ascending or descending order, point to the *Text Filters* option to display a values list, or click one of the values filters at the bottom of the menu. The shortcut menu can also be displayed by selecting specific text within a field entry and then right-clicking the selection.

Figure 7.2 Selection Button Drop-Down List

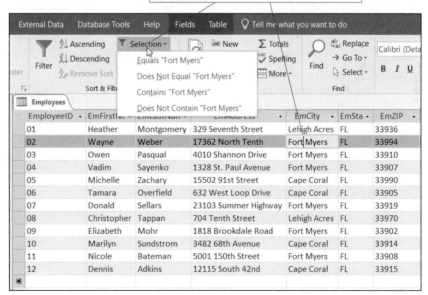

Figure 7.3 Filtering Shortcut Menu

> Right-click in a field entry to display a shortcut menu with sorting and filtering options.

BanquetReservations					
ResDate ▾	ResFirstNam ▾	ResLastNan ▾	ResTelephone ▾	Event ▾	EmployeeID ▾
6/9/2021	Joanne	Blair	(239) 555-7783	Birthday	03
6/17/2021	Jason	Haley	(239) 555-6641	Birthday	06
6/22/2021	Heidi	Thompson	(941) 555-3215	Birthday	01
6/30/2021	Kirsten	Simpson	(941) 555-4425	Birthday	02
6/16/2021	Aaron	Williams	(239) 555-3821	Bar mitzvah	04
6/29/2021	Robin	Gehring	(239) 555-0126	Bar m ✂ Cut	
6/10/2021	Tim	Drysdale	(941) 555-0098	Bat m 📋 Copy	
6/8/2021	Bridget	Kohn	(239) 551-1299	Othe 📋 Paste	
6/10/2021	Gabrielle	Johnson	(239) 555-1882	Othe A↓ Sort A to Z	
6/15/2021	Tristan	Strauss	(941) 555-7746	Othe Z↓ Sort Z to A	
6/17/2021	Lillian	Krakosky	(239) 555-8890	Othe	
6/29/2021	David	Fitzgerald	(941) 555-3792	Othe Clear filter from Event	
6/1/2021	Terrance	Schaefer	(239) 555-6239	Wedo Text Filters ▸	
6/10/2021	Cliff	Osborne	(239) 555-7823	Wedo Equals "Bar mitzvah"	
6/2/2021	David	Hooper	(941) 555-2338	Wedo Does Not Equal "Bar mitzvah"	
6/23/2021	Anthony	Wiegand	(239) 555-7853	Wedo Contains "Bar mitzvah"	
6/30/2021	Shane	Rozier	(239) 555-1033	Wedo Does Not Contain "Bar mitzvah"	
6/1/2021	Andrea	Wyatt	(239) 555-4282	Wedo	
6/15/2021	Janis	Semala	(239) 555-0476	Wedding reception	06

Activity 1c Filtering Records by Selection

Part 3 of 4

1. With **7-Skyline** open, open the Inventory table.
2. Filter only those records with a supplier number of 6 by completing the following steps:
   a. Click in the first entry containing *6* in the *SupplierID* field.
   b. Click the Selection button in the Sort & Filter group on the Home tab and then click *Equals "6"* at the drop-down list.
   c. Print the filtered table by pressing Ctrl + P and then clicking OK at the Print dialog box.
   d. Click the Toggle Filter button in the Sort & Filter group.

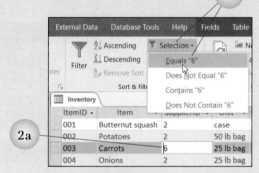

3. Filter any records in the *Item* field containing the word *peppers* by completing the following steps:
   a. Click in an entry in the *Item* field containing the text *Green peppers*.
   b. Using the mouse, select the word *peppers*.
   c. Click the Selection button and then click *Contains "peppers"* at the drop-down list.
   d. Print the filtered table by pressing Ctrl + P and then clicking OK at the Print dialog box.
4. Close the Inventory table without saving the changes.

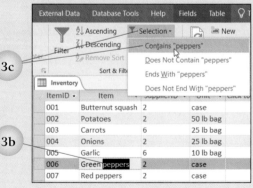

5. Open the BanquetReservations query.
6. Filter the records so that all but those containing *Birthday* in the *Event* field display by completing the following steps:
   a. Right-click in the first *Birthday* entry in the *Event* field.
   b. Click *Does Not Equal "Birthday"* at the shortcut menu.

c. Print the filtered query.
d. Click the Toggle Filter button.
7. Filter the records so that only those containing the word *mitzvah* in the *Event* field display by completing the following steps:
   a. Click in an entry in the *Event* field containing the entry *Bar mitzvah*.
   b. Using the mouse, select the word *mitzvah*.
   c. Right-click in the selected word and then click *Contains "mitzvah"* at the shortcut menu.
   d. Print the filtered query.
8. Close the BanquetReservations query without saving the changes.

Check Your Work

Using the *Filter By Form* Option

Advanced

Quick Steps

Use *Filter By Form* Option
1. Click Advanced button.
2. Click *Filter By Form*.
3. Click in empty field below column to filter.
4. Click arrow.
5. Click item to filter.

The Advanced button in the Sort & Filter group on the Home tab contains additional options for filtering records, including *Filter By Form*. Click this option and a blank record displays in a Filter by Form window in the work area. In the Filter by Form window, the Look for tab and the Or tab display at the bottom of the form. The Look for tab is active by default and tells Access to look for whatever data is inserted in a field. Click in the empty field below the column and an arrow displays at the right side of the field. Click the arrow and then click the item by which to filter. Click the Toggle Filter button to display the desired records. Add an additional value to a filter by clicking the Or tab at the bottom of the form.

1. With **7-Skyline** open, open the Banquets table.
2. Filter records by a specific employee ID number by completing the following steps:
   a. Click the Advanced button in the Sort & Filter group on the Home tab and then click *Filter By Form* at the drop-down list.

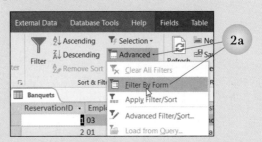

   b. At the Filter by Form window, click in the empty field below the *EmployeeID* column heading.
   c. Click the arrow at the right side of the field and then click *03* at the drop-down list.
   d. Click the Toggle Filter button in the Sort & Filter group.
3. Print the filtered table by completing the following steps:
   a. Click the File tab, click the *Print* option, and then click the *Print Preview* option.
   b. Change the orientation to landscape and the left and right margins to 0.5 inch.
   c. Click the Print button and then click OK at the Print dialog box.
   d. Click the Close Print Preview button.
4. Close the Banquets table without saving the changes.
5. Open the Inventory table.
6. Filter records by the supplier number 2 or 7 by completing the following steps:
   a. Click the Advanced button and then click *Filter By Form* at the drop-down list.
   b. At the Filter by Form window, click in the empty field below the *SupplierID* column heading.
   c. Click the arrow at the right side of the field and then click *2* at the drop-down list.
   d. Click the Or tab at the bottom of the form.
   e. If necessary, click in the empty field below the *SupplierID* column heading.

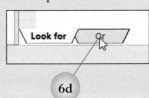

   f. Click the arrow at the right side of the field and then click *7* at the drop-down list.
   g. Click the Toggle Filter button.
   h. Print the filtered table.
   i. Click the Toggle Filter button to redisplay all records in the table.
   j. Click the Advanced button and then click *Clear All Filters* at the drop-down list.
7. Close the Inventory table without saving the changes.

Check Your Work

Tutorial

Viewing Object
Dependencies

Quick Steps

View Object
Dependencies

1. Open database.
2. Click object in
   Navigation pane.
3. Click Database Tools
   tab.
4. Click Object
   Dependencies
   button.

Object
Dependencies

Viewing Object Dependencies

The structure of a database is comprised of table, query, form, and report objects. Tables are related to other tables by the relationships that have been created. Queries, forms, and reports draw the source data from the records in the tables to which they have been associated, and forms and reports can include subforms and subreports, which further expand the associations between objects. A database with a large number of interdependent objects is more complex to work with than a simpler database. Viewing a list of the objects within a database and viewing the dependencies between objects can be beneficial to ensure an object is not deleted or otherwise modified, causing an unforeseen effect on another object.

View the structure of a database—including tables, queries, forms, and reports, as well as relationships—at the Object Dependencies task pane. Display this task pane by opening the database, clicking an object in the Navigation pane, clicking the Database Tools tab, and then clicking the Object Dependencies button in the Relationships group. The Object Dependencies task pane, shown in Figure 7.4, displays the objects in the Skyline database that depend on the Banquets table.

Figure 7.4 Object Dependencies Task Pane

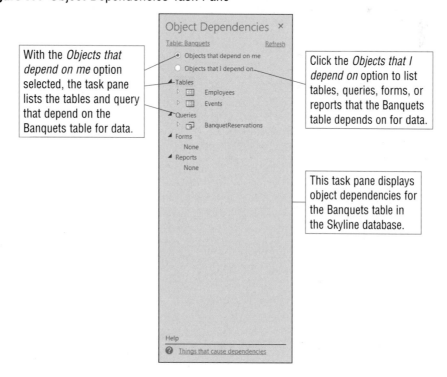

With the *Objects that depend on me* option selected, the task pane lists the tables and query that depend on the Banquets table for data.

Click the *Objects that I depend on* option to list tables, queries, forms, or reports that the Banquets table depends on for data.

This task pane displays object dependencies for the Banquets table in the Skyline database.

By default, *Objects that depend on me* is selected in the Object Dependencies task pane and the list box displays the names of the objects for which the selected object is the source. Next to each object in the task pane list box is an expand button (a right-pointing, white triangle). Clicking the expand button next to an object shows the other objects that depend on it. For example, if a query is based on the Banquets and Events tables and the query is used to generate a report, clicking the expand button next to the query name will show the report name. Clicking an object name in the Object Dependencies task pane opens the object in Design view.

Activity 2a Viewing Object Dependencies

Part 1 of 4

1. With **7-Skyline** open, display the structure of the database by completing the following steps:
   a. Click *Banquets* in the Tables group in the Navigation pane.
   b. Click the Database Tools tab and then click the Object Dependencies button in the Relationships group. (This displays the Object Dependencies task pane. By default, *Objects that depend on me* is selected and the task pane lists the names of the objects for which the Banquets table is the source.)

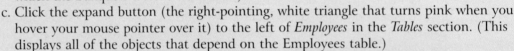

   c. Click the expand button (the right-pointing, white triangle that turns pink when you hover your mouse pointer over it) to the left of *Employees* in the *Tables* section. (This displays all of the objects that depend on the Employees table.)
   d. Click the *Objects that I depend on* option near the top of the Object Dependencies task pane.

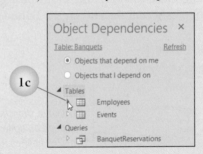
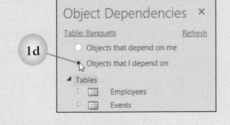

   e. Click *Events* in the Tables group in the Navigation pane. (Make sure to click *Events* in the Navigation pane and not the Object Dependencies task pane.)
   f. Click the <u>Refresh</u> hyperlink in the upper right corner of the Object Dependencies task pane.
   g. Click the *Objects that depend on me* option near the top of the Object Dependencies task pane.
2. Close the Object Dependencies task pane.

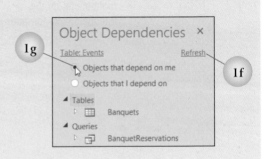

Using Options at the Info Backstage Area

The Info backstage area contains options for compacting and repairing a database, encrypting a database with a password, and viewing and customizing database properties. Display the Info backstage area, shown in Figure 7.5, by opening a database and then clicking the File tab.

Compacting and Repairing a Database

Compact & Repair Database

Quick Steps

Compact and Repair Database
1. Open database.
2. Click File tab.
3. Click Compact & Repair Database button.

Compacting and Repairing a Database

To optimize the performance of a database, compact and repair it on a regular basis. When working in a database on an ongoing basis, data in it can become fragmented, causing the amount of space the database takes on the storage medium or in the folder to be larger than necessary. To compact and repair a database, open the database, click the File tab and then click the Compact & Repair Database button or click the Compact and Repair Database button in the Tools group on the Database Tools tab.

A database can be compacted and repaired each time it is closed. To do this, click the File tab and then click *Options*. At the Access Options dialog box, click the *Current Database* option in the left panel. Click the *Compact on Close* check box to insert a check mark and then click OK to close the dialog box. Before compacting and repairing a database in a multi-user environment, make sure that no other user has the database open.

Figure 7.5 Info Backstage Area

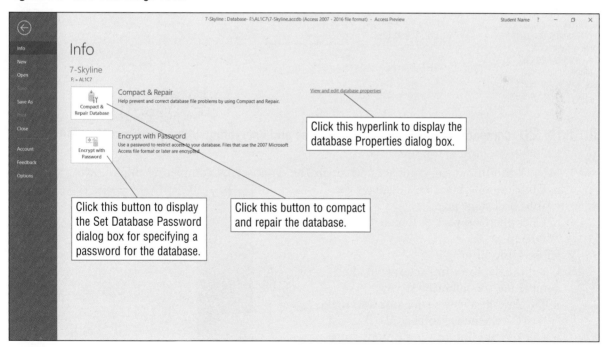

Tutorial

Encrypting a
Database with a
Password

 Encrypt with
Password

Decrypt
Database

Encrypting a Database with a Password

To prevent unauthorized access to a database, encrypt the database with a password to ensure that it can be opened only by someone who knows the password. Be careful when encrypting a database with a password because if the password is lost, the database will not open.

To encrypt a database with a password, the database must be opened in Exclusive mode. To do this, display the Open dialog box, navigate to the folder containing the database, and then click the database to select it. Click the Open button arrow (in the lower right corner of the dialog box) and then click *Open Exclusive* at the drop-down list. When the database opens, click the File tab and then click the Encrypt with Password button at the Info backstage area. This displays the Set Database Password dialog box, as shown in Figure 7.6. At this dialog box, type a password in the *Password* text box, press the Tab key, and then type the password again. The typed text will display as asterisks. Click OK to close the Set Database Password dialog box.

To remove a password from a database, open the database in Exclusive mode, click the File tab, and then click the Decrypt Database button. At the Unset Database Password dialog box, type the password and then click OK.

Quick Steps

Open Database in Exclusive Mode
1. Display Open dialog box.
2. Click database.
3. Click Open button arrow.
4. Click *Open Exclusive*.

Encrypt Database with Password
1. Open database in Exclusive mode.
2. Click File tab.
3. Click Encrypt with Password button.
4. Type password, press Tab key, and type password again.
5. Click OK.

Hint When encrypting a database with a password, use a password that combines uppercase and lowercase letters, numbers, and symbols.

Figure 7.6 Set Database Password Dialog Box

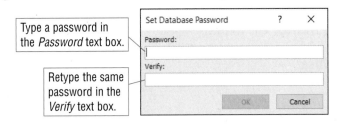

Type a password in the *Password* text box.

Retype the same password in the *Verify* text box.

Activity 2b **Compacting and Repairing a Database and Encrypting with a Password** Part 2 of 4

1. With **7-Skyline** open, compact and repair the database by completing the following steps:
   a. Click the File tab. (This displays the Info backstage area.)
   b. Click the Compact & Repair Database button.
2. Close **7-Skyline**.
3. Open the database in Exclusive mode by completing the following steps:
   a. Display the Open dialog box and make AL1C7 the active folder.
   b. Click **7-Skyline** in the Content pane to select it.

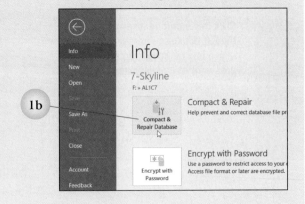

c. Click the Open button arrow (in the lower right corner of the dialog box) and then click *Open Exclusive* at the drop-down list.

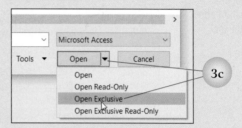

4. Encrypt the database with a password by completing the following steps:
   a. Click the File tab.
   b. At the Info backstage area, click the Encrypt with Password button.
   c. At the Set Database Password dialog box, type your first and last names in all lowercase letters with no space, press the Tab key, and then type your first and last names again in lowercase letters.
   d. Click OK to close the dialog box.
   e. If a message displays with information about encrypting with a block cipher, click OK.

5. Close **7-Skyline**.
6. Display the Open dialog box with AL1C7 the active folder and then open **7-Skyline** in Exclusive mode.
7. At the Password Required dialog box, type your password and then click OK.
8. Remove the password by completing the following steps:
   a. Click the File tab.
   b. Click the Decrypt Database button.
   c. At the Unset Database Password dialog box, type your first and last names in lowercase letters and then press the Enter key.

Viewing and Customizing Database Properties

Each database has associated properties, such as the type of file; its location; and when it was created, accessed, and modified. These properties can be viewed and modified at the Properties dialog box. To view properties for the currently open database, click the File tab to display the Info backstage area and then click the View and edit database properties hyperlink at the right side of the backstage area. This displays the Properties dialog box, similar to what is shown in Figure 7.7.

The Properties dialog box for an open database contains tabs with information about the database. With the General tab selected, the dialog box provides information about the database type, size, and location. Click the Summary tab to see fields such as *Title, Subject, Author, Category, Keywords,* and *Comments.* Some fields contain data and others are blank. Text can be inserted, edited, or deleted in the fields. Move the insertion point to a field by clicking in the field or by pressing the Tab key until the insertion point is positioned in the field.

Quick Steps
View Database Properties
1. Click File tab.
2. Click View and edit database properties.

Figure 7.7 Properties Dialog Box

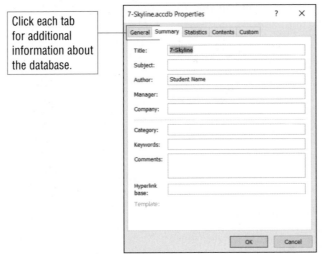

Click each tab for additional information about the database.

Click the Statistics tab to view information such as the dates the database was created, modified, accessed, and printed. With the Content tab selected, the *Document contents* section lists the objects in the database, including tables, queries, forms, reports, macros, and modules.

Use options at the Properties dialog box with the Custom tab selected to add custom properties to the database. For example, a property can be added to include the date the database was completed, information on the department in which the database was created, and much more. The list box below the *Name* option box displays the predesigned properties provided by Access. Choose a predesigned property from this list box or create a custom property.

To choose a predesigned property, click a predesigned property in the list box, specify what type of property it is (such as value, date, number, yes/no), and then type a value. For example, to specify the department in which the database was created, click *Department* in the list box, make sure the *Type* displays as *Text*, click in the *Value* text box, and then type the name of the department.

Activity 2c Viewing and Customizing Database Properties

Part 3 of 4

1. With **7-Skyline** open, click the File tab and then click the <u>View and edit database properties</u> hyperlink at the right side of the backstage area.

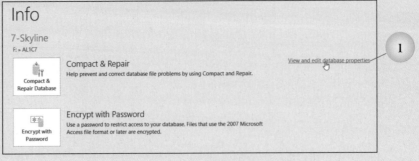

2. At the 7-Skyline.accdb Properties dialog box, click the General tab and then read the information in the dialog box.
3. Click the Summary tab and then type the following text in the specified text boxes:

Title	7-Skyline database
Subject	Restaurant and banquet facilities
Author	(*Type your first and last names.*)
Category	restaurant
Keywords	restaurant, banquet, event, Fort Myers
Comments	This database contains information on Skyline Restaurant employees, banquets, inventory, and orders.

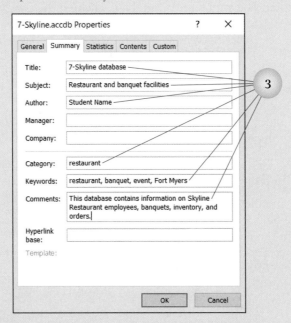

4. Click the Statistics tab and read the information in the dialog box.
5. Click the Contents tab and notice that the *Document contents* section of the dialog box displays the objects in the database.
6. Click the Custom tab and then create custom properties by completing the following steps:
   a. Click the *Date completed* option in the *Name* list box.
   b. Click the *Type* option box arrow and then click *Date* at the drop-down list.
   c. Click in the *Value* text box and then type the current date in this format: *dd/mm/yyyy*.
   d. Click the Add button.
   e. With the insertion point positioned in the *Name* text box, type Course.
   f. Click the *Type* option box arrow and then click *Text* at the drop-down list.
   g. Click in the *Value* text box, type your current course number, and then press the Enter key.
   h. Click OK to close the dialog box.
7. Click the Back button to return to the database.

Check Your Work

Tutorial

Saving a Database
and Database
Object in Different
Formats

Saving a Database and Database Object in Different Formats

Access 2003 and earlier versions save a database with a different file extension. To save an Access database in Access 2003 or earlier, open the database, click the File tab, and then click the *Save As* option. This displays the Save As backstage area, as shown in Figure 7.8. Click the *Access 2002-2003 Database* option in the *Save Database As* section and then click the Save As button at the bottom of the *Save Database As* section. This displays the Save As dialog box with the *Save as type* option set to *Microsoft Access Database (2002-2003)*. At this dialog box, click the Save button.

With an object open in a database, clicking the *Save Object As* option in the *File Types* section of the Save As backstage area displays options for saving the object. Click the *Save Object As* option to save the selected object in the database or click the *PDF or XPS* option to save the object as a PDF or XPS file. The letters *PDF* stand for *portable document format*, a file format developed by Adobe Systems that captures all of the elements of a file as an electronic image. An XPS file is a Microsoft file format for publishing content in an easily viewable format. The letters *XPS* stand for *XML paper specification* and the letters *XML* stand for *extensible markup language*, which is a set of rules for encoding files electronically.

Quick Steps

Save Database in Earlier Version
1. Open database.
2. Click File tab.
3. Click *Save As* option.
4. Click version in *Save Database As* section.
5. Click Save As button.
6. Click Save button.

Hint An Access 2007 or later version cannot be opened with an earlier version of Access.

Saving an Object as a PDF or XPS File

Quick Steps

Save Object in PDF or XPS File Format
1. Open object.
2. Click File tab.
3. Click *Save As* option.
4. Click *Save Object As* option.
5. Click *PDF or XPS* option.
6. Click Save As button.

To save an object as a PDF or XPS file, open the object, click the File tab, and then click the *Save As* option. At the Save As backstage area, click the *Save Object As* option in the *File Types* section, click the *PDF or XPS* option in the *Save the current database object* section, and then click the Save As button. This displays the Publish as PDF or XPS dialog box with the *Save as type* option set at *PDF*. Click the Publish button and the object is saved as a PDF file. To specify that the object should open in Adobe Acrobat Reader, click the *Open file after publishing* check box to insert a check box. With this check box active, the object will open in Adobe Acrobat Reader when the Publish button is clicked.

Figure 7.8 Save As Backstage Area with *Save Database As* Option Selected

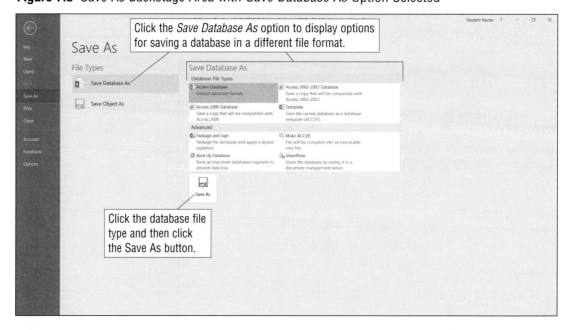

A PDF file can be opened in Adobe Acrobat Reader, Microsoft Edge, or Microsoft Word. (A PDF file may also open in other available web browsers such as Google Chrome.) An XPS file can be opened in Internet Explorer or XPS Viewer. One method for opening a PDF or XPS file is to open File Explorer, navigate to the folder containing the file, right-click on the file, and then point to *Open with*. This displays a side menu with the programs that will open the file.

Tutorial

Backing Up a
Database

Backing Up a Database

Databases often contain important company information, and loss of this information can cause major problems. Backing up a database is important to minimize the chances of losing critical company data and is especially important when several people update and manage a database.

Quick Steps

Back Up Database
1. Click File tab.
2. Click *Save As* option.
3. Click *Back Up Database* option.
4. Click Save As button.
5. Click Save button.

To back up a database, open the database, click the File tab, and then click the *Save As* option. At the Save As backstage area, click the *Back Up Database* option in the *Advanced* section and then click the Save As button. This displays the Save As dialog box with a default database file name, which is the original database name followed by the current date, in the *File name* text box. Click the Save button to save the backup database while keeping the original database open.

Activity 2d Saving an Object as a PDF File, Saving a Database in a Previous Version, and Backing Up a Database

Part 4 of 4

1. With **7-Skyline** open, save the Orders table as a PDF file by completing the following steps:
   a. Open the Orders table.
   b. Click the File tab and then click the *Save As* option.
   c. At the Save As backstage area, click the *Save Object As* option in the *File Types* section.
   d. Click the *PDF or XPS* option in the *Save the current database object* section.
   e. Click the Save As button.

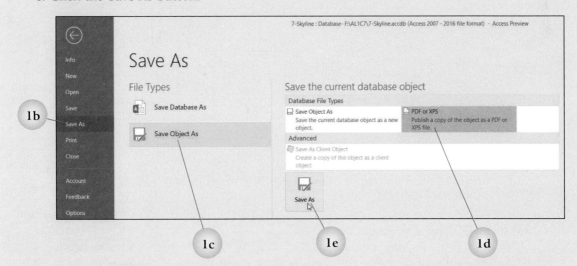

   f. At the Publish as PDF or XPS dialog box, make sure your AL1C7 folder is active and then click the *Open file after publishing* check box to insert a check mark. (Skip this step if the check box already contains a check mark.)
   g. Click the Publish button.

h. When the Orders table opens in Adobe Acrobat Reader, scroll through the file and then close the file by clicking the Close button in the upper right corner of the screen.

2. Close the Orders table.

3. Save the database in a previous version of Access by completing the following steps:
   a. Click the File tab and then click the *Save As* option.
   b. At the Save As backstage area, click the *Access 2002-2003 Database* option in the *Save Database As* section.
   c. Click the Save As button.

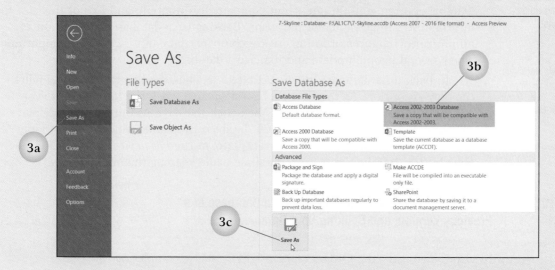

d. At the Save As dialog box, make sure your AL1C7 folder is active, type 7-Skyline-2002-2003format, and then click the Save button. (This saves the database with the file extension *.mdb*.)

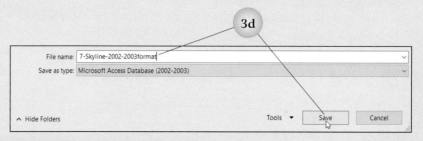

e. Notice that the Title bar displays the database file name *7-Skyline-2002-2003format : Database (Access 2002 - 2003 file format)*.

4. Close the database.

5. Open **7-Skyline**.

6. Create a backup of the database by completing the following steps:
   a. Click the File tab and then click the *Save As* option.
   b. At the Save As backstage area, click the *Back Up Database* option in the *Advanced* section and then click the Save As button.
   c. At the Save As dialog box, notice that the database name in the *File name* text box displays the original file name followed by the current date (year, month, day).
   d. Make sure your AL1C7 folder is active and then click the Save button. (This saves the backup copy of the database to your folder and the original database remains open.)

7. Close **7-Skyline**.

Chapter 7 | Modifying, Filtering, and Viewing Data

Chapter Summary

- A set of restrictions called a filter can be put on records in a table or form. A filter is used to select specific field values.

- Filter records with the Filter button in the Sort & Filter group on the Home tab.

- Click the Toggle Filter button in the Sort & Filter group to switch back and forth between data and filtered data.

- Remove a filter by clicking the Filter button in the Sort & Filter group and then clicking the *Clear filter from xxx* (where *xxx* is the name of the field).

- Another method for removing a filter is to click the Advanced button in the Sort & Filter group and then click *Clear All Filters*.

- Display a list of filter values by clicking the Filter button and then pointing to *Text Filters* (if the data is text), *Number Filters* (if the data is numbers), or *Date Filters* (if the data is dates).

- Filter by selection by clicking the Selection button in the Sort & Filter group.

- Right-click in a field entry to display a shortcut menu with filtering options.

- Filter by form by clicking the Advanced button in the Sort & Filter group and then clicking *Filter By Form* at the drop-down list. This displays a blank record with two tabs: Look for and Or.

- View the structure of a database and relationships between objects at the Object Dependencies task pane. Display this task pane by clicking the Database Tools tab and then clicking the Object Dependencies button in the Relationships group.

- Click the Compact & Repair Database button at the Info backstage area or in the Tools group on the Database Tools tab to optimize database performance.

- To prevent unauthorized access to a database, encrypt the database with a password. To encrypt a database, the database must be opened in Exclusive mode using the Open button drop-down list in the Open dialog box. While in Exclusive mode, encrypt a database with a password using the Encrypt with Password button at the Info backstage area.

- To view properties for the current database, click the View and edit database properties hyperlink at the Info backstage area. The Properties dialog box displays with a number of tabs containing information about the database.

- Save a database in Access 2003 or earlier using options in the *Save Database As* section of the Save As backstage area.

- To save a database object as a PDF or XPS file, display the Save As backstage area, click the *Save Object As* option, click the *PDF or XPS* option, and then click the Save As button.

- Back up a database to maintain critical data. Back up a database with the *Back Up Database* option at the Save As backstage area.

Commands Review

FEATURE	RIBBON TAB, GROUP/OPTION	BUTTON, OPTION
filter	Home, Sort & Filter	
filter by form	Home, Sort & Filter	, *Filter By Form*
filter by selection	Home, Sort & Filter	
Info backstage area	File, *Info*	
Object Dependencies task pane	Database Tools, Relationships	
remove filter	Home, Sort & Filter	, *Clear filter from xxx* OR , *Clear All Filters*
toggle filter	Home, Sort & Filter	

Microsoft®

Access®

Exporting and Importing Data

Performance Objectives

Upon successful completion of Chapter 8, you will be able to:

1   Export Access data to Excel

2   Export Access data to Word

3   Merge Access data with a Word document

4   Export an Access object to a PDF or XPS file

5   Import data to a new table

6   Link data to a new table

7   Use the Office Clipboard

Microsoft Office is a suite of applications that allows for easy data exchange. In this chapter, you will learn how to export data from Access to Excel and Word, merge Access data with a Word document, export an Access object to a PDF or XPS file, import and link Excel data to a new table, and copy and paste data between applications and programs.

 Data Files

Before beginning chapter work, copy the AL1C8 folder to your storage medium and then make AL1C8 the active folder.

The online course includes additional training and assessment resources.

You will export a table and query to Excel, export a table and report to Word, and export an Access object to a PDF file. You will also merge data in an Access table and query with a Word document.

Exporting Data

One of the advantages of using the Microsoft Office suite is the ability to exchange data between programs. Access, like other programs in the suite, offers a feature to export data from Access into Excel and/or Word. The Export group on the External Data tab contains buttons for exporting a table, query, form, or report to other programs, such as Excel and Word.

Tutorial

Exporting Access
Data to Excel

Excel

♀ *Hint* Store data in Access and analyze it using Excel.

Quick Steps

**Export Access Data
to Excel**

1. Click table, query, or
   form.
2. Click External Data
   tab.
3. Click Excel button in
   Export group.
4. Make changes
   at Export - Excel
   Spreadsheet dialog
   box.
5. Click OK.

Insert a check mark
in this check box
to export all object
formatting and layout.

Insert a check mark
in this check box to
open the file in the
destination program.

Exporting Access Data to Excel

Use the Excel button in the Export group on the External Data tab to export data in a table, query, or form to an Excel workbook. Click the object containing the data to be exported to Excel, click the External Data tab, and then click the Excel button in the Export group. The Export - Excel Spreadsheet dialog box displays, as shown in Figure 8.1.

At the dialog box, Access uses the name of the object as the Excel workbook name. This can be changed by selecting the current name and then typing a new name. The file format also can be changed with the *File format* option. Click the

Figure 8.1 Export - Excel Spreadsheet Dialog Box

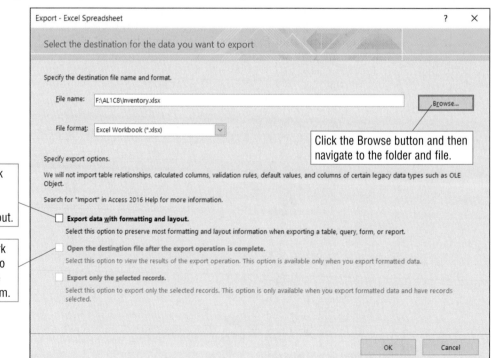

Hint Data exported from Access to Excel is saved as an Excel workbook with the .xlsx file extension.

Hint You can export only one database object at a time.

Export data with formatting and layout check box to insert a check mark. This exports all data formatting to the Excel workbook. To open Excel with the exported data, click the *Open the destination file after the export operation is complete* option to insert a check mark.

When all changes have been made, click OK. This opens Excel with the data in a workbook. Make changes to the workbook and then save, print, and close the workbook. When Excel is closed, Access displays with a dialog box, asking if the export step should be saved. At this dialog box, insert a check mark in the *Save export steps* check box to save the export steps or leave the check box blank and then click the Close button.

Activity 1a Exporting a Table and Query to Excel

Part 1 of 5

1. Open **8-Hilltop** from your AL1C8 folder and enable the content.
2. Save the Inventory table as an Excel workbook by completing the following steps:
   a. Click *Inventory* in the Tables group in the Navigation pane.
   b. Click the External Data tab and then click the Excel button in the Export group.
   c. At the Export - Excel Spreadsheet dialog box, click the Browse button.
   d. At the File Save dialog box, navigate to your AL1C8 folder and then click the Save button.
   e. Click the *Export data with formatting and layout* option to insert a check mark in the check box.
   f. Click the *Open the destination file after the export operation is complete* option to insert a check mark in the check box.
   g. Click OK.

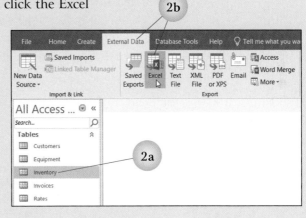

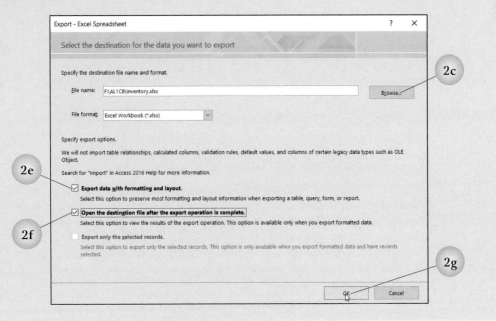

h. When the data displays on the screen in Excel as a worksheet, select the range A2:A11 and then click the Center button in the Alignment group on the Home tab.

i. Select the range D2:F11 and then click the Center button.

j. Click the Save button on the Quick Access Toolbar.

k. Print the worksheet by pressing Ctrl + P and then clicking the Print button at the Print backstage area.

l. Close the worksheet and then close Excel.

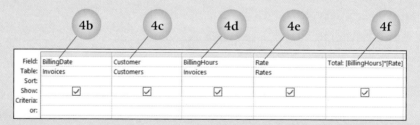

3. In Access, click the Close button to close the dialog box.

4. Design a query that extracts records from three tables with the following specifications:
   a. Add the Invoices, Customers, and Rates tables to the query window.
   b. Insert the *BillingDate* field from the Invoices table field list box in the first field in the *Field* row.
   c. Insert the *Customer* field from the Customers table field list box in the second field in the *Field* row.
   d. Insert the *BillingHours* field from the Invoices table field list box in the third field in the *Field* row.
   e. Insert the *Rate* field from the Rates table field list box in the fourth field in the *Field* row.
   f. Click in the fifth field in the *Field* row, type Total: [BillingHours]*[Rate], and then press the Enter key.

Field:	BillingDate	Customer	BillingHours	Rate	Total: [BillingHours]*[Rate]
Table:	Invoices	Customers	Invoices	Rates	
Sort:					
Show:	☑	☑	☑	☑	☑
Criteria:					
or:					

g. Run the query.

h. If necessary, adjust the column width of the *Customer* field to display all records.

i. Save the query, typing CustomerInvoices as the query name.

j. Close the query.

5. Export the CustomerInvoices query to Excel by completing the following steps:
   a. Click *CustomerInvoices* in the Queries group in the Navigation pane.
   b. Click the External Data tab and then click the Excel button.
   c. At the Export - Excel Spreadsheet dialog box, click the *Export data with formatting and layout* option to insert a check mark in the check box.
   d. Click the *Open the destination file after the export operation is complete* option to insert a check mark in the check box.
   e. Click OK.

f. When the data displays on the screen in Excel as a worksheet, select the range C2:C31 and then click the Center button in the Alignment group on the Home tab.

g. Click the Save button on the Quick Access Toolbar.

h. Print the worksheet by pressing Ctrl + P and then clicking the Print button at the Print backstage area.

i. Close the worksheet and then close Excel.

6. In Access, click the Close button to close the dialog box.

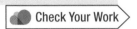

Exporting Access Data to Word

Exporting Access
Data to Word

 More

Quick Steps

Export Data to Word
1. Click object in Navigation pane.
2. Click External Data tab.
3. Click More button in Export group.
4. Click *Word*.
5. Make changes at Export - RTF File dialog box.
6. Click OK.

Export data from Access to Word in a similar manner as exporting to Excel. To export data to Word, click the object in the Navigation pane, click the External Data tab, click the More button in the Export group, and then click *Word* at the drop-down list. At the Export - RTF File dialog box, make changes and then click OK. Word automatically opens and the data displays in a Word document that is saved automatically with the same name as the database object. The difference is that the file extension *.rtf* is added to the name. An RTF file is saved in rich-text format, which preserves formatting such as fonts and styles. A document saved with the .rtf extension can be exported in Word and other Windows word processing or desktop publishing programs.

Activity 1b Exporting a Table and Report to Word   Part 2 of 5

1. With **8-Hilltop** open, click *Invoices* in the Tables group in the Navigation pane.
2. Click the External Data tab, click the More button in the Export group, and then click *Word* at the drop-down list.

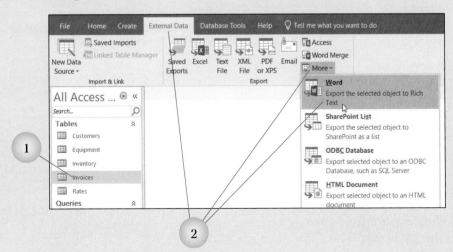

3. At the Export - RTF File dialog box, click the Browse button.
4. At the File Save dialog box, make sure your AL1C8 folder is active and then click the Save button.
5. At the Export - RTF File dialog box, click the *Open the destination file after the export operation is complete* check box to insert a check mark.

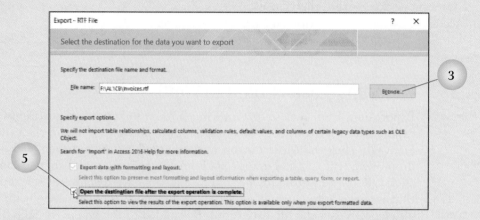

6. Click OK.
7. With **Invoices** open in Word, print the file by pressing Ctrl + P and then clicking the Print button at the Print backstage area.
8. Close **Invoices** and then close Word.
9. In Access, click the Close button to close the dialog box.
10. Create a report with the Report Wizard by completing the following steps:
   a. Click the Create tab and then click the Report Wizard button in the Reports group.
   b. At the first Report Wizard dialog box, insert the following fields in the *Selected Fields* list box:
      From the Customers table:
         Customer
      From the Equipment table:
         Equipment
      From the Invoices table:
         BillingDate
         BillingHours

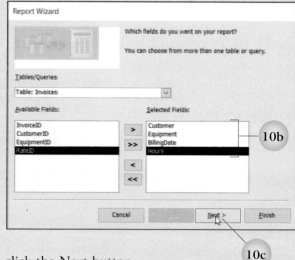

   c. After inserting the fields, click the Next button.
   d. At the second Report Wizard dialog box, make sure *by Customers* is selected in the list box at the left and then click the Next button.
   e. At the third Report Wizard dialog box, click the Next button.
   f. At the fourth Report Wizard dialog box, click the Next button.
   g. At the fifth Report Wizard dialog box, click *Block* in the *Layout* section and then click the Next button.
   h. At the sixth and final Report Wizard dialog box, select the current name in the *What title do you want for your report?* text box, type CustomerReport, and then click the Finish button.
   i. When the report displays in Print Preview, click the Print button at the left of the Print Preview tab and then click OK at the Print dialog box.
   j. Save and then close the CustomerReport report.

11. Export the CustomerReport report to Word by completing the following steps:
    a. Click *CustomerReport* in the Reports group in the Navigation pane.
    b. Click the External Data tab, click the More button, and then click *Word* at the drop-down list.
    c. At the Export - RTF File dialog box, click the *Open the destination file after the export operation is complete* option to insert a check mark in the check box and then click OK.
    d. When the data displays on the screen in Word, print the document by pressing Ctrl + P and then clicking the Print button at the Print backstage area.
    e. Save and then close the CustomerReport document.
    f. Close Word.
12. In Access, click the Close button to close the dialog box.

Check Your Work

Merging Access Data with a Word Document

Merging Access
Data with a Word
Document

Data from an Access table or query can be merged with a Word document. When merging data, the data in the Access table is considered the data source and the Word document is considered the main document. When the merge is completed, the merged document displays in Word.

 Word
Merge

To merge data in a table, click the table in the Navigation pane, click the External Data tab, and then click the Word Merge button. When merging Access data, either type the text in the main document or merge Access data with an existing Word document.

Activity 1c Merging Access Table Data with a Word Document   Part 3 of 5

1. With **8-Hilltop** open, click *Customers* in the Tables group in the Navigation pane.
2. Click the External Data tab.
3. Click the Word Merge button in the Export group.

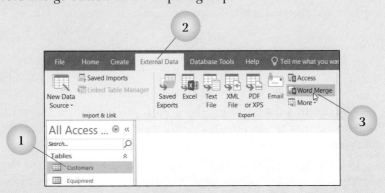

4. At the Microsoft Word Mail Merge Wizard dialog box, make sure *Link your data to an existing Microsoft Word document* is selected and then click OK.
5. At the Select Microsoft Word Document dialog box, make sure your AL1C8 folder is active and then double-click the document named **8-HilltopLetter**.
6. Click the Word button on the taskbar.
7. Click the Maximize button in the *8-HilltopLetter* Title bar and then close the Mail Merge task pane.

8. Press the Down Arrow key six times (not the Enter key) and then type the current date.
9. Press the Down Arrow key four times and then insert fields for merging from the Customers table by completing the following steps:
   a. If necesssary, click the Mailings tab.
   b. Click the Insert Merge Field button arrow in the Write & Insert Fields group and then click *Customer* in the drop-down list. (This inserts the «Customer» field in the document.)
   c. Press the Enter key, click the Insert Merge Field button arrow, and then click *CuStreetAddress* in the drop-down list.
   d. Press the Enter key, click the Insert Merge Field button arrow, and then click *CuCity* in the drop-down list.
   e. Type a comma (,) and then press the spacebar.
   f. Click the Insert Merge Field button arrow and then click *CuState* in the drop-down list.
   g. Press the spacebar, click the Insert Merge Field button arrow, and then click *CuZipCode* in the drop-down list.
   h. Replace the letters *XX* at the bottom of the letter with your initials.
   i. Click the Finish & Merge button in the Finish group and then click *Edit Individual Documents* in the drop-down list.
   j. At the Merge to New Document dialog box, make sure *All* is selected and then click OK.
   k. When the merge is completed, save the new document and name it **8-HilltopLtrs** in your AL1C8 folder.
10. Print just the first two pages (two letters) of **8-HilltopLtrs**.
11. Close **8-HilltopLtrs** and then close **8-HilltopLetter** without saving the changes.
12. Close Word.
13. Close **8-Hilltop**.

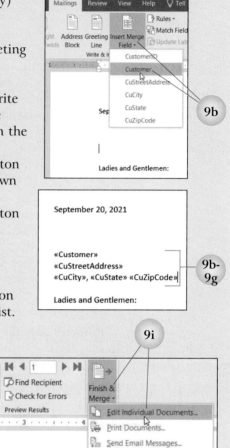

 Check Your Work

Quick Steps

Merge Data with Word
1. Click table or query.
2. Click External Data tab.
3. Click Word Merge button.
4. Make choices at each dialog box.

Address Block

A query in a database can be used to merge with a Word document. In Activity 1c, a table was merged with an existing Word document. A table or query also can be merged and then the Word document typed.

In Activity 1c, a number of merge fields were inserted for the inside address of a letter. Another method for inserting fields for the inside address is to insert the «AddressBlock» field, which inserts all of the fields required for the inside address. The «AddressBlock» field is an example of a composite field, which groups a number of fields together. Insert the «AddressBlock» composite field by clicking the Address Block button in the Write & Insert Fields group on the Mailings tab. Clicking the button displays the Insert Address Block dialog box with a preview of how the fields will be inserted in the document to create the inside address. The dialog box also contains buttons and options for customizing the fields. Click OK and the «AddressBlock» field is inserted in the document.

In Activity 1c, the «AddressBlock» composite field could not be used because the *Customer* field was not recognized by Word as a field for the inside address. In Activity 1d, a query will be created that contains the *FirstName* and *LastName* fields, which Word recognizes and uses for the «AddressBlock» composite field.

Activity 1d Performing a Query and then Merging with a Word Document   Part 4 of 5

1. Open **8-CopperState** from your AL1C8 folder and enable the content.
2. Perform a query with the Query Wizard and modify the query by completing the following steps:
   a. Click the Create tab and then click the Query Wizard button in the Queries group.
   b. At the New Query dialog box, make sure *Simple Query Wizard* is selected and then click OK.
   c. At the first Simple Query Wizard dialog box, click the *Tables/Queries* option box arrow and then click *Table: Clients*.
   d. Click the All Fields button to insert all of the fields in the *Selected Fields* list box.
   e. Click the Next button.
   f. At the second Simple Query Wizard dialog box, make the following changes:
      1) Select the current name in the *What title do you want for your query?* text box and then type ClientsPhoenix.
      2) Click the *Modify the query design* option.
      3) Click the Finish button.
   g. At the query window, click in the *City* field in the *Criteria* row, type Phoenix, and then press the Enter key.
   h. Click the Run button in the Results group. (Only clients living in Phoenix will display.)
   i. Save and then close the query.
3. Click *ClientsPhoenix* in the Queries group in the Navigation pane.
4. Click the External Data tab and then click the Word Merge button in the Export group.
5. At the Microsoft Word Mail Merge Wizard dialog box, click the *Create a new document and then link the data to it* option and then click OK.
6. Click the Word button on the taskbar.
7. Click the Maximize button in the Document1 Title bar and then close the Mail Merge task pane.

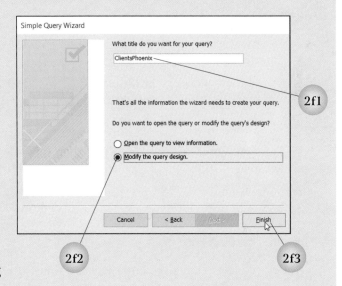

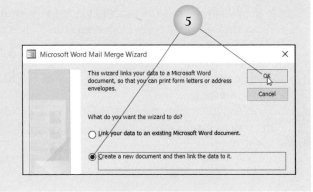

8. Complete the following steps to type text and insert the «AddressBlock» composite field in the blank Word document:

   a. Click the Home tab and then click the *No Spacing* style in the Styles group.

   b. Press the Enter key six times.

   c. Type the current date.

   d. Press the Enter key four times.

   e. Click the Mailings tab.

   f. Insert the «AddressBlock» composite field by clicking the Address Block button in the Write & Insert Fields group on the Mailings tab and then clicking OK at the Insert Address Block dialog box. (This inserts the «AddressBlock» composite field in the document.)

   g. Press the Enter key two times and then type the salutation Ladies and Gentlemen:.

   h. Press the Enter key two times and then type the following paragraphs of text (press the Enter key two times after typing the first paragraph):

> At the Grant Street West office of Copper State Insurance, we have hired two additional insurance representatives as well as one support staff member to ensure that we meet all your insurance needs. To accommodate the new staff, we have moved to a larger office just a few blocks away. Our new address is 3450 Grant Street West, Suite 110, Phoenix, AZ 85003. Our telephone number, (602) 555-6300, has remained the same.
>
> If you have any questions or concerns about your insurance policies or want to discuss adding or changing current coverage, please stop by or give us a call. We are committed to providing our clients with the most comprehensive automobile insurance coverage in the county.

   i. Press the Enter key two times and then type the following complimentary close at the left margin (press the Enter key four times after typing *Sincerely,*):

> Sincerely,
>
>
>
> Lou Galloway
> Manager (Press the Enter key two times after typing *Manager*.)
>
> XX (Type your initials instead of XX.)
> 8-CSLtrs

   j. Click the Finish & Merge button in the Finish group on the Mailings tab and then click *Edit Individual Documents* in the drop-down menu.

   k. At the Merge to New Document dialog box, make sure *All* is selected, and then click OK.

   l. When the merge is complete, save the new document in your AL1C8 folder and name it **8-CSLtrs**.

9. Print the first two pages (two letters) of **8-CSLtrs**.

10. Close **8-CSLtrs**.

11. Save the main document as **8-CSMainDoc** in your AL1C8 folder and then close the document.

12. Close Word.

Check Your Work

Exporting an Access Object to a PDF or XPS File

Quick Steps

**Export Access Object
to PDF or XPS File**
1. Click object in
   Navigation pane.
2. Click External Data
   tab.
3. Click PDF or XPS
   button.
4. Navigate to folder.
5. Click Publish button.

 PDF or XPS

With the PDF or XPS button in the Export group on the External Data tab, an Access object can be exported to a PDF or XPS file. As explained in Chapter 7, the letters *PDF* stand for *portable document format*, which is a file format that captures all of the elements of a file as an electronic image. The letters *XPS* stand for *XML paper specification* and the letters *XML* stand for *extensible markup language*, which is a set of rules for encoding files electronically. PDF and XPS files both have a fixed layout, so they can be shared or displayed for viewing without worry that content will be modified or formatting will be lost.

To export an Access object to a PDF or XPS file, click the object, click the External Data tab, and then click the PDF or XPS button in the Export group. This displays the Publish as PDF or XPS dialog box with the *PDF* option selected in the *Save as type* option box. To save the Access object in XPS file format, click the *Save as type* option box and then click *XPS Document* at the drop-down list. At the Save As dialog box, type a name in the *File name* text box and then click the Publish button.

To open a PDF or XPS file in a web browser, open the browser and then press Ctrl + O to display the Open dialog box. At the Open dialog box, change the *Files of type* to *All Files*, navigate to the folder containing the file, and then double-click the file.

Activity 1e Exporting an Access Object to a PDF File   Part 5 of 5

1. With **8-CopperState** open, export the Coverage table to a PDF file by completing the following steps:
   a. Click *Coverage* in the Tables group in the Navigation pane.
   b. Click the External Data tab.
   c. Click the PDF or XPS button in the Export group.
   d. At the Publish as PDF or XPS dialog box, navigate to your AL1C8 folder and make sure the *Save as type* option box displays with *PDF*. (If not, click the option box and then click *PDF* at the drop-down list.)
   e. Click the *Open file after publishing* check box to insert a check mark and then click the Publish button.
   f. When the Coverage table data displays in Adobe Acrobat Reader, scroll through the file to see how it looks.
   g. Print the PDF file by clicking the Print button on the menu bar and then clicking the Print button at the Print dialog box.
   h. Close Adobe Acrobat Reader by clicking the Close button in the upper right corner of the window.
2. In Access, click the Close button to close the dialog box.

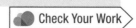 Check Your Work

Importing and Linking Data to a New Table

In addition to exporting Access data to Excel or Word, data from other applications can be imported into an Access table. For example, data from an Excel worksheet can be imported to create a new table in a database. Data in the original application is not connected to the data imported into an Access table. If changes are made to data in the original application, those changes are not reflected in the Access table. To connect the imported data with the data in the original application, link the data.

Tutorial

Importing Data to a
New Table

New Data
Source

Quick Steps

**Import Data into
New Table**
1. Click External Data tab.
2. Click New Data
   Source button.
3. Click location and
   application.
4. Click Browse button.
5. Double-click file name.
6. Click OK.
7. Make choices at each
   wizard dialog box.

Importing Data to a New Table

To import data, click the External Data tab, click the New Data Source button in the Import & Link group, and then click a location and application from which to retrieve data. At the Import dialog box that displays, click the Browse button, double-click the file name, and then click OK. This activates the Import Wizard and displays the first wizard dialog box. The appearance of the dialog box varies depending on the file selected. Complete the steps of the Import Wizard, specifying information such as the range of data, whether the first row contains column headings, whether to store the data in a new table or existing table, the primary key, and the name of the table.

Activity 2a Importing an Excel Worksheet to an Access Table **Part 1 of 2**

1. With **8-CopperState** open, import an Excel worksheet into a new table in the database by completing the following steps:
   a. Click the External Data tab.
   b. Click the New Data Source button in the Import & Link group, point to the *From File* option at the-drop down menu, and then click the *Excel* option at the side menu.

c. At the Get External Data - Excel Spreadsheet dialog box, click the Browse button and then make your AL1C8 folder active.

d. Double-click **8-Policies** in the list box.

e. Click OK at the Get External Data - Excel Spreadsheet dialog box.

f. At the first Import Spreadsheet Wizard dialog box, make sure the *First Row Contains Column Headings* check box contains a check mark and then click the Next button.

g. At the second Import Spreadsheet Wizard dialog box, click the Next button.

h. At the third Import Spreadsheet Wizard dialog box, click the *Choose my own primary key* option (which inserts *PolicyID* in the option box to the right of the option) and then click the Next button.

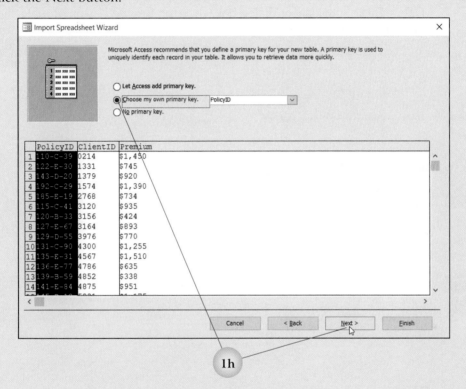

1h

i. At the fourth Import Spreadsheet Wizard dialog box, type Policies in the *Import to Table* text box and then click the Finish button.

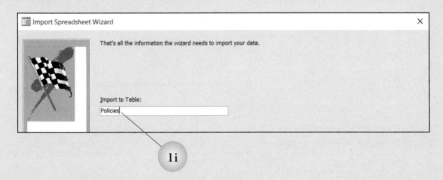

1i

j. At the Get External Data - Excel Spreadsheet dialog box, click the Close button.

2. Open the new Policies table in Datasheet view.

3. Print and then close the Policies table.

Tutorial

Linking Data to a
New Table

Quick Steps

**Link Data to Excel
Worksheet**
1. Click External Data
   tab.
2. Click New Data
   Source button.
3. Point to *From File*.
4. Click *Excel*.
5. Click Browse button.
6. Double-click file
   name.
7. Click *Link to a data
   source by creating a
   linked table*.
8. Click OK.
9. Make choices at
   each wizard dialog
   box.

Linking Data to an Excel Worksheet

Imported data is not connected to the source program. If the data will only be used in Access, import it. However, to update the data in a program other than Access, link the data. Changes made to linked data in the source program file are reflected in the destination program file. For example, an Excel worksheet can be linked with an Access table and, when changes are made to the Excel worksheet, the changes are reflected in the Access table.

To link Excel data to a new table, click the External Data tab, click the New Data Source button in the Import & Link group, point to *From File* at the drop-down list, and then click the *Excel* option at the side menu. At the Get External Data - Excel Spreadsheet dialog box, click the Browse button, double-click the file name, click the *Link to a data source by creating a linked table* option, and then click OK. This activates the Link Wizard and displays the first wizard dialog box. Complete the steps of the Link Wizard, specifying the same basic information as the Import Wizard.

Activity 2b **Linking an Excel Worksheet to an Access Table**   Part 2 of 2

1. With **8-CopperState** open, click the External Data tab, click the New Data Source button in the Import & Link group, point to *From File* at the drop-down list, and then click *Excel* at the side menu.
2. At the Get External Data - Excel Spreadsheet dialog box, click the Browse button, make sure your AL1C8 folder is active, and then double-click **8-Policies**.
3. At the Get External Data - Excel Spreadsheet dialog box, click the *Link to the data source by creating a linked table* option and then click OK.

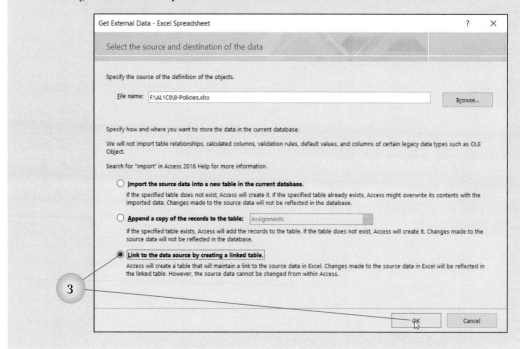

4. At the first Link Spreadsheet Wizard dialog box, make sure the *First Row Contains Column Headings* option contains a check mark and then click the Next button.
5. At the second Link Spreadsheet Wizard dialog box, type LinkedPolicies in the *Linked Table Name* text box and then click the Finish button.
6. At the message stating the linking is finished, click OK.
7. Open the new LinkedPolicies table in Datasheet view.
8. Close the LinkedPolicies table.
9. Open Excel, open the **8-Policies** workbook, click the Enable Editing button, if necessary, and then make the following changes:
   a. Change the amount *$745* in cell C3 to *$850*.
   b. Add the following information in the specified cells:

   A26: 190-C-28
   B26: 3120
   C26: 685
10. Save, print, and then close **8-Policies**.
11. Close Excel.
12. With Access as the active program and **8-CopperState** open, open the LinkedPolicies table. Notice the changes you just made in Excel are reflected in the table.
13. Close the LinkedPolicies table.
14. Close **8-CopperState**.

Check Your Work

Activity 3 **Collect Data in Word and Paste It into an Access Table**   **1 Part**

You will open a Word document containing Hilltop customer names and addresses and then copy the data and paste it into an Access table.

Tutorial

Using the Office Clipboard

Using the Office Clipboard

Quick Steps
Display Clipboard Task Pane
Click Clipboard group task pane launcher.

Use the Office Clipboard to collect and paste multiple items. Up to 24 different items can be collected and pasted in Access or other applications in the Office suite. To copy and paste multiple items, display the Clipboard task pane, shown in Figure 8.2, by clicking the Clipboard group task pane launcher on the Home tab.

Select the data or object to be copied and then click the Copy button in the Clipboard group on the Home tab. Continue selecting text or items and clicking the Copy button. To insert an item from the Clipboard task pane to a field in an Access table, make the destination field active and then click the button in the task pane representing the item. If the copied item is text, the first 50 characters display in the Clipboard task pane. After items have been inserted, click the Clear All button to remove any remaining items from the Clipboard task pane.

Data can be copied from one object to another in an Access database or from a file in another application to an Access database. In Activity 3, data from a Word document will be copied and then pasted into an Access table. Data also can be collected from other applications, such as PowerPoint and Excel.

Figure 8.2 Office Clipboard Task Pane

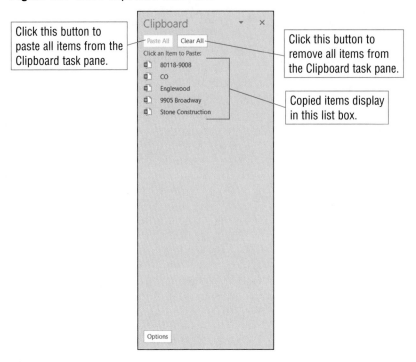

Click this button to paste all items from the Clipboard task pane.

Click this button to remove all items from the Clipboard task pane.

Copied items display in this list box.

Activity 3 Collecting Data in Word and Pasting It into an Access Table

Part 1 of 1

1. Open **8-Hilltop** and then open the Customers table.
2. Copy data from Word and paste it into the Customers table by completing the following steps:
   a. Open Word, make AL1C8 the active folder, and then open **8-HilltopCustomers**.
   b. If necessary, click the Enable Editing button, and then make sure the Home tab is active.
   c. Click the Clipboard group task pane launcher to display the Clipboard task pane. (If the Clipboard contains any items, click the Clear All button.)
   d. Select the first company name, *Stone Construction*, and then click the Copy button in the Clipboard group.

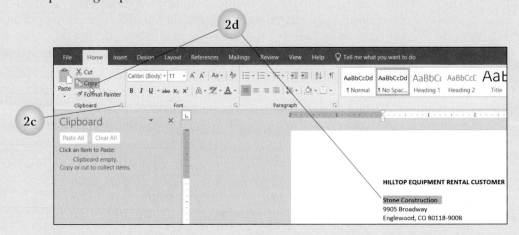

   e. Select the street address, *9905 Broadway*, and then click the Copy button.
   f. Select the city, *Englewood* (selecting only the city and not the comma after the city), and then click the Copy button.

g. Select the state, *CO* (selecting only the two letters and not the space after the letters), and then click the Copy button.

h. Select the zip code, *80118-9008*, and then click the Copy button.

i. Click the Access button on the taskbar. (Make sure the Customers table is open and displays in Datasheet view.)

j. Click in the first empty field in the *CustomerID* field column and then type 178.

k. Display the Clipboard task pane by clicking the Home tab and then clicking the Clipboard group task pane launcher.

l. Close the Navigation pane by clicking the Shutter Bar Open/Close Button.

m. Click in the first empty field in the *Customer* field column and then click *Stone Construction* in the Clipboard task pane.

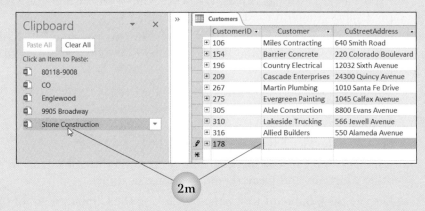

2m

n. Click in the *CuStreetAddress* field and then click *9905 Broadway* in the Clipboard task pane.

o. Click in the *CuCity* field and then click *Englewood* in the Clipboard task pane.

p. Click in the *CuState* field and then click *CO* in the Clipboard task pane.

q. Click in the *CuZipCode* field, make sure the insertion point is positioned at the left side of the field, and then click *80118-9008* in the Clipboard task pane.

r. Click the Clear All button in the Clipboard task pane. (This removes all entries from the Clipboard.)

2r

3. Complete steps similar to those in 2d through 2q to copy the information for Laughlin Products and paste it into the Customers table. (The customer ID number is 255.)

4. Click the Clear All button in the Clipboard task pane.

5. Close the Clipboard task pane by clicking the Close button (X) in the upper right corner of the task pane.

6. Save, print, and then close the Customers table.

7. Open the Navigation pane by clicking the Shutter Bar Open/Close Button.

8. Make Word the active program, close **8-HilltopCustomers** without saving changes, and then close Word.

9. Close **8-Hilltop**.

 Check Your Work

Chapter Summary

- Use the Excel button in the Export group on the External Data tab to export data in a table, query, or form to an Excel worksheet.
- Export data in a table, query, form, or report to a Word document by clicking the More button and then clicking *Word* at the drop-down list. Access exports the data to an RTF (rich-text format) file.
- Access data can be merged with a Word document. The Access data is the data source and the Word document is the main document. To merge data, click the table or query, click the External Data tab, and then click the Word Merge button in the Export group.
- Export an Access object to a PDF or XPS file with the PDF or XPS button in the Export group on the External Data tab.
- Click the New Data Source button in the Import group on the External Data tab, point to *From File*, and then click *Excel* to import Excel data to an Access table.
- Imported data can be linked so that changes made to the data in the source program file are reflected in the destination source file.
- To link imported data, click the *Link to the data source by creating a linked table* option at the Get External Data dialog box.
- Use the Clipboard task pane to collect up to 24 different items in Access or other programs and paste them in various locations.
- Display the Clipboard task pane by clicking the Clipboard group task pane launcher on the Home tab.

Commands Review

FEATURE	RIBBON TAB, GROUP	BUTTON, OPTION
Clipboard task pane	Home, Clipboard	
export object to Excel	External Data, Export	
export object to PDF or XPS	External Data, Export	
export object to Word	External Data, Export	, *Word*
import data	External Data, Import & Link	
merge Access data with Word	External Data, Export	

Index

A

Access
  exporting object to PDF or
    XPS file, 233
  merging data in Word
    document, 229–230
  opening, 4
  screen elements, 5–6
Access Help Window, 127–128
Add a group button, 186
Add a sort button, 187
Add Existing Fields button, 157
Address Block button, 230
Advanced button, 209
aggregate function, designing
    queries with, 89–92
Align Left button, 118
Align Right button, 118
Alternate Row Color button, 118
And criteria, queries designed
    with, 76–80
Ascending button, 117, 139, 177
Attachment data type, 103
AutoNumber data type, 22, 23,
    103
  assigning, 37
  changing, 28

B

Background button, 182
Background Color button, 118
Background Image button, 151
backing up database, 219–220
backstage area, getting help in,
    128–129
Back Up Database option, 219
Bold button, 118
Build button, 106
Builder button, 87

C

Calculated data type, 103
calculator
  inserting in form, 160–162

inserting in report, 190–191
Caption text box, 26
Cascade Update Related Fields,
    41, 50
Center button, 118
Clear All button, 237–238
Close button, 7
Close Print Preview button, 17
closing
  database, 7–9
  objects, 7–9
collapse indicator, 55
Colors button, 150
columns. *See also* fields
  hiding, unhiding, freezing
    and unfreezing,
    14–16
  width changes, 14–16
compacting, database, 213–215
Compact & Repair Database
    button, 213–215
conditional formatting
  of forms, 153–157
  of reports, 182–186
Conditional Formatting button,
    153, 182
Control Margins button, 147
control objects
  arranging in form, 147–150
  defined, 142
  deleting, 143
  inserting, 143–144
  modifying
    in forms, 143–144
    in reports, 176–177
  moving form table, 147
  sizing, 143
Control Padding button, 147
Control Source property box,
    160–161
Copy button, 237
Create button, 4
Create tab, 62, 136
criterion statement
  adding, 63–64
  criteria examples, 64

queries with *Or* and *And*
    criteria, 76–80
crosstab query, 92–95
Crosstab Query Wizard, 92
Currency data type, 23, 103
Current Record box, 10
customizing
  database properties, 215–217
  forms, 142–169
  reports
    layout view, 182–191
    in print preview, 177–178

D

data
  collecting data in Word and
    pasting in Access
    table, 238–239
  exporting
    to Excel, 224–227
    to Word, 227–229
  filtering
    by Filter button, 202–203
    by *Filter By Form* option,
      209–210
    in query and report,
      205–206
    records in table and form,
      203–205
    removing a filter, 203
    by selection, 207–209
    by shortcut menu, 207–
      208
    on specific values, 205
  finding and replacing, 123–
    126
  formatting in table data,
    118–119
  grouping and sorting in
    reports, 186–189
  importing into new table,
    234–235
  inserting in form header, 143
  linking, to Excel worksheet,
    236–237

Interior Photo Credits

Microsoft® Access® Level 2

Unit 1

Advanced Tables, Relationships, Queries, and Forms

Microsoft®
Access®

Designing the Structure of Tables

Performance Objectives

Upon successful completion of Chapter 1, you will be able to:

1   Design the structure of a table

2   Select field data type based on an analysis of the source data

3   Disallow blank field values

4   Allow and disallow zero-length strings in fields

5   Create custom formats for Short Text, Number, and Date/Time data type fields

6   Restrict data entry using input masks

7   Enable rich text formatting in a Long Text data type field

8   Maintain a history of changes for a Long Text data type field

9   Define and use an Attachment data type field with multiple attachments

Designing tables in Access is the most important task when creating a database because all the other objects are based on tables. All queries, forms, and reports rely on tables as the sources of their data. Designing a new database involves planning the number of tables needed, the fields to be included in each table, and the methods to be used to check and/or validate new data as it is entered. In this chapter, you will learn the basic steps involved in planning a new database by analyzing existing data. You will also learn to select appropriate data types and use field properties to control, restrict, or otherwise validate data.

In order to complete the activities in this chapter, you will already need to know the steps in creating a new table, including changing the data type and field size and assigning the primary key field. Readers are also assumed to know the meanings of the terms *field*, *record*, *table*, and *database*.

 Data Files

Before beginning chapter work, copy the AL2C1 folder to your storage medium and then make AL2C1 the active folder.

The online course includes additional training and assessment resources.

Use sample data to decide how to structure a new database to track computer service work orders using best practices for table design and then create the tables.

Tutorial

Review: Creating a Table in Design View

Designing Tables and Fields for a New Database

Most databases encountered in the workplace have been created by database designers. Even so, understanding the process involved in creating a new database will help users make sense of how objects are organized and related. Creating a new database from scratch involves careful planning.

Database designers spend considerable time analyzing existing data and asking questions of users and managers. Designers want to know how data will be used so they can identify the forms, queries, and reports that will need to be generated. Often designers begin by modeling a required report from the database to see the data used to populate the report. The designer then compiles a data dictionary, which is a list of fields as well as the attributes of each field. The designer uses the data dictionary to map out the number of required tables.

A sample work order for RSR Computer Services is analyzed in Activity 1. RSR started as a small computer service company and the owners used Excel worksheets to enter information from service records and to produce revenue reports. The company's success has created the need for a relational database to track customer information. The owners want to be able to generate queries and reports that will help them in decision making. Examine the data in the sample work order shown in Figure 1.1. The work order that the technicians have been filling out at the customer site will be used as the input source document for the database.

Figure 1.1 Sample Work Order for RSR Computer Services

Designers analyze all the input documents and output requirements to capture the entire set of data elements that needs to be created. Once all the data has been identified, the designer maps out the number of tables required to hold it. During the process of mapping out the tables and fields to be associated with each table, the designer follows these guidelines and techniques:

- Consider each table an entity that describes a single person, place, object, event, or other subject. Each table should store facts related only to that entity.

- Segment the data until it is in its smallest unit. For example, in the work order shown in Figure 1.1, the customer's name and address should be split into separate fields for first name, last name, street address, city, state, and zip code. Using this approach provides maximum flexibility for generating other objects and allows the user to sort or filter by any individual data element.

- Do not include fields that can be calculated using data from other fields. For example, the total labor and total due amounts in the work order can be calculated using other elements of numeric data.

- Identify fields that can be used to answer questions from the data. Queries and reports can be designed to extract information based on the results of a conditional expression (sometimes referred to as Boolean logic). For example, in the work order in Figure 1.1, the technician indicates whether the customer has a service contract. Providing a field that stores a *Yes* or *No* (true or false) condition for the service contract data element allows the business to generate reports of customers that have subscribed to a service contract (true condition) and customers that have not subscribed to a service contract (false condition).

- Identify a field in each table that will hold the data that uniquely identifies each record. This field becomes the primary key field. If the data that the database is designed to organize do not reveal a logical unique identifier, it is possible to use the ID field that Access automatically generates with the AutoNumber data type for each record as the primary key field.

- Identify each table that will relate to another table and the field that will be used to join the two tables when the relationships are created. Identifying relationships at this stage helps determine whether a field needs to be added to a related table to allow the tables to be joined.

- Keep in mind that relational databases are built on the concept that data redundancy should be avoided, except for fields that will be used to join tables in a relationship. (The term *data redundancy* means that data in one table is repeated in another table.) Repeating fields in multiple tables wastes storage space, promotes inefficiency and inconsistency, and increases the likelihood of errors being made when adding, updating, and deleting field values.

The database design process may seem time consuming, but creating a well-designed database will save time later. A poorly designed database often contains logical and structural errors that require redefining data or objects after live data has been entered.

Diagramming a Database

Recall from Level 1, Chapter 1 that designers often create a visual representation of the structure of a database in a diagram similar to the one shown in Figure 1.2. In the database diagram, each table is represented in a box with the table name at the top. Within each box, the fields that will be stored in the table are listed with the field names that will be used when the table is created. The

Figure 1.2 Diagram of Table Structure for RSR Computer Services Database

Customers	ServiceContracts	Technicians
*CustID	*CustID	*TechID
FName	SCNo	SSN
LName	StartDate	FName
Street	EndDate	LName
City	FeePd	Street
State		City
ZIP	**WorkOrders**	State
HPhone	*WO	ZIP
CPhone	CustID	HPhone
ServCont	TechID	CPhone
	WODate	
	Descr	
	ServDate	
	Hours	
	Rate	
	Parts	
	Comments	

Hint Words such as *Name* and *Date* are reserved words in Access and cannot be used as field names. A prompt will appear when trying to save a table containing field names that use reserved words.

primary key field is denoted with an asterisk. Tables that will be joined are connected with lines at their common fields. The database represented in Figure 1.2 will be built in the remainder of this chapter and the relationships will be created in Chapter 2.

Notice that many of the field names in the diagram are abbreviated. Although a field name can contain up to 64 characters, using field names that are short enough to be understood is recommended; they are easier to manage and to type into expressions. For abbreviated field names, the Caption property is used to display descriptive headings that contain spaces and/or longer words when viewing the data in a datasheet, form, or report.

Also notice that none of the field names contains spaces. Spaces are allowed in field names but most database designers avoid using them. Instead, designers indicate a space between words by changing the case, using an underscore character (_), or using a hyphen (-).

Assigning Data Types

Designers assign each field a data type based on the types of entries that will be allowed into the field and the operations that will be used to manipulate the data. Selecting the appropriate data type is important because restrictions will be placed on a field based on its data type. For example, in a field designated with the Number data type, only numbers, a period to represent a decimal point, and a plus or minus symbol (+ or −) can be entered into the field in a datasheet or form. Table 1.1 identifies the available data types.

Table 1.1 Data Types

Data Type	Description
Short Text	Alphanumeric data up to 255 characters—for example, a name, an address, or a value such as a telephone number or social security number that is used as an identifier and not for calculating.
Long Text	Alphanumeric data longer than 255 characters; up to 65,535 characters can be stored in a field, although only 64,000 can be displayed. These fields are used to store longer passages of text in a record. Rich text formatting can be added such as bold, italic, or font color.
Number	Positive or negative values that can be used in calculations; not to be used for monetary amounts (see Currency).
Large Number	Non-monetary, numeric values used to calculate large numbers. Not backwards compatible with previous Access versions.
Date/Time	Used to ensure dates and times are entered and sorted properly.
Currency	Values that involve money; Access will not round off during calculations.
AutoNumber	Used to automatically number records sequentially (increments of 1); each new record is numbered as it is typed.
Yes/No	Data in the field is restricted to conditional logic of *Yes* or *No*, *True* or *False*, *On* or *Off*.
OLE Object	Used to embed or link objects in other Office applications.
Hyperlink	Used to store a hyperlink, such as a URL.
Attachment	Used to add file attachments to a record such as a Word document or Excel workbook.
Calculated	Used to display the Expression Builder dialog box, where an expression is entered to calculate the value of the calculated column.
Lookup Wizard	Used to enter data in a field from another existing table or to display a list of values in a drop-down list from which the user chooses.

Using the Field Size Property to Restrict Field Length

By default, a Short Text data type field is set to a width of 255 characters in the Field Size property. Access uses only the amount of space needed for the data entered, even when the field size allows for more characters. Even so, it can be helpful to change this property to a smaller value.

One reason to change the Field Size property to a smaller value is that it restricts the length of the data that can be entered into the field. For example, assume that RSR Computer Services has developed a four-character numbering system for customer numbers. Setting the field size for the *CustID* field to four characters will ensure that no one enters a longer customer number by accident. Access will disallow any character typed after the fourth character.

Figure 1.3 shows the table structure diagram for the RSR Computer Services database expanded to include the data type and Field Size property for each field. Use this diagram to create the tables in Activity 1a.

Figure 1.3 Expanded Table Structure Diagram with Data Types and Field Sizes for Activity 1a

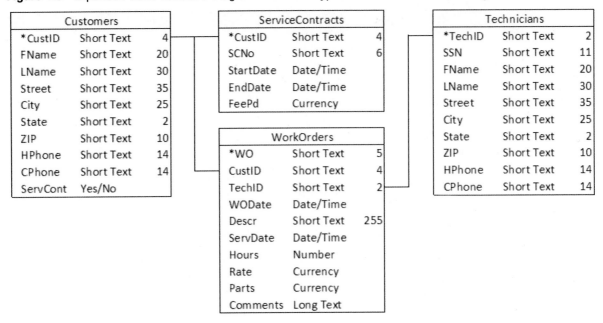

Customers		
*CustID	Short Text	4
FName	Short Text	20
LName	Short Text	30
Street	Short Text	35
City	Short Text	25
State	Short Text	2
ZIP	Short Text	10
HPhone	Short Text	14
CPhone	Short Text	14
ServCont	Yes/No	

ServiceContracts		
*CustID	Short Text	4
SCNo	Short Text	6
StartDate	Date/Time	
EndDate	Date/Time	
FeePd	Currency	

WorkOrders		
*WO	Short Text	5
CustID	Short Text	4
TechID	Short Text	2
WODate	Date/Time	
Descr	Short Text	255
ServDate	Date/Time	
Hours	Number	
Rate	Currency	
Parts	Currency	
Comments	Long Text	

Technicians		
*TechID	Short Text	2
SSN	Short Text	11
FName	Short Text	20
LName	Short Text	30
Street	Short Text	35
City	Short Text	25
State	Short Text	2
ZIP	Short Text	10
HPhone	Short Text	14
CPhone	Short Text	14

Activity 1a Creating Tables in Design View

Part 1 of 6

1. Start Access.
2. At the Access 365 opening screen, complete the following steps to create a new database to store the work orders for RSR Computer Services:
   a. Click the *Blank database* template.
   b. At the Blank database window, click the Browse button and navigate to the AL2C1 folder on your storage medium. Select the current text in the *File Name* text box, type 1-RSRCompServ, and then click OK.
   c. Click the Create button below the *File Name* text box.

   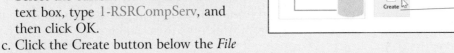

3. Close the Table1 blank table datasheet that displays. Design view will be used to access all the field properties available for fields.
4. Click the Create tab and then click the Table Design button in the Tables group.
5. Create the fields shown in the Customers table in Figure 1.3, including the data type and field size settings, by completing the following steps:
   a. Type CustID and then press the Tab key. Short Text is the default data type.
   b. Double-click *255* in the *Field Size* property box in the *Field Properties* section and then type 4.
   c. Click in the blank *Field Name* text box directly under the *CustID* field.
   d. Enter the rest of the field names, data types, and field sizes for the Customers table shown in Figure 1.3 by completing steps similar to those in Steps 5a–c. For the *ServCont* data type, type y or click the *Data Type* option box arrow and then select *Yes/No*.

6. Select the *CustID* field and then click the Primary Key button in the Tools group.
7. Click the Save button on the Quick Access Toolbar, type Customers in the *Table Name* text box, and then click OK.

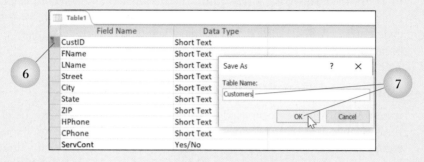

8. Close the table.
9. Create the ServiceContracts, WorkOrders, and Technicians tables shown in Figure 1.3 by completing steps similar to those in Steps 4–8. Assign the primary key field in each table using the fields denoted with asterisks in Figure 1.3 (on page 8).
10. Make sure all the tables are closed.

Check Your Work

Tutorial

Controlling Data Entry

Controlling Data Display and Data Entry Using Field Properties

What properties are available for a field depend on the field's data type. For example, a Yes/No data type field has 7 properties, whereas a Short Text data type field has 14 and a Number data type field has 11.

Use the options available in the *Field Properties* section in Design view to place restrictions on or control data accepted into a field and ensure that data is entered and displayed consistently. Field properties carry over when objects such as queries, forms, and reports are created if the properties are defined before the other objects are created. Taking the time to define the properties when the table is created reduces the number of changes needed if the properties are modified later.

When entering field properties, it is important to select the appropriate field and enter the properties correctly. If properties are applied to the wrong field or entered incorrectly, Access may not be able to save the table or enter the data.

Adding Captions

In Level 1, captions were entered using the Caption property in the Name & Caption dialog box when a new table was created using Datasheet view. The same property appears in Design view in the *Field Properties* section. Recall that the Caption property allows the user to enter a more descriptive title for a field if the field name has been truncated or abbreviated.

The words in field name captions should be separated using spaces, rather than the underscore or hyphen characters used in the field names themselves. In the absence of a caption, Access displays the field name in datasheets, queries, forms, and reports.

Add Caption to Existing Field
1. Open table in Design view.
2. Select field.
3. Click in *Caption* property box.
4. Type descriptive text.
5. Save table.

Require Data in Field
1. Open table in Design view.
2. Select field.
3. Double-click in *Required* property box.
4. Save table.

Requiring Data in a Field

Hint Set the *Required* field to *Yes* and *Allow Zero Length* to *No* to make sure a field value (and not a space) will be entered when the record is added.

Use the Required property to make sure that a certain field is never left empty when a new record is added. By default, the Required property is set to *No*. Change this value to *Yes* to make sure data is typed into the field when a new record is added. For example, setting the Required property to *Yes* will force all new records to have zip code entries. A field defined as a primary key field already has this property set to *Yes* because a primary key field cannot be left empty.

Using and Disallowing Zero-Length Strings

Quick Steps

Disallow Zero-Length String in Field
1. Open table in Design view.
2. Activate a field.
3. Double-click in *Allow Zero Length* property box.
4. Save table.

Hint Press the spacebar to insert a zero-length string.

A zero-length field can be used to indicate that a value will not be entered into the field because the field does not apply to the current record. When a new record is entered and a field is left blank, Access records a null value in the field. For example, assume that a new record for a customer is being entered and the cell phone number is not known; leave the field empty with the intention of updating it later. Access will record a null value in the field. Alternatively, assume that if there is no home phone number; enter a zero-length string in the field to indicate that no field value applies to this record.

To enter a zero-length string, type two double quotation marks with no space between them (""). It is impossible to distinguish between a field with a null value and a field with a zero-length string when viewing the field in a datasheet, query, form, or report; both display as blanks. To help distinguish between the two, create a control in a form or report that returns a user-defined message in a blank field. For example, display the word *Unknown* in a field with a null value and the phrase *Not applicable* in a field with a zero-length string.

By default, Short Text, Long Text, and Hyperlink data type fields allow zero-length strings. Change the Allow Zero Length property to *No* to disallow zero-length strings.

Activity 1b Modifying Field Properties to Add Captions and Disallow Blank Values in a Field

Part 2 of 6

1. With **1-RSRCompServ** open, add captions to the fields in the Customers table by completing the following steps:
   a. Right-click *Customers* in the Tables group in the Navigation pane and then click *Design View* at the shortcut menu.
   b. With *CustID* the active field, click in the *Caption* property box in the *Field Properties* section and then type Customer ID.
   c. Click in the *FName* field row to activate the field, click in the *Caption* property box in the *Field Properties* section, and then type First Name.
   d. Add captions to the following fields by completing the step similar to Step 1c:

LName	Last Name
HPhone	Home Phone
CPhone	Cell Phone
ServCont	Service Contract?

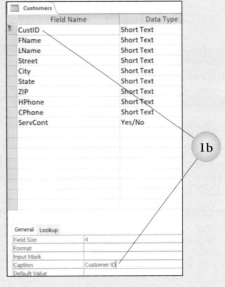

e. Click the Save button on the Quick Access Toolbar.

f. Click the View button (not the button arrow) to switch to Datasheet view and then select all the columns in the datasheet. If necessary, click the Shutter Bar Open/Close button (the two left-pointing chevrons at the top of the Navigation pane) to minimize the Navigation pane.

g. Click the More button in the Records group on the Home tab, click *Field Width* at the drop-down list, and then click the Best Fit button at the Column Width dialog box to adjust the column widths to fit the longest entries.

h. To deselect the columns, click in the *Customer ID* field in the first row of the datasheet.

1h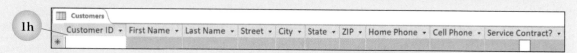

2. Switch to Design View and then click the Shutter Bar Open/Close button (the two right-pointing chevrons) to redisplay the Navigation pane if the pane was minimized in Step 1f.

3. Ensure that no record is entered without an entry in the *ZIP* field and disallow blank values in the field, including zero-length strings, by completing the following steps:

   a. Click in the *ZIP* field row to select the field.

   b. Click in the *Required* property box in the *Field Properties* section (which displays *No*), click the option box arrow that appears, and then click *Yes* at the drop-down list.

3b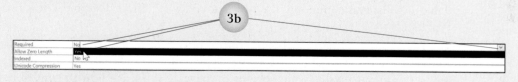

   c. Double-click in the *Allow Zero Length* property box (which displays *Yes*) to change the *Yes* to *No*.

   d. Save the changes to the table design.

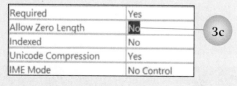
3c

4. Use a new record to test the restrictions on the *ZIP* field by completing the following steps:

   a. Switch to Datasheet view.

   b. Add the following data in the fields indicated:

Customer ID	1000
First Name	Jade
Last Name	Fleming
Street	12109 Woodward Avenue
City	Detroit
State	MI

   c. At the *ZIP* field, press the Tab key to move past the field, leaving it blank.

   d. Type 313-555-0214 in the *Home Phone* field.

   e. Type 313-555-3485 in the *Cell Phone* field.

   f. Press the spacebar in the *Service Contract?* field to insert a check mark in the check box.

   g. Press the Enter key. Access displays an error message because the record cannot be saved without an entry in the *ZIP* field.

   h. Click OK at the Microsoft Access message box.

   i. Click in the *ZIP* field, type 48203-3579, and then press the Enter key four times to move to the *Customer ID* field in the second row of the datasheet.

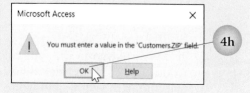
4h

5. Double-click the right boundaries of the *Street* and *ZIP* columns to adjust the column widths so that the entire field values can be read.

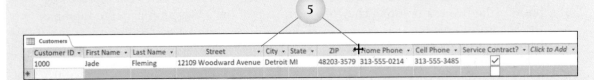

6. Close the Customers table. Click Yes when prompted to save changes to the layout of the table.

 Tutorial

Creating a Custom Format for a Short Text Field

Creating a Custom Format for a Short Text Data Type Field

The Format property controls how data is displayed in the field in the datasheet, query, form, or report. What formats are available depends on the data type of the field. Predefined formats are available for some data types and can be selected from a drop-down list in the *Format* property box. No predefined formats exist for Short Text and Long Text data type fields. If no predefined format exists or the predefined format options are not suitable, then a custom format can be created. Table 1.2 displays commonly used format codes for Short Text and Long Text data type fields.

The Format property does not control how data is entered into the field. Rather, formatting a field controls the display of accepted field values. Refer to the section on input masks (which begins on page 17) to learn how to control new data as it is being entered.

Ö̈uick Steps

Format Short Text Field
1. Open table in Design view.
2. Activate field.
3. Click in *Format* property box.
4. Type format codes.
5. Save table.

Table 1.2 Format Codes for Short Text and Long Text Data Type Fields

Code	Description	Format Property Example
@	Used as a placeholder, one symbol for each character position. An unused position in a field value is replaced with a blank space to the left of the text entered into the field.	@@@@ Field value entered is *123*. Access displays one blank space followed by *123*, left-aligned in the field.
!	Placeholder positions are filled with characters from left to right instead of the default right to left sequence.	!@@@@ Field value entered is *123*. Access displays *123* left-aligned in the field with one blank space after the *3*.
>	All text is converted to uppercase.	> Field value entered is *mi*. Access displays *MI* in the field.
<	All text is converted to lowercase.	< Field value entered is *Jones@EMCP.NET*. Access displays *jones@emcp.net* in the field.
[color]	Text is displayed in the font color specified. Available colors are black, blue, cyan, green, magenta, red, yellow, and white.	[red]@@@@@-@@@@ Field value entered is *482033579*. Access displays *48203-3579*.

1. With **1-RSRCompServ** open, format the *State* field in the Customers table to ensure that all the text is displayed in uppercase letters by completing the following steps:
    a. Right-click *Customers* in the Tables group in the Navigation pane and then click *Design View* at the shortcut menu.
    b. Click in the *State* field row to activate the field.
    c. Click in the *Format* property box and then type >.
    d. Save the table.

2. Format the *ZIP* field to fill it with characters from left to right, display the text in red, and provide for the five-plus-four-character US zip code (with the two sets of characters, each separated by a hyphen) by completing the following steps:
    a. Click in the *ZIP* field row to activate the field.
    b. Click in the *Format* property box and then type ![red]@@@@@-@@@@.
    c. Save the table.

3. Test the custom formats in the *State* and *ZIP* fields using a new record by completing the following steps:
    a. Switch to Datasheet view.
    b. Add the following data in a new record. Type the text for the *State* field as indicated in lowercase text. Notice that Access automatically converts the lowercase text to uppercase when moving to the next field. As you type the *ZIP* field text, notice that it displays in red. Since no field values are entered for the last four characters of the *ZIP* field, Access displays blank spaces in these positions.

Customer ID	1005
First Name	Cayla
Last Name	Fahri
Street	12793 Riverdale Avenue
City	Detroit
State	mi
ZIP	48223
Home Phone	313-555-6845
Cell Phone	313-555-4187
Service Contract?	Press the spacebar for *Yes*.

4. Look at the data in the *ZIP* field for the first record. This data was entered before the *ZIP* field was formatted. Since a hyphen was typed when the data was entered in Activity 1b and the field is now formatted to automatically add the hyphen, two hyphen characters appear in the existing record. Update the record by editing the *ZIP* field value for record 1 to remove the extra hyphen.

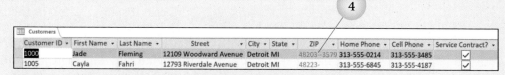

5. Display the datasheet in Print Preview. Change to landscape orientation. Set the top margin to 1 inch and the bottom, left, and right margins to 0.25 inch. Print the datasheet and then close Print Preview.

6. Close the Customers table.

Creating a Custom Format for a Numeric Data Type Field

Access provides predefined formats for Number, AutoNumber, and Currency data type fields that include options for displaying a fixed number of digits past the decimal point, a comma separator, and a currency symbol. Table 1.3 displays format codes that can be used to create custom formats. Use the placeholders shown in Table 1.3 in combination with other characters (such as the dollar symbol, comma, and period) to create the desired custom numeric format.

Specify up to four formats for a numeric data type field to include different options for displaying positive values, negative values, zero (0), and null values. Examine the following custom format code:

#,###.00;-#,###.00[red];0.00;"Unknown"

Notice that the four sections are separated with semicolons (;). The first section, *#,###.00*, defines the format for positive values. It includes a comma in thousands values and two digits past the decimal point; zeros are used if no decimal value is entered. The second section, *-#,###.00[red]*, defines negative values. It includes the same placeholders as used with positive values but starts the field with a minus symbol and displays the numbers in red. The color code can be placed before or after the codes for the numbers. The third section, *0.00*, instructs Access to show *0.00* in the field if a zero is entered. Finally, a field value that is a null value (the spacebar is pressed) will display the text *Unknown* (italic used here for emphasis only) in the field. Text to be directly entered into the field is enclosed in quotation marks.

Table 1.3 Format Codes for Numeric Data Type Fields

Code	Description	Format Property Example
#	Used as a placeholder to display a number.	*#.##* Field value entered is *123.45*. Access displays *123.45* in the field. Notice that the number of placeholder positions does not restrict the data entered into the field.
0	Used as a placeholder to display a number. Access displays a *0* in place of a position for which no value is entered.	*000.00* Field value entered is *55.4*. Access displays *055.40* in the field.
%	Value is multiplied times 100 and a percent symbol is added.	*#.0%* Field value entered is *.1246*. Access displays *12.5%* in the field. Field value entered is *.1242*. Access displays *12.4%* in the field. Having only one digit past the decimal point causes rounding up or down to occur.

1. With **1-RSRCompServ** open, format the *Rate* field in the WorkOrders table with a
   custom format that displays positive numbers with two decimal places and blue text and
   null values with the text *Not Available* by completing the following steps:
   a. Open the WorkOrders table in Design view.
   b. Make the *Rate* field active.
   c. Click in the *Format* property box, delete
      the current entry (*Currency*), and then type
      #.00[blue];;;"Not Available". Notice that three
      semicolons are typed after the first custom format
      option, *#.00[blue]*. When a custom format for
      negative or zero values is not needed, include the
      semicolon to indicate that there is no format setting.
      Since an hourly rate is never a negative or zero
      value, do not use custom formats in these situations.
   d. Save the table.

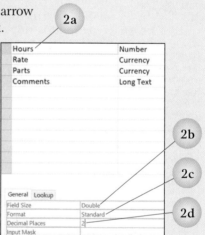

2. Change the *Hours* field size to one that displays fractional numbers and format the field
   using a predefined format by completing the following steps:
   a. Make the *Hours* field active.
   b. Click in the *Field Size* property box, click the option box arrow
      that appears, and then click *Double* at the drop-down list.
      The default setting for a Number data type field is
      Long Integer, which stores only whole numbers. (This
      means that a decimal value entered into the field is
      rounded.) Changing the *Field Size* property box to
      Double allows decimal values to be stored.
   c. Click in the *Format* property box, click the option
      box arrow that appears, and then click *Standard* at
      the drop-down list.
   d. Click in the *Decimal Places* property box, click the
      option box arrow that appears, and then click *2* at
      the pop-up list.
   e. Save the table.
3. Switch to Datasheet view.
4. Add the following data in a new record to test the custom format and the predefined
   format. Notice that when you move past the *Rate* field, the value is displayed in blue.
   Compare your data entry to the screen image on the next page and notice the new format
   for the *Hours* and *Rate* fields.

WO	65012
CustID	1000
TechID	11
WODate	09-07-2021
Descr	Biannual desktop computer cleaning and maintenance
ServDate	09-07-2021
Hours	1.25
Rate	50
Parts	35.15
Comments	Keys are sticking; cleaning did not resolve. Customer is considering buying a new keyboard.

5. Review the WorkOrders table and note the formatting.

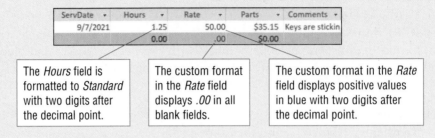

The *Hours* field is formatted to *Standard* with two digits after the decimal point.

The custom format in the *Rate* field displays *.00* in all blank fields.

The custom format in the *Rate* field displays positive values in blue with two digits after the decimal point.

6. Close the WorkOrders table.

 Tutorial

Creating a Custom Format for a Date/Time Field

Creating a Custom Format for a Date/Time Field

Access provides predefined formats for fields with a Date/Time data type. These formats provide a variety of combinations of month, day, and year options for dates and hour and minute display options for time. If the predefined display formats are not suitable, custom formats can be created using a combination of the codes described in Table 1.4, along with symbols (such as hyphens and slashes) between parts of the date. If a format option for a Date/Time data type field is not chosen, Access displays the date in the format *m/d/yyyy*. For example, in Activity 1d, the date entered into the *WODate* field displays as *9/7/2021*.

A custom format for a Date/Time data type field can contain two sections separated by a semicolon. The first section specifies the format for displaying dates. To add a format for displaying times, type a semicolon and then add the format codes.

Quick Steps

Format Date/Time Field

1. Open table in Design view.
2. Activate field.
3. Click in *Format* property box.
4. Type format codes or select from predefined list.
5. Save table.

Table 1.4 Format Codes for Date/Time Data Type Fields

Code	Description
d or dd	displays the day of the month as one digit (*d*) or two digits (*dd*)
ddd or dddd	displays the day of the week abbreviated (*ddd*) or in full (*dddd*)
m or mm	displays the month as one digit (*m*) or two digits (*mm*)
mmm or mmmm	displays the month abbreviated (*mmm*) or in full (*mmmm*)
yy or yyyy	displays the year as the last two digits (*yy*) or all four digits (*yyyy*)
h or hh	displays the hour as one digit (*h*) or two digits (*hh*)
n or nn	displays the minutes as one digit (*n*) or two digits (*nn*)
s or ss	displays the seconds as one digit (*s*) or two digits (*ss*)
AM/PM	displays 12-hour clock values followed by *AM* or *PM*

1. With **1-RSRCompServ** open, format the *WODate* field in the WorkOrders table with a custom format by completing the following steps:
   a. Open the WorkOrders table in Design view.
   b. Make *WODate* the active field.
   c. Click in the *Format* property box, type ddd, mmm dd yyyy, and then press the Enter key. This format will display dates beginning with the day of the week in abbreviated form, followed by a comma, the month in abbreviated form, the day of the month as two digits, and the year as four digits. Spaces separate the sections of the date. Notice that Access puts quotation marks around the comma.

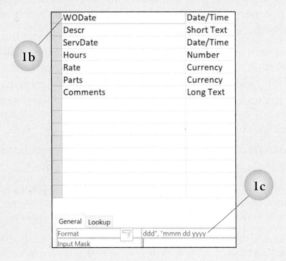

   d. Save the table.
2. Switch to Datasheet view.
3. If necessary, adjust the column width of the *WODate* field to allow reading the entire entry.

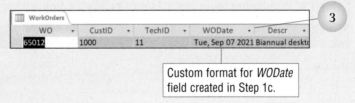

Custom format for *WODate* field created in Step 1c.

4. Switch to Design view.
5. Format the *ServDate* field using the same custom format used for *WODate* by completing steps similar to those in Steps 1b–1c.
6. Save the table and then switch to Datasheet view.
7. Double-click the right column boundary to adjust the column width of the *ServDate* field and view the custom date format.

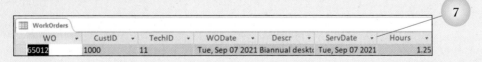

8. Close the WorkOrders table. Click Yes when prompted to save changes to the table layout.

Controlling Data Entry Using an Input Mask

Whereas a format controls how the data is displayed, an input mask controls the type of data and pattern in which the data is entered into a field. In Activity 1e, a date format was created to display the date as *Tue, Sep 7 2021*. When confirming dates with clients, it is good practice to provide the day of the week, as well as the day of the month. To type the date consistently and efficiently into the *ServDate* or *WODate* field, an input mask will be created in Activity 1f to enter the date by typing *sep072021*. Access knows that this date is a Tuesday and displays *Tue, Sep 7 2021*.

Recall from Level 1, Chapter 4, that Access provides the Input Mask Wizard, which can be used to create an input mask for a text or date field. Commonly used input masks are predefined within the wizard for telephone numbers, social security numbers, zip codes, dates, and times. To create an input mask without the wizard, use the codes described in Table 1.5.

An input mask can contain up to three sections, with each section separated by a semicolon. The first section contains the input mask codes for the data entry in the field. The second section instructs Access to store or not store the display characters used in the field, such as hyphens, slashes, or brackets). A zero indicates that Access should store the characters. Leaving the second section blank means the display characters will not be stored. The third section specifies the placeholder character to display in the field when it becomes active for data entry.

The following is an example of an input mask to store a four-digit customer identification number with a pound symbol (#) as the placeholder: *0000;;#*. The first section, *0000*, contains the four required digits for the customer identification number. Since the mask contains no display characters (hyphens, slashes, etc.), the second section is blank. The pound symbol after the second semicolon is the placeholder character.

In addition to the symbols in Table 1.5, use the format code > to force characters to be uppercase or the format code < to force characters to be lowercase. Decimal points, hyphens, slashes, and other punctuation symbols can also be used.

Table 1.5 Commonly Used Input Mask Codes

Code	Description
0	Required digit.
9	Optional digit.
#	Digit, space, plus or minus symbol. If no data is typed at this position, Access leaves a blank space.
L	Required letter.
?	Optional letter.
A	Required letter or digit.
a	Optional letter or digit.
&	Required character or space.
C	Optional character or space.
!	Field is filled from left to right instead of right to left.
\	Character is displayed that immediately follows in the field.

1. With **1-RSRCompServ** open, create a custom input mask for the work order numbers in the WorkOrders table by completing the following steps:
   a. Open the WorkOrders table in Design view.
   b. With *WO* the active field, click in the *Input Mask* property box and then type 00000;;_. This mask requires a five-digit work order number to be entered. The underscore is used as the placeholder character that displays when the field becomes active.

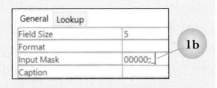

   c. Save the table.
2. Create an input mask to require the two date fields to be entered as three characters for the month, followed by two digits for the day and four digits for the year by completing the following steps:
   a. Make *WODate* the active field.
   b. Click in the *Input Mask* property box and then type >L<LL\ 00\ 0000;0;_. This mask requires three letters for the month; the first letter is converted to uppercase and the remaining two letters are converted to lowercase. The backslash symbol (\ ) followed by the space instructs Access to display a space after the month as data is entered. Two digits are required for the day, followed by another space and then four digits for the year. The *0* after the first semicolon instructs Access to store the display characters. At the end of the mask, the underscore character is used again as the placeholder character.

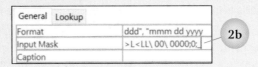

   c. Make *ServDate* the active field, click in the *Input Mask* property box, and then type >L<LL\ 00\ 0000;0;_.
   d. Save the table.
3. Switch to Datasheet view.
4. Test the input masks using a new record by completing the following steps:
   a. Click the New button in the Records group on the Home tab.
   b. Type 6501. Notice that as soon as the first character is typed, the placeholders appear in the field.
   c. Press the Tab key to move to the next field in the datasheet. Since the mask contains five zeros (indicating five required digits), Access displays a message box stating that the value entered is not appropriate for the input mask.
   d. Click OK at the Microsoft Access message box.

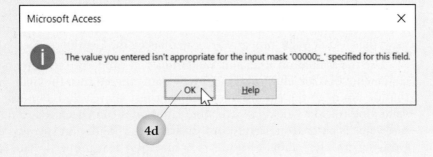

e. Type 3 in the last position in the *WO* field and then press the Tab key to move to the next field.

f. Type 1000 in the *CustID* field and then press the Tab key.

g. Type 10 in the *TechID* field and then press the Tab key.

h. Type sep072021 in the *WODate* field and then press the Tab key. Notice that the placeholder characters and spaces appear as soon as the first letter is typed. Notice also that the first character is converted to uppercase and that the spaces do not need to be typed; Access moves automatically to the next position after the month and day are typed.

i. Type Replace keyboard in the *Descr* field and then press the Tab key.

j. Type sep102021 in the *ServDate* field and then press the Tab key.

k. Complete the remainder of the record as follows:

Hours   .5
Rate   45.50
Parts   67.25
Comments   Serial Number AWQ-982358

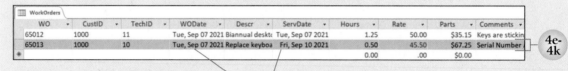

Notice that once the date is accepted into the field, the custom *Format* property controls how the date is presented in the datasheet; the abbreviated day of the week is at the beginning of the field and spaces are between the month, day, and year.

5. Display the datasheet in Print Preview. Change the orientation to landscape. Set the top and bottom margins to 1 inch and the left and right margins to 0.25 inch. Print the datasheet and then close Print Preview.

6. Close the WorkOrders table. Click Yes when prompted to save changes to the table layout.

Tutorial

Review: Modifying Field Properties in Design View

Other field properties that promote accuracy when entering data and that should be considered when designing database tables include the Default Value, Validation Rule, and Validation Text properties. Use the Default Value property to populate the field in a new record with the field value that is used most often. For example, when most employees live in the same city and state, use default values for these fields to ensure consistent spelling and capitalization within the table. The text appears automatically in the fields when new records are added to the table. The user can accept the default value by pressing the Tab key or the Enter key to move past the field or type new data in the field. In Level 1, default values were created using the Default Value button in the Properties group on the Table Tools Fields tab. In Design view, the Default Value property is located below the Caption property.

Tutorial

Review: Applying a Validation Rule in Design View

Use the Validation Rule and Validation Text properties to enter conditional statements that are checked against new data entered into the field. Invalid entries that do not meet the conditional statement test are rejected. For example, a validation rule on a field used to store labor rates can check that a minimum labor rate value is entered in each record. In Level 1, a validation rule was added in Design view. The Validation Rule and Validation Text properties are located just above the Required property.

| Activity 2 | **Work with Long Text and Attachment Data Type Fields** | 2 Parts |

You will edit properties for a Long Text data type field, apply rich text formatting to text, and attach files to records using an Attachment data type field.

Tutorial

Working with a
Long Text Field

Quick Steps

Enable Rich Text Formatting in Long Text Field
1. Open table in Design view.
2. Select field defined as Long Text.
3. Double-click in *Text Format* property box.
4. Save table.

Track Changes in Long Text Data Type Field
1. Open table in Design view.
2. Select field defined as Long Text.
3. Double-click in *Append Only* property box.
4. Save table.

Working with a Long Text Data Type Field

By default, Access formats a Long Text data type field as plain text. However, formatting attributes can be applied to text by enabling rich text formatting. Enabling rich text formatting in a Long Text data type field allows the user to change the font, apply bold or italic formatting, or add font color to text, among other formatting options. To enable rich text formatting, change the Text Format property to *Rich Text*.

The Append Only property for a Long Text data type field is set to *No* by default. Change the property to *Yes* to track changes made to the field value in the datasheet. Scroll down the General tab in the *Field Properties* section to locate the Append Only property. When this property is set to *Yes*, Access maintains a history of additions to the field, which can be viewed in the datasheet. Changing the Append Only property to *No* causes Access to delete any existing history.

In Activity 2a, the text format of the *Comments* field is changed from *Plain Text* to *Rich Text* and then bold red formatting is applied to the serial number of the new keyboard. These formatting changes make it easier to find important information in the *Comments* field—in this case, the serial number. Changing the Append Only property from *No* to *Yes* allows the user to keep track of the dates for any comments entered into the field.

| Activity 2a | **Working with Rich Text Formatting and Maintaining a History** |
| | **of Changes in a Long Text Data Type Field** | Part 1 of 2 |

1. With **1-RSRCompServ** open, enable rich text formatting and turn on tracking of history in a field defined as a Long Text data type field by completing the following steps:
   a. Open the WorkOrders table in Design view.
   b. Make *Comments* the active field.
   c. Double-click in the *Text Format* property box (displays *Plain Text*). It should now read *Rich Text*.
   d. At the Microsoft Access message box indicating that the field will be converted to Rich Text, click Yes.

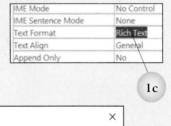

1c

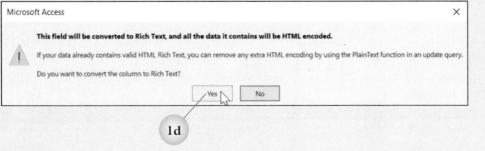

1d

e. If necessary, scroll down the General tab in the *Field Properties* section to locate the *Append Only* property box.

f. Double-click in the *Append Only* property box to change *No* to *Yes*.

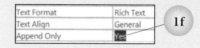

g. Save the table.

2. Switch to Datasheet view.

3. Minimize the Navigation pane and then adjust all the column widths except the *Descr* and *Comments* columns to best fit.

4. Change the column width of the *Comments* field to 25 characters.

5. Select the serial number text *AWQ-982358* in the second record in the *Comments* field and then apply bold formatting and the standard red font color using the buttons in the Text Formatting group on the Home tab. Click at the end of the serial number to deselect the text.

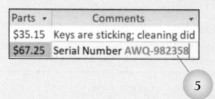

6. Click in the *Comments* field in the first record. Press the End key to move the insertion point to the end of the existing text. Press the spacebar, update the record by typing Microsoft wireless keyboard was recommended., and then press the Enter key to save the changes and move to the next row.

7. Right-click in the *Comments* field of the first record and then click *Show column history* at the shortcut menu.

8. Click OK after reading the text in the History for Comments dialog box.

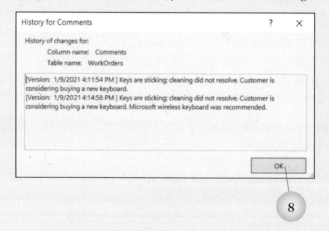

9. Click in the *Comments* field of the first record. Press the End key to move the insertion point to the end of the current text. Press the spacebar, type See work order 65013 for replacement keyboard request., and then press the Enter key.

10. Right-click in the *Comments* field of the first record and then click *Show column history* at the shortcut menu.

11. Click OK after reading the text in the History for Comments dialog box.

12. Display the datasheet in Print Preview. Change to landscape orientation. Set the top and bottom margins to 1 inch and the left and right margins to 0.25 inch. Print the datasheet and then close Print Preview.

13. Close the WorkOrders table. Click Yes if prompted to save changes to the table layout and then redisplay the Navigation pane.

History for Comments   ?   ×

History of changes for:
Column name: Comments
Table name: WorkOrders

[Version: 1/9/2021 4:11:54 PM] Keys are sticking; cleaning did not resolve. Customer is considering buying a new keyboard.
[Version: 1/9/2021 4:14:58 PM] Keys are sticking; cleaning did not resolve. Customer is considering buying a new keyboard. Microsoft wireless keyboard was recommended.
[Version: 1/9/2021 4:21:25 PM] Keys are sticking; cleaning did not resolve. Customer is considering buying a new keyboard. Microsoft wireless keyboard was recommended. See work order 65013 for replacement keyboard request.

OK

(11)

> Check Your Work

> Tutorial

Creating an Attachment Field

Creating an Attachment Data Type Field and Attaching Files to Records

Use an Attachment data type field to store several files in a single field attached to a record. The attachments can be opened within Access and viewed in the program from which the document originated. For example, attach a Word document to a field in a record. Opening the attached file in the Access table causes Microsoft Word to start and the document to display. A file that is attached to a record cannot be larger than 256 megabytes (MB).

An Attachment data type field displays with a paper clip in Datasheet view. To manage the attached files, double-click the paper clip to open the Attachments dialog box, shown in Figure 1.4. A field that is created with an Attachment data type cannot be changed. Multiple files can be attached to a record, as long as the combined size of all the files does not exceed 2 gigabytes (GB).

Any file created within the Microsoft Office suite can be attached to a record or an image file (.bmp, .jpg, .gif, .png), a log file (.log), a text file (.txt), or a compressed file (.zip). Some files, such as any file ending with *.com* or *.exe*, are considered potential security risks and therefore blocked by Access.

Figure 1.4 Attachments Dialog Box

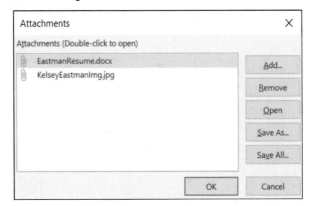

Attachments   ×

Attachments (Double-click to open)

EastmanResume.docx
KelseyEastmanImg.jpg

Add...
Remove
Open
Save As...
Save All...

OK   Cancel

Quick Steps

Create Attachment Field
1. Open table in Design view.
2. Click in first blank field row.
3. Type field name.
4. Click in *Data Type* column.
5. Click option box arrow.
6. Click *Attachment*.
7. Save table.

Attach Files to Record
1. Open table in Datasheet view.
2. Double-click paper clip in record.
3. Click Add button.
4. Navigate to drive and/or folder location.
5. Double-click file name.
6. Click OK.

View Attached File
1. Open table in Datasheet view.
2. Double-click paper clip in record.
3. Double-click file name.
4. View file contents.
5. Exit source program.
6. Click OK.

Saving an Attached File to Another Location

To save a copy of an attachment outside Access, select the file and then click the Save As button in the Attachments dialog box. At the Save Attachment dialog box, navigate to the drive and/or folder in which the duplicate copy is to be saved, click the Save button, and then click OK to close the Attachments dialog box.

Editing or Removing an Attached File

Access 365 provides two options for editing an attachment. The first option is to save the attached file to another location, as noted above. After the necessary changes are made, save the file, remove the original attachment from the database, and then attach the edited file. The second option is to open the attachment directly in Access and enable the content. After making the necessary changes in the source program (for example, Word), close the application and save the changes, click OK in the Attachments dialog box and respond to the message asking if you would like to save the updates to the database.

To remove a file attached to a record in a database, open the Attachments dialog box in the record containing the file attachment, click the file name for the file to be removed, click the Remove button, and then click OK to close the Attachments dialog box.

Activity 2b Creating an Attachment Data Type Field, Attaching Files to a Record, and Viewing the Contents of Attached Files
Part 2 of 2

1. With **1-RSRCompServ** open, create a new field in the Technicians table to store file attachments by completing the following steps:
   a. Open the Technicians table in Design view.
   b. Click in the blank row below *CPhone*, type Attachments, and then press the Tab key.
   c. Click the option box arrow in the *Data Type* column and then click *Attachment* at the drop-down list.
   d. Save the table.
2. Switch to Datasheet view.
3. Add the following data in the first row of the datasheet:

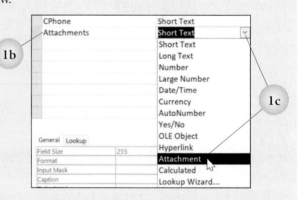

TechID	10
SSN	000-43-5789
FName	Kelsey
LName	Eastman
Street	550 Montclair Street
City	Detroit
State	MI
ZIP	48214-3274
HPhone	"" (Recall that double quotation marks indicate a zero-length field.)
CPhone	313-555-6315

4. Attach two files to the record for Kelsey Eastman by completing the following steps:

a. Double-click the paper clip in the first row of the datasheet. An Attachment data type field displays a paper clip in each record in a column and has a paper clip in the field name row. The number in brackets next to the paper clip indicates the number of files attached to the record.

b. At the Attachments dialog box, click the Add button.

c. At the Choose File dialog box, navigate to the AL2C1 folder on your storage medium.

d. Click the file named *EastmanResume.*

e. Press and hold down the Ctrl key, click the file named *KelseyEastmanImg*, and then release the Ctrl key.

f. Click the Open button.

g. Click OK. Access closes the Attachments dialog box and displays *(2)* next to the paper clip in the first record.

5. Open the attached files by completing the following steps:

a. Double-click the paper clip in the first row of the datasheet to open the Attachments dialog box.

b. Double-click *EastmanResume.docx* in the *Attachments* list box to open the Word document.

c. Read the resume in Microsoft Word and then exit Word.

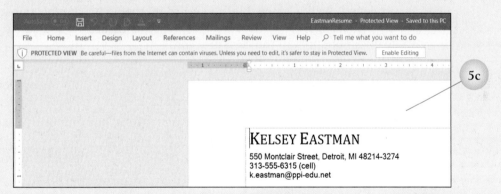

d. Double-click *KelseyEastmanImg.jpg* in the *Attachments* list box to open the picture file.

e. View the picture and then exit the photo viewer program.

f. Click OK to close the Attachments dialog box.

6. Adjust all the column widths to best fit.

7. Display the datasheet in Print Preview. Change the orientation to landscape, print the datasheet, and then close Print Preview.

8. Close the Technicians table. Click Yes when prompted to save changes to the table layout.

9. Close **1-RSRCompServ**.

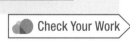

Chapter Summary

- Database designers plan the tables for a new database by analyzing sample data, input documents, and output requirements to generate the entire set of data elements needed.

- Once all the data has been identified, the designer maps out the number of tables required.

- Each table holds data only for a single entity (topic). Data is split into the smallest unit that will be manipulated.

- Designers also consider the relationships that will be created. Doing so in the planning stage helps determine if a field needs to be added to a table to join tables.

- Data redundancy should be avoided. This means that fields should not be repeated in another table, except those fields needed to join tables in a relationship.

- A diagram of a database portrays the database tables, providing field names, data types, field sizes, and notation of the primary key fields.

- A field is assigned a data type by selecting a type appropriate for the kind of data that will be accepted into the field.

- Changing the Field Size property is one way to restrict entries in the field to a maximum length. This prevents having longer entries added to the field by accident.

- Add field properties before creating queries, forms, and reports as the properties carry over to these objects.

- Change the Required property to *Yes* to make sure data is typed into the field when a new record is added to the table.

- A zero-length field is entered into a record by typing two double quotation marks with no space between them. This method is used to indicate that a field value does not apply to the current record.

- Disallow zero-length strings by changing the Allow Zero Length property to *No*.

- The Format property controls the display of data in a field. Custom formats can be created by typing the appropriate format codes in the *Format* property box.

- A custom numeric format can contain four sections: one for positive values, one for negative values, one for zero, and one for null values.

- Use an input mask to control the type and pattern of data entered into the field.

- Create a custom input mask for a Short Text or Date/Time data type field by typing the appropriate input mask codes in the *Input Mask* property box.

- A Long Text data type field can be formatted using rich text formatting options in the Text Formatting group on the Home tab. To enable rich text formatting, change the Text Format property to *Rich Text*.

- Change the Append Only property of a Long Text data type field to *Yes* to track changes made to field values.

- An Attachment data type field can be used to store files in a single field attached to a record.

- Double-click the paper clip in the Attachment data type field for a record to add, view, save, or remove a file attachment.

Commands Review

FEATURE	RIBBON TAB, GROUP	BUTTON	KEYBOARD SHORTCUT
create table in Design view	Create, Tables		
minimize Navigation pane			F11
redisplay Navigation pane			F11
switch to Datasheet view from Design view	Table Tools Design, Views		
switch to Design view from Datasheet view	Home, Views		

Microsoft® Access®

Building Relationships and Lookup Fields

Performance Objectives

Upon successful completion of Chapter 2, you will be able to:

1 Create and edit relationships between tables, including one-to-many, one-to-one, and many-to-many relationships

2 Define a table with a multiple-field primary key field

3 Create and modify a lookup field to populate records with data from another table

4 Create a lookup field that allows having multiple values in records

5 Create single-field and multiple-field indexes

6 Define the term *normalization*

7 Determine if a table is in first, second, or third normal form

Once the table design has been completed, the next step is to establish relationships and relationship options between tables. This involves analyzing the type of relationship that exists between two tables. Some database designers draw a relationship diagram to depict the primary table and the matching record frequency of the related table. In this chapter, you will create and edit relationships and lookup fields, multiple-field primary key fields, multiple-value fields, and indexes. The concept of database normalization and three forms of normalization will be introduced to complete the examination of database design fundamentals.

Data Files

Before beginning chapter work, copy the AL2C2 folder to your storage medium and then make AL2C2 the active folder.

The online course includes additional training and assessment resources.

Activity 1 **Create and Edit Relationships**   **4 Parts**

You will create relationships and edit relationship options for the tables designed to track work orders for RSR Computer Services and create and print a relationship report.

Building Relationships

Hint Are you unsure whether two tables should be related? Consider whether data will need to be extracted from both tables in the same query, form, or report. If yes, then the tables should be joined in a relationship.

After determining which tables to relate to one another, the next step in designing a database is to examine the types of relationships that exist between the tables. A relationship is based on an association between two tables. For example, in the database created in Chapter 1 for RSR Computer Services, there is an association between the Customers table and the WorkOrders table. A customer is associated with all of his or her work orders involving computer maintenance requests, and each work order is associated with the individual customer who requested the service.

When building relationships, consider the associations between tables and how these associations affect the data that will be entered into the tables. In the database diagram presented in Chapter 1, relationships were shown with lines connecting the common field name between tables. In this chapter, consider the type of relationship that should exist between the tables and the relationship options to use to place restrictions on data entry. Access provides for three types of relationships: one-to-many, one-to-one, and many-to-many. Access Level 1, Chapter 2 addressed one-to-many and one-to-one relationships. This chapter begins by reviewing these two relationship types and then discusses how to establish a many-to-many relationship.

 Tutorial

Review: Creating a One-to-Many Relationship

Establishing a One-to-Many Relationship

In the computer service database in Chapter 1, the relationship between the Customers table and WorkOrders table exists because a work order involves computer maintenance for a specific customer. The customer is identified by the customer number stored in the Customers table. In the Customers table, only one record exists per customer. In the WorkOrders table, the same customer number can be associated with several work orders. This means that the relationship between the Customers table and WorkOrders table is a one-to-many relationship. This is the most common type of relationship created in Access.

 Relationships

A common field is needed to join the Customers table and WorkOrders table, so the *CustID* field is included in both tables. In the Customers table, *CustID* is the primary key field because it contains a unique identification number for each customer. In the WorkOrders table, *CustID* cannot be the primary key field because the same customer can be associated with several work orders. In the WorkOrders table, *CustID* is a type of field referred to as a *foreign key* field. A foreign key field is included in a table for the purpose of creating a relationship to a field that is a primary key field in another table. The Customers-to-WorkOrders one-to-many relationship is illustrated in Figure 2.1.

Figure 2.1 One-to-Many Relationship between the Customers Table and WorkOrders Table

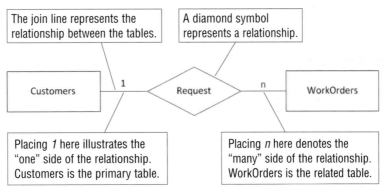

The join line represents the relationship between the tables.

A diamond symbol represents a relationship.

Customers | 1 | Request | n | WorkOrders

Placing *1* here illustrates the "one" side of the relationship. Customers is the primary table.

Placing *n* here denotes the "many" side of the relationship. WorkOrders is the related table.

Quick Steps

Create One-to-Many Relationship
1. Click Database Tools tab.
2. Click Relationships button.
3. Add tables from Show Table dialog box.
4. Close Show Table dialog box.
5. Drag primary key field name from primary table to foreign key field name in related table.
6. Click Create button.

When diagramming a database, designers may choose to show relationships in a separate illustration. In the diagram shown in Figure 2.1, table names are displayed in rectangles and connected with lines to a diamond symbol, which represents a relationship. Inside the diamond, a word (usually a verb) describes the action that relates the two tables. For example, in the relationship shown in Figure 2.1, the word *Request* is used to show that "Customers *request* WorkOrders." On the line connecting the rectangle to the diamond symbol (called the *join line*), a *1* is placed next to the primary table, or the "one" side of the relationship, and an *n* is placed next to the related table, or the "many" side of the relationship.

Activity 1a Creating a One-to-Many Relationship   Part 1 of 4

1. Open **2-RSRCompServ**. This database has the same structure as the database created in Chapter 1. However, additional field properties have been defined and several records have been added to each table to provide data for testing relationships and lookup fields.
2. If a security warning appears in the message bar to indicate that some active content has been disabled, click the Enable Content button.
3. Create a one-to-many relationship between the Customers table and WorkOrders table by completing the following steps:
   a. Click the Database Tools tab.
   b. Click the Relationships button in the Relationships group.
   c. At the Show Table dialog box with the Tables tab active and *Customers* selected in the list box, press and hold down the Ctrl key, click *WorkOrders*, click the Add button, and then release the Ctrl key.
   d. Click the Close button to close the Show Table dialog box.
   e. Resize both table field list boxes by dragging down the bottom borders of the boxes until all the field names are visible.

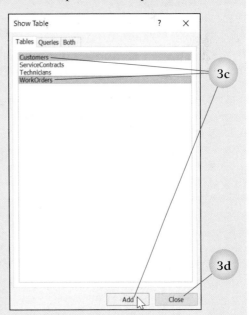

f. Drag the *CustID* field from the Customers table field list box to the *CustID* field in the WorkOrders table field list box. Be careful to drag the common field name from the primary table (Customers) to the related table (WorkOrders) and not vice versa.

g. At the Edit Relationships dialog box, notice that *One-To-Many* appears in the *Relationship Type* section. Access detected this type of relationship because the field used to join the tables is a primary key field in only one of the tables. Establishing this relationship makes *CustID* a foreign key field in the WorkOrders table. Always check that the correct table and field names are shown in the *Table/Query* and *Related Table/Query* option boxes. If the table name and/or the common field name is not shown correctly, click the Cancel button. Errors can occur if the mouse is dragged starting or ending at the wrong table or field. If this happens, click the Cancel button, return to Step 3f and then try again.

h. Click the Create button.

4. Click the Close button in the Relationships group on the Relationship Tools Design tab.

5. Click Yes at the message box asking if you want to save changes to the layout of the Relationships window.

The bottom border of each table field list box is resized to show all the field names in Step 3e.

Another one-to-many relationship exists between the Technicians table and WorkOrders table. A technician is associated with each work order assigned to that technician and a work order is associated with the technician that carried out the service request. The Technicians-to-WorkOrders relationship diagram is shown in Figure 2.2.

Figure 2.2 One-to-Many Relationship between the Technicians Table and WorkOrders Table

1. With **2-RSRCompServ** open, display the Relationships window by clicking the Database Tools tab and then clicking the Relationships button.
2. Click the Show Table button in the Relationships group on the Relationship Tools Design tab.
3. Double-click *Technicians* in the list box at the Show Table dialog box with the Tables tab selected. Move the Show Table dialog box, if necessary, to verify that the Technicians table has been added and then click the Close button.
4. Drag the bottom and right borders of the Technicians table field list box until all the field names are fully visible.
5. Drag the *TechID* field from the Technicians table field list box to the *TechID* field in the WorkOrders table field list box.
6. Check that the correct table and field names appear in the *Table/Query* and *Related Table/Query* option boxes. If necessary, click the Cancel button and repeat Step 5.
7. Click the Create button.
8. Click the Close button on the Relationship Tools Design tab.
9. Click Yes at the message box asking if you want to save changes to the layout of the Relationships window.

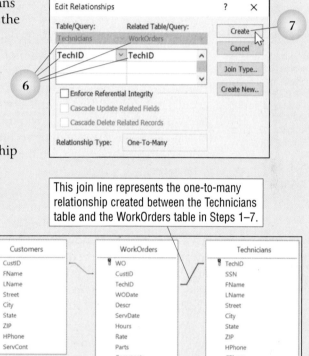

This join line represents the one-to-many relationship created between the Technicians table and the WorkOrders table in Steps 1–7.

Editing a Relationship

At the Edit Relationships dialog box, shown in Figure 2.3, select relationship options and/or specify the type of join to create. To open the Edit Relationships dialog box, click the Database Tools tab, click the Relationships button, click the join line for the relationship to be edited, and then click the Edit Relationships button.

Once the Edit Relationships dialog box is open, select any of the options. The *Cascade Update Related Fields* and *Cascade Delete Related Records* options do not become active unless referential integrity is turned on. To turn on referential integrity, click the *Enforce Referential Integrity* check box to insert a check mark.

Figure 2.3 Edit Relationships Dialog Box

Selecting *Enforce Referential Integrity* places restrictions on data entry. In this example, it means that a technician cannot be assigned to a work order if the technician does not exist in the Technicians table.

Hint To enable referential integrity, the primary key and foreign key fields must be the same data type.

If an error message displays when you are attempting to enforce referential integrity, open each table in Design view and compare the data types for the fields used to join the tables.

Activating referential integrity in a one-to-many relationship is a good way of ensuring that orphan records do not occur. An orphan record is a record in a related table for which no "parent" record exists in the primary table. Assigning a technician to a work order in the WorkOrders table when there is no matching technician record in the Technicians table results in creating an orphan record in the WorkOrders table. Once referential integrity has been turned on, Access checks for the existence of a matching record in the primary table as each new record is added to the related table. If no match is found, Access does not allow the record to be saved.

For example, suppose that referential integrity has not been activated and a typing mistake is made that causes the accidental entry of an unassigned *TechID* to a work order. If the customer later has a question for the technician about the service, no one will know which technician to contact. However, if referential integrity is activated before the record is saved, Access will check the Technicians table and verify that the *TechID* exists. If the *TechID* does not exist, the error message shown in Figure 2.4 will display. Click OK at the message and then enter a *TechID* that exists in the Technicians table.

When the *Enforce Referential Integrity* check box is clicked to insert a check mark, the *Cascade Update Related Fields* and *Cascade Delete Related Records* check boxes become available. When a check mark is inserted in the *Cascade Update Related Fields* check box, Access automatically updates all the occurrences of the same data in the foreign key field in the related table when a change is made to the primary key field in the primary table. When a check mark is inserted in the *Cascade Delete Related Records* check box, deleting a record from the primary table for which related records exist in the related table results in all the related records being automatically deleted.

Join types and situations in which changing the join type is warranted will be discussed in Chapter 3.

Figure 2.4 Referential Integrity Error Message

1. With **2-RSRCompServ** open, edit the one-to-many relationship between the Customers table and WorkOrders table to enforce referential integrity and add both cascade options by completing the following steps:
   a. Open the Relationships window.
   b. Click to select the join line between the Customers table and WorkOrders table.
   c. Click the Edit Relationships button in the Tools group on the Relationship Tools Design tab.

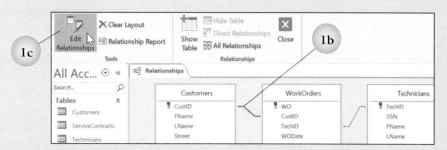

   d. At the Edit Relationships dialog box, click to insert check marks in the *Enforce Referential Integrity* check box, the *Cascade Update Related Fields* check box, and the *Cascade Delete Related Records* check box.
   e. Click OK. The *1* at the primary table (the "one") side of the join line and the infinity symbol (∞) at the related table (the "many" side) of the join line indicate that referential integrity has been activated.

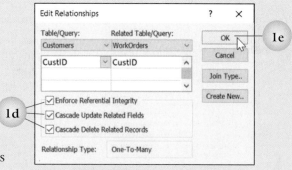

2. Edit the one-to-many relationship between the Technicians table and WorkOrders table to enforce referential integrity and add both cascade options by completing the following steps:
   a. Double-click the join line between the Technicians table and WorkOrders table in the Relationships window. (Or right-click the join line and then click *Edit Relationship* at the shortcut menu.)
   b. At the Edit Relationships dialog box, click to insert check marks in the *Enforce Referential Integrity* check box, the *Cascade Update Related Fields* check box, and the *Cascade Delete Related Records* check box.
   c. Click OK.
3. Notice with referential integrity turned on, the 1 and the infinity symbol (∞) display on the join lines.
4. Close the Relationships window.

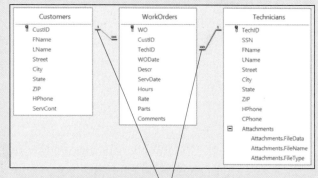

With referential integrity turned on, the 1 and the infinity symbol (∞) display on the join lines.

Check Your Work

Tutorial

Review: Creating
a One-to-One
Relationship

Quick Steps

**Create One-to-One
Relationship**
1. Click Database Tools
   tab.
2. Click Relationships
   button.
3. Add tables from
   Show Table dialog
   box.
4. Close Show Table
   dialog box.
5. Drag primary key
   field name from
   primary table to
   primary key field
   name in related
   table.
6. Select relationship
   options.
7. Click Create button.

Relationship
Report

Creating a One-to-One Relationship

In the database for RSR Computer Services, a table is used to store service contract information for each customer. That table, ServiceContracts, is associated with the Customers table. Only one record exists for a customer in the Customers table and each customer subscribes to only one service contract in the ServiceContracts table. This means that the two tables have a one-to-one relationship, as shown in Figure 2.5.

When a new customer is added to the database, the customer's name and contact information are entered into the Customers table first and then the service contract information (including start date, end date, and fee paid) is entered into the ServiceContracts table. When creating the relationship, drag the primary key field name from the table in which the data is entered into the database first (Customers table) to the primary key field name from the table in which the data is entered second (ServiceContracts table). When the relationship is established, the Customers table is placed in the *Table/Query* option box and the ServiceContracts table is placed in the *Related Table/Query* option box; otherwise, when data is being entered, an error message will appear, as shown in Figure 2.4.

To print the relationship, first create the Relationship Report by clicking the Relationship Report button in the Tools group. When Access displays the report in Print Preview, click the Print button on the Print Preview tab.

Figure 2.5   One-to-One Relationship between the Customers Table and ServiceContracts Table

Activity 1d **Creating a One-to-One Relationship**   Part 4 of 4

1. With **2-RSRCompServ** open, create a one-to-one relationship between the Customers table and ServiceContracts table by completing the following steps:
   a. Open the Relationships window.
   b. Click the Show Table button.
   c. Double-click *ServiceContracts* in the list box at the Show Table dialog box with the Tables tab selected and then click the Close button.
   d. Drag the *CustID* field from the Customers table field list box to the *CustID* field in the ServiceContracts table field list box.
   e. At the Edit Relationships dialog box, check that the correct table and field names appear in the *Table/Query* and *Related Table/Query* option boxes. If necessary, click the Cancel button and repeat Step 1d.
   f. Notice that *One-To-One* appears in the *Relationship Type* section. Access detected this type of relationship because the field used to join the tables is a primary key field in both tables.
   g. Click to insert check marks in the *Enforce Referential Integrity* check box, the *Cascade Update Related Fields* check box, and the *Cascade Delete Related Records* check box.
   h. Click the Create button.

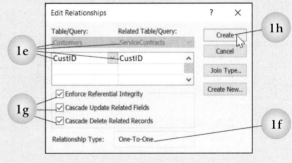

2. Drag the Title bar of the ServiceContracts table field list box to the approximate location in the Relationships window shown below. This makes it easier to view the join line and the *1* at each end of the join line between the Customers and ServiceContracts table field list boxes.

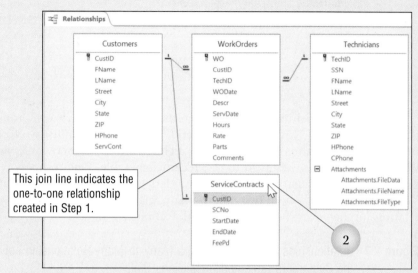

This join line indicates the one-to-one relationship created in Step 1.

3. Create a relationship report by clicking the Relationship Report button in the Tools group on the Relationship Tools Design tab.
4. Access displays the report in Print Preview. Click the Print button in the Print group on the Print Preview tab and then click OK at the Print dialog box.
5. Close the relationship report for **2-RSRCompServ** by clicking the Close Relationship button on the far right side of the screen. At the Microsoft Access message box, click Yes to save the report and click OK at the Save As dialog box to accept the default name.
6. Close the Relationships window.

Check Your Work

Tutorial

Creating a Many-to-Many Relationship

Creating a Many-to-Many Relationship

Consider the association between the Customers table and Technicians table in the RSR Computer Services database. Over time, any individual customer can have computer service work done by many different technicians and any individual technician can perform computer service work at many different customer locations. In other words, a record in the Customers table can be matched to many records in the Technicians table and a record in the Technicians table can be matched to many records in the Customers table. This is an example of a many-to-many relationship. A diagram of the many-to-many relationship between the Customers table and Technicians table is shown in Figure 2.6.

A many-to-many relationship is problematic because it creates duplicate records. If the same customer number is associated with many technicians and vice versa, many duplicates will occur in the two tables and Access may experience data conflicts when trying to identify unique records. To resolve the duplication and create unique entries, a third table is used to associate, or link, the many-to-many tables. That table is called a *junction table* and contains the primary key fields from both tables in the many-to-many relationship as its foreign key fields. Using the junction table, two one-to-many relationships are created.

Figure 2.6 Many-to-Many Relationship between the Customers Table and Technicians Table

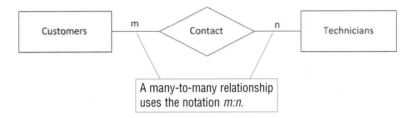

In the Relationships window shown in Figure 2.7, the WorkOrders table is the junction table. Notice that the WorkOrders table contains two foreign key fields: *CustID*, which is the primary key field in the Customers table, and *TechID*, which is the primary key field in the Technicians table. One-to-many relationships exist between the Customers and WorkOrders tables and the Technicians and WorkOrders tables. These two one-to-many relationships create a many-to-many relationship between the Customers and Technicians tables.

Figure 2.7 Relationships Window Showing a Many-to-Many Relationship between the Customers Table and Technicians Table

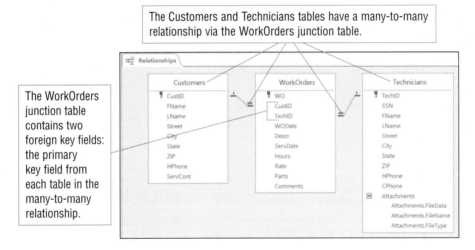

Activity 2 Create a Table with a Multiple-Field Primary Key Field and Lookup Fields

5 Parts

You will create a new table that requires three fields to uniquely identify each record. To restrict data entry in the table, you will create fields that each display a list from which the user selects the field value(s).

Tutorial

Defining a Multiple-Field Primary Key Field

Quick Steps

Define Multiple-Field Primary Key Field

1. Open table in Design view.
2. Select first field.
3. Press and hold down Shift (adjacent row) or Ctrl (nonadjacent row) and select second field.
4. Click Primary Key button.
5. Save table.

Primary Key

Hint Delete a primary key field designation by opening the table in Design view, activating the primary key field, and then clicking the Primary Key button.

Defining a Multiple-Field Primary Key Field

In most tables, one field is designated the primary key field. However, in some situations, a single field may not be guaranteed to hold unique data. Look at the fields in the table shown in Figure 2.8. This is a new table that will be created in the RSR Computer Services database to store computer profiles for RSR customers. The company stores the profiles as a service to clients, who sometimes forget their login information. Technicians can also access the profile data when troubleshooting at a customer's site.

A customer may have more than one computer in his or her home or office and each computer may have a different profile for each username. The *CustID* field will not serve as the primary key field if the customer has more than one record in the Profiles table. However, a combination of the three fields *CustID, CompID,* and *Username* will uniquely identify each record. In this table, three fields will be defined as primary key fields. A primary key field that is made up of two or more fields is called a *composite key* field.

Figure 2.8 Activity 2a Profiles Table

Profiles		
*CustID	Short Text	4
*CompID	Short Text	2
*Username	Short Text	15
Password	Short Text	15
Remote	Yes/No	

Activity 2a Creating a New Table with a Multiple-Field Primary Key

Part 1 of 5

1. With **2-RSRCompServ** open, create a new table to store customer profiles by completing the following steps:
   a. Click the Create tab and then click the Table Design button in the Tables group.
   b. Type the field names, assign the data types, and change the field sizes according to the data structure shown in Figure 2.8.

2. Hover the mouse pointer over the field selector bar (the blank column left of the field names) next to *CustID* until the pointer changes to a black right-pointing arrow and then click to select the field.

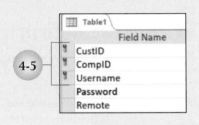

3. Press and hold down the Shift key, click in the field selector bar next to *Username*, and then release the Shift key. The three adjacent fields *CustID*, *CompID*, and *Username* are now selected.
4. Click the Primary Key button in the Tools group on the Table Tools Design tab. Access displays the primary key icon next to each field name.
5. Click in any data type field to deselect the first three rows.
6. Save the table with the name *Profiles*.
7. Close the table.

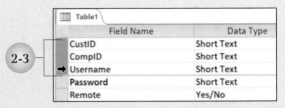
Check Your Work

 Tutorial

Creating a Field to Look Up Values in Another Table

 Tutorial

Modifying Lookup List Properties

Creating a Field to Look Up Values in Another Table

In Level 1, Chapter 4, the Lookup Wizard was used to create a lookup field where the values in the list were typed. The Lookup Wizard can also be used to create a lookup field to look up values found in records from another table. Lookup fields allow the user to enter data by pointing and clicking, rather than by typing the entry in the field.

A lookup field that draws its data from a field in another table can be useful in several ways. Data can be restricted to items within the list, which prevents orphan records, data entry errors, and spelling inconsistencies from occurring. The lookup field can also provide more information to help the user select the correct option. For example, suppose that a lookup field requires the user to select a customer's identification number. If the lookup field displays a drop-down list of identification numbers, identifying the correct number will be difficult. However, if the lookup field displays the identification number along with the customer's name, the correct entry will be easy to identify. When the user chooses the field entry based on the name, Access automatically enters the correct identification number.

When using the Lookup Wizard to create a lookup field, make sure to do so before the relationships are created. If a relationship already exists between the table for the lookup field and the source data table, Access will display a message stating that the relationship needs to be deleted before the Lookup Wizard can run.

Use the Lookup tab in the *Field Properties* section to change the *Limit to List* property from *No* to *Yes*. A user will not be able to type in an entry that is not in the list.

Quick Steps

Create Lookup Field to Another Table
1. Open table in Design view.
2. Click in column of lookup field.
3. Click option box arrow.
4. Click *Lookup Wizard*.
5. Click Next.
6. Choose table and click Next.
7. Choose fields to display in column.
8. Click Next.
9. Choose field by which to sort.
10. Click Next.
11. If necessary, expand column widths.
12. Clear *Hide key column*.
13. Click Next.
14. Choose field value to store in table.
15. Click Next.
16. Click Finish.
17. Click Yes.

1. With **2-RSRCompServ** open, open the Profiles table in Design view.
2. Create a lookup field to select and enter a customer's identification number from a list of customers in the Customers table by completing the following steps:
   a. With *CustID* the active field, click in the *Data Type* column, click the option box arrow, and then click *Lookup Wizard* at the drop-down list.
   b. At the first Lookup Wizard dialog box with *I want the lookup field to get the values from another table or query* selected, click the Next button.
   c. At the second Lookup Wizard dialog box with *Table: Customers* already selected in the *Which table or query should provide the values for your lookup field?* list box, click the Next button.

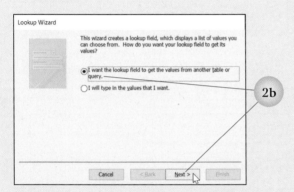
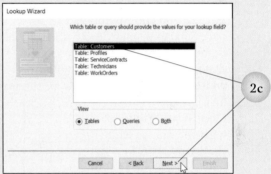

   d. At the third Lookup Wizard dialog box, double-click *FName* in the *Available Fields* list box to move the field to the *Selected Fields* list box.
   e. Double-click *LName* in the *Available Fields* list box to move the field to the *Selected Fields* list box and then click the Next button.

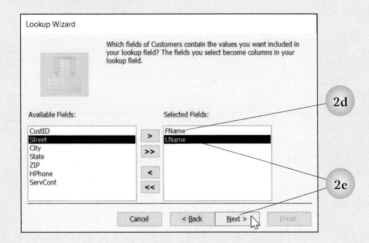

f. At the fourth Lookup Wizard dialog box, click the first sort option box arrow, and then click *LName* at the drop-down list. Notice that up to four sort keys can be defined to sort the lookup list and that an Ascending button appears next to each *Sort* option box. You can change the sort order from Ascending to Descending by clicking the Ascending button. Click the Next button.

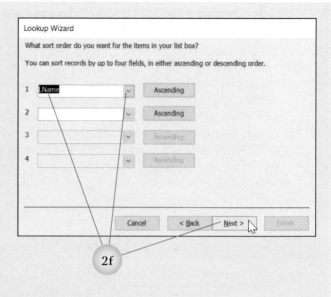

2f

g. At the fifth Lookup Wizard dialog box, expand the column widths if necessary to display all the data. Scroll down the list of entries in the dialog box. Notice that the column widths are sufficient to show all the text.

h. To view the customer identification numbers with the names while the list is open in a record, click the *Hide key column (recommended)* check box to remove the check mark. Removing the check mark displays the *CustID* field values as the first column in the lookup list.

2h

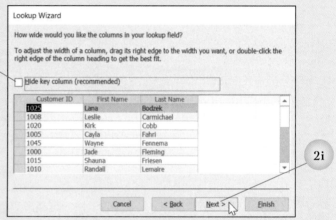

2i

i. Click the Next button.

j. At the sixth Lookup Wizard dialog box, choose the field value to be stored in the table when an entry is selected in the drop-down list. With *CustID* already selected in the *Available Fields* list box, click the Next button.

k. At the last Lookup Wizard dialog box, click the Finish button to accept the existing field name for the lookup field of *CustID*.

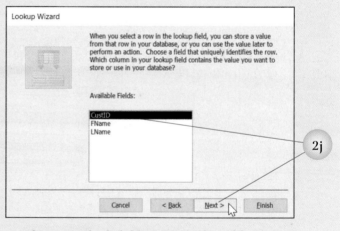

2j

l. At the Lookup Wizard message box stating that the table must be saved before relationships can be created, click Yes to save the table. Access automatically creates a relationship between the Customers table and Profiles table based on the *CustID* field used to create the lookup field.

3. Close the Profiles table.

1. With **2-RSRCompServ** open, open the Profiles table in Design view.
2. Type the following text in the *Caption* property box for each of the following fields:

    CustID   Customer ID
    CompID   Computer ID
    Remote   Remote Access?

3. Modify the lookup field properties to restrict entries in new records to items within the list by completing the following steps:
    a. Make *CustID* the active field.
    b. Click the Lookup tab in the *Field Properties* section.
    c. Look at the entries in all the property boxes for the Lookup tab. These entries were created by the Lookup Wizard.
    d. Double-click in the *Limit To List* property box to change *No* to *Yes*. This means that the field will accept data only from existing customer records. A user will not be able to type in an entry that is not in the list.

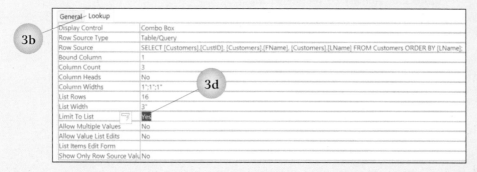

    e. Save the table.
4. Switch to Datasheet view.
5. With *Customer ID* in the first row of the datasheet as the active field, click the option box arrow in the field and then click *Jade Fleming* at the drop-down list. Notice that Access inserts *1000* as the field value in the first column since that is the customer number associated with the selected name.
6. Type the remaining data as indicated:

    Computer ID   D1
    Username   jade
    Password   P$ck7
    Remote Access? (Leave blank to indicate *No*.)

7. Best fit the width of each column.
8. Print and then close the Profiles datasheet. Click Yes when prompted to save changes to the table layout.

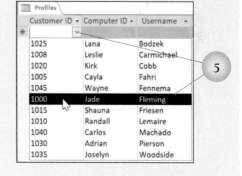

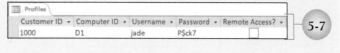

Check Your Work

Tutorial

Creating a Field That Allows Multiple Values

Hint Multiple-value fields should be used when more than one choice from a small list needs to be stored. This makes it unnecessary to create an advanced database design.

Quick Steps

Create Multiple-Value Lookup Field

1. Open table in Design view.
2. Start Lookup Wizard for desired field.
3. Create list by typing values or binding data to field in another table.
4. At last Lookup Wizard dialog box, insert check mark in *Allow Multiple Values* check box.
5. Click Finish.
6. Click Yes.

Creating a Field That Allows Multiple Values

The industry certifications that a technician has achieved could be organized by creating a separate field for each certification. A more efficient approach is to create a single field that displays a list of certifications with check boxes to indicate whether they have been attained.

Look at the fields in the table structure shown in Figure 2.9. For each technician, a list in the *Certifications* field is opened and the check box next to the applicable certification title is checked if the technician has obtained that specific certification. In the *OperatingSys* field, another list can be used to keep track of the operating systems for which the technician is considered an expert.

Use the Lookup Wizard to create a field to store multiple values. Choose to look up the values in a field in another table or create a custom value list. At the last Lookup Wizard dialog box, make sure to click the *Allow Multiple Values* check box to insert a check mark.

Figure 2.9 Activity 2d TechSkills Table

TechSkills		
*TechID	Short Text	2
Certifications	Short Text	20
OperatingSys	Short Text	20
NetworkSpc	Yes/No	
WebDesign	Yes/No	
Programming	Yes/No	

Activity 2d Creating Fields That Allow Multiple Values in a New Table — Part 4 of 5

1. With **2-RSRCompServ** open, create a new table to store technician competencies by completing the following steps:
   a. Create a new table using Design view.
   b. Type the field names and assign the data types according to the data structure shown in Figure 2.9.
   c. Assign the field denoted with an asterisk in Figure 2.9 as the primary key field.
   d. Save the table with the name *TechSkills*.
2. Create a lookup field to select a technician from a list of names in the Technicians table by completing the following steps:
   a. Click in the *Data Type* column for the *TechID* field, click the option box arrow that appears, and then click *Lookup Wizard*.
   b. Click the Next button at the first Lookup Wizard dialog box.
   c. Click *Table: Technicians* and then click the Next button.
   d. Double-click *FName* in the *Available Fields* list box to move the field to the *Selected Fields* list box.
   e. Double-click *LName* in the *Available Fields* list box and then click the Next button.
   f. Sort by *LName* and then click the Next button.

g. With a check mark in the *Hide key column (recommended)* check box, click the Next button to accept the current column widths. (In this lookup example, you are electing not to show the technician's ID field value. Although you will view and select by name, Access stores the primary key field's value in the table. *TechID* is considered the bound field and *FName* and *LName* are considered display fields).

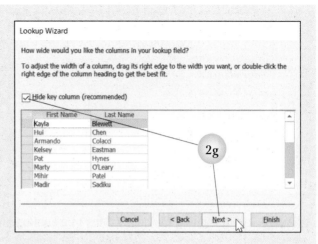

h. Click the Finish button and then click Yes to save the table.

3. Create a lookup field that allows multiple values for certification information by completing the following steps:

a. Click in the *Data Type* column for the *Certifications* field, click the option box arrow that appears, and then click *Lookup Wizard*.

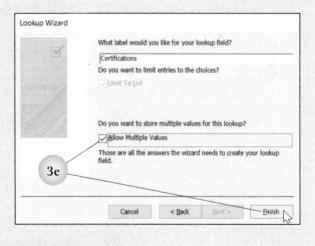

b. Click *I will type in the values that I want* and then click the Next button.

c. At the second Lookup Wizard dialog box, type the following entries in *Col1* pressing the Tab key to move down to the next entry:

    CCNA Cloud
    CCNA Wireless
    Comp TIA A+
    Microsoft MCT
    MOS Master

d. Click the Next button.

e. At the last Lookup Wizard dialog box, click the *Allow Multiple Values* check box to insert a check mark and then click the Finish button.

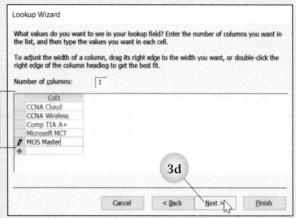

f. At the message box indicating that once the field is set to store multiple values, the action cannot be undone, click Yes to change the *Certifications* field to multiple values.

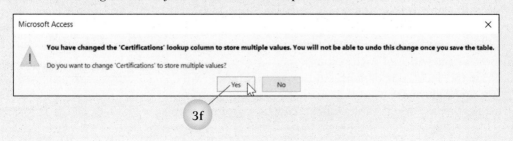

4. Complete steps similar to those in Steps 3a–3f to create a lookup list to store multiple values in the *OperatingSys* field using the following value list:

> Windows 10
> Windows 8
> Windows 7
> Linux
> Unix

5. Save and then close the TechSkills table.

Activity 2e Assigning Multiple Values in a Lookup List

Part 5 of 5

1. With **2-RSRCompServ** open, open the TechSkills table in Design view.
2. Type the following text in the *Caption* property box for each field as indicated:

TechID	Technician ID
OperatingSys	Operating Systems
NetworkSpc	Network Specialist?
WebDesign	Design Web Sites?
Programming	Programming?

3. Save the table and then switch to Datasheet view.
4. Add a new record to the table by completing the following steps:
   a. With the insertion point positioned in the *Technician ID* column, click the option box arrow and then click *Kelsey Eastman* at the drop-down list. Notice that Access displays the technician's first name in the column. *FName* is considered a display field for this column but the identification number associated with the name *Kelsey Eastman* is stored in the table.

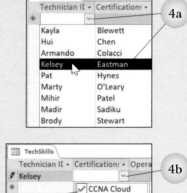

   b. Press the Tab key and then click the option box arrow in the *Certifications* column.
   c. Since *Certifications* is a multiple-value field, the drop-down list displays with check boxes next to all the items. Click the *CCNA Cloud* check box and the *Microsoft MCT* check box to insert check marks and then click OK.
   d. Press the Tab key and then click the option box arrow in the *Operating System* column.
   e. Click the *Windows 10, Windows 8*, and *Linux* check boxes to insert check marks and then click OK.

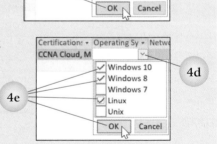

   f. Press the Tab key and then press the spacebar to insert a check mark in the *Network Specialist?* check box.
   g. Press the Tab key three times to finish the record, leaving the check boxes blank in the *Design Web Sites?* and *Programming?* columns.
5. Best fit the width of each column.
6. Print the TechSkills table in landscape orientation with left and right margins of 0.25 inch.
7. Close the TechSkills table. Click Yes when prompted to save changes to the layout.

Tutorial

Creating an Index

Quick Steps

Create Single-Field Index
1. Open table in Design view.
2. Make field active.
3. Click in *Indexed* property box.
4. Click option box arrow.
5. Click *Yes (Duplicates OK)* or *Yes (No Duplicates)*.
6. Save table.

Create Multiple-Field Index
1. Open table in Design view.
2. Click Indexes button.
3. Click in first blank row in *Index Name* column.
4. Type name for index.
5. Press Tab.
6. Click option box arrow in *Field Name* column.
7. Click field.
8. If necessary, change sort order.
9. Click in *Field Name* column in next row.
10. Click option box arrow.
11. Click field.
12. If necessary, change sort order.
13. Repeat Steps 9–12 until finished.
14. Close Indexes window.

 Indexes

♀ **Hint** An index cannot be generated for fields with a data type of OLE Object or Attachment.

Creating an Index

An index is a list created by Access containing pointers that direct Access to the locations of specific records in a table. A database index is very similar to an index found at the end of a book. Search the book's index for a keyword associated with the topic being searched, and the index gives the page number(s) in the book that contains information on that topic. Using a book's index allows information to be retrieved quickly and efficiently. Although the information in an Access index cannot be seen, it operates in much the same way, reducing the amount of time it takes to find a particular record.

Access automatically generates an index for a field designated the primary key field in a table. In a database with a large number of records, it is useful to identify fields other than the primary key field that are often sorted or searched and to create indexes for these fields to speed up sorting and searching. For example, in the Customers table in the RSR Computer Services database, creating an index for the *LName* field is a good idea because the table data will frequently be sorted by a customer's last name.

An index can be created to restrict the data in a field to unique values. This creates a field similar to a primary key field in that Access will not allow two records to hold the same data. For example, an email field in a table that is frequently searched is a good candidate for an index. To avoid data entry errors in a field that should contain unique values (and is not the primary key field), set up the index so it will not accept duplicates.

Create a multiple-field index if a large table is frequently sorted by two or more fields at the same time. In Table Design view, click the Indexes button to open the Indexes window, as shown in Figure 2.10. Create an index for a combination of up to 10 fields.

Figure 2.10 Indexes Window

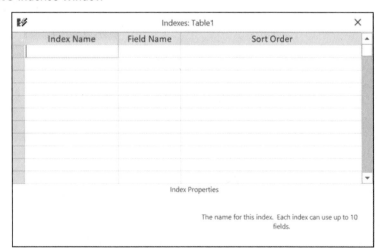

1. With **2-RSRCompServ** open, open the Customers table in Design view.
2. Create a single-field index for the *ZIP* field by completing the following steps:
   a. Make *ZIP* the active field.
   b. Double-click in the *Indexed* property box to change *No* to *Yes (Duplicates OK)*.
   c. Save the table.
3. Create a multiple-field index for the *LName* and *FName* fields by completing the following steps:
   a. Click the Indexes button in the Show/Hide group on the Table Tools Design tab.
   b. At the Indexes: Customers window, click in the first blank row in the *Index Name* column (below *ZIP*) and then type Names.
   c. Press the Tab key, click the option box arrow that appears in the *Field Name* column, and then click *LName* at the drop-down list. The sort order for *LName* defaults to *Ascending*.

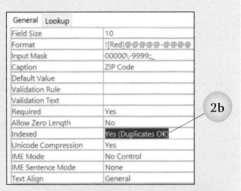

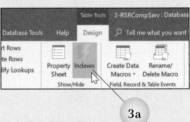

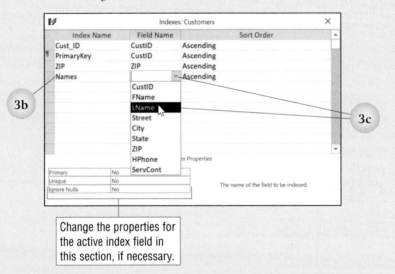

Change the properties for the active index field in this section, if necessary.

   d. In the *Field Name* column, click in the row below *LName*, click the option box arrow that appears, and then click *FName*.
   e. Close the Indexes: Customers window.
   f. Save the table.
4. Close the Customers table.
5. Close **2-RSRCompServ**.

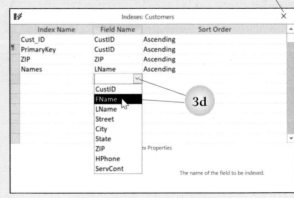

Tutorial

Normalizing a
Database

Normalizing a Database

Normalizing a database involves reviewing the database structure to ensure that the tables are set up to eliminate redundancy. If data redundancy is discovered, the process of normalization often involves splitting fields into smaller units and/or breaking down larger tables into smaller tables and creating relationships to remove repeating groups of data. Three normalization states are tested: first normal form, second normal form, and third normal form.

Checking First Normal Form

A table meets first normal form when it does not contain any fields that can be broken into smaller units and when it does not have similar information stored in several fields.

For example, a table that contains a single field called *TechnicianName* that stores the technician's first and last names in the same column violates first normal form. To correct the structure, split *TechnicianName* into two fields, such as *TechLastName* and *TechFirstName*.

A table that has multiple fields and in which each field contains similar data—such as *Week1*, *Week2*, *Week3*, and *Week4*—also violates first normal form. To correct this structure, delete the four week fields and replace them with a single field named *WeekNumber*.

Checking Second Normal Form

Meeting second normal form is of concern only for a table that has a multiple-field primary key field (composite key field). A table with a composite key field meets second normal form when it is in first normal form and when all of its fields are dependent on all the fields that form the primary key field.

Hint If a field cannot relate to the individual fields of a composite key field, then it should be included in a different table.

For example, assume that a table is defined with two fields that form the primary key field: *CustID* and *ComputerID*. Also assume that a field in the same table is titled *EmailAdd*. The contents of the *EmailAdd* field are dependent on the customer only (not the computer). Since *EmailAdd* is not dependent on both *CustID* and *ComputerID*, the table is not in second normal form. To correct the structure, delete the *EmailAdd* field. (The *EmailAdd* field belongs in another table in the database.)

Checking Third Normal Form

Meeting third normal form applies to a table that has a single primary key field and is in first normal form. If a field exists in the table for which the field value is not dependent on the field value of the primary key field, the table is not in third normal form.

For example, assume that a table is defined with a single primary key field titled *TechnicianID*. Also assume that fields in the same table are titled *PayCode* and *PayRate*. Finally, assume that a technician's pay rate is dependent on the pay code assigned to him or her. Since the pay rate is dependent on the field value in the pay code field and not on the technician's identification number, the table is not in third normal form. To convert the table to third normal form, delete the *PayRate* field from the table. (The *PayRate* field belongs in another table in the database.)

Chapter Summary

- When building relationships, consider the frequency of matching data in the common field in both tables to determine if the relationship is one-to-many, one-to-one, or many-to-many.

- The most common type of relationship is one-to-many. It involves joining tables by dragging the primary key field from the "one" table to the foreign key field in the "many" table.

- In a one-to-many relationship, only one record for a matching field value exists in the primary table, while many records for the same value can exist in the related table.

- A relationship diagram depicts the two tables joined in the relationship and the type of relationship between them.

- Enforce referential integrity to place restrictions on new data entered into a related table at the Edit Relationships dialog box. A record is not allowed in the related table if a matching record does not already exist in the primary table.

- Selecting the *Cascade Update Related Fields* option automatically updates all the occurrences of the same data in the foreign key field in the related table when a change is made to the primary key field in the primary table.

- Selecting the *Cascade Delete Related Records* option automatically deletes related records when a record is deleted from the primary table.

- In a one-to-one relationship, only one record exists that has a matching value in the joined field in both tables.

- In a many-to-many relationship, many records can exist that have matching values in the joined fields in both tables.

- A junction table is used to create a many-to-many relationship. A junction table contains a minimum of two fields, which are the primary key fields from the two tables in the many-to-many relationship.

- Using a junction table, two one-to-many relationships are joined to form a many-to-many relationship.

- In some tables, two or more fields are used to create the primary key field if a single field is not guaranteed to hold unique data.

- A primary key field that is made up of two or more fields is called a *composite key* field.

- A lookup field displays a drop-down list in a field, in which the user points and clicks to enter the field value. The list can be generated from records in a related table or by typing in a value list.

- Once a lookup field has been created, options at the Lookup tab in the *Field Properties* section in Table Design view can be selected to modify individual properties.

- Clicking *Yes* at the Limit To List property box allows entries in the field to be restricted to items within the lookup list.

- A field that allows selecting multiple entries from a drop-down list can be created by clicking *Allow Multiple Values* at the final Lookup Wizard dialog box.

- Access displays check boxes next to all the items in the drop-down list if the field has been set to allow multiple values.

- An index is a list generated by Access containing pointers that direct Access to the locations of specific records in a table.

- Access automatically generates an index for a field designated the primary key field in a table.
- To create a single-field table index, change the Indexed property to *Yes (Duplicates OK)* or *Yes (No Duplicates)*.
- To create a multiple-field index, open the Indexes window.
- Normalizing a database involves reviewing the database structure to ensure that the tables are set up to eliminate redundancy. Three normalization states are tested: first normal form, second normal form, and third normal form.

Commands Review

FEATURE	RIBBON TAB, GROUP	BUTTON
Edit Relationships dialog box	Relationship Tools Design, Tools	
indexes window	Table Tools Design, Show/Hide	
primary key field	Table Tools Design, Tools	
relationships report	Relationship Tools Design, Tools	
Relationships window	Database Tools, Relationships OR Table Tools Design, Relationships	
Show Table dialog box	Relationship Tools Design, Relationships	

Microsoft®

Access®

Advanced Query Techniques

Performance Objectives

Upon successful completion of Chapter 3, you will be able to:

1 Save a filter as a query

2 Create and run a parameter query to prompt for criteria

3 Create an inner join, left outer join, and right outer join to modify query results

4 Add tables to and remove tables from a query

5 Create a self-join query to match two fields in the same table

6 Assign an alias to a table name

7 Create a query that includes a subquery

8 Create a query that uses conditional logic

9 Select records using a multiple-value field in a query

10 Create a new table using a make table query

11 Remove records from a table using a delete query

12 Add records to a table using an append query

13 Modify records using an update query

In this chapter, you will create, save, and run queries that incorporate advanced query features, such as saving a filter as a query, prompting for criteria on single and multiple fields, modifying join properties to view alternative query results, and using action queries to perform operations on groups of records. In addition, you will create an alias for a table and incorporate subqueries to manage multiple calculations.

 Data Files

Before beginning chapter work, copy the AL2C3 folder to your storage medium and then make AL2C3 the active folder.

The online course includes additional training and assessment resources.

You will create queries to select records by saving the criteria for a filter and creating a query that prompts the user for the criteria when the query is run.

Extracting Records Using Select Queries

A select query is the type of query most often used in Access. It extracts records that meet specified criteria from a single table or multiple tables. The subset of records that a query returns can be edited, viewed, and/or printed. In Query Design view, the criteria used to select records are entered by typing expressions in the *Criteria* row for the required field(s). Access also provides other methods for specifying query criteria.

Saving a Filter as a Query

Quick Steps

Save Filter as Query
1. Open table.
2. Filter table as needed.
3. Click Advanced button.
4. Click *Filter By Form*.
5. Click Advanced button.
6. Click *Save As Query*.
7. Type query name.
8. Click OK.

Advanced

Saving a Filter as a Query

A filter is used in a datasheet or form to temporarily hide records that do not meet specified criteria. For example, filter a WorkOrders datasheet to display only those work orders completed on a specified date. The subset of records can then be edited, viewed, and/or printed. Use the Filter by Form feature to filter a datasheet by multiple criteria using a blank datasheet. A filter is active until it is removed or the datasheet or form is closed. A filter can be saved when the table is closed but when the worksheet is reopened, all the records redisplay. Click the Toggle Filter button to reapply the filter.

If these steps are being repeated often, save the filtered datasheet as a query. Some users are more comfortable using filters to select records than with typing criteria expressions in Query Design view. To save a filter as a query, click the Advanced button in the Sort & Filter group on the Home tab and then click *Filter By Form* at the drop-down list to display the criteria. Click the Advanced button again and then click *Save As Query* at the drop-down list. Type a query name at the Save As Query dialog box and then press the Enter key or click OK. Saving a filter as a query means that all the columns in the table display in the query results datasheet. Use the Hide Fields feature to remove a field(s) from the results.

Activity 1a Saving a Filter as a Query   Part 1 of 3

1. Open **3-RSRCompServ** and enable the content.
2. Use the Filter by Form feature to display only those service calls that required two or more hours of labor by technicians billed at $50.00 per hour by completing the following steps:
   a. Open the WorkOrders table in Datasheet view.
   b. Minimize the Navigation pane.
   c. Hide the *Comments* field by right-clicking the *Comments* column heading in the datasheet and then clicking *Hide Fields* at the shortcut menu.

d. Click the Advanced button in the Sort & Filter group on the Home tab and then click *Filter By Form* at the drop-down list.

e. Click in the empty record in the *Hours* column. Type >=2 and then press the Tab key.

f. With the insertion point positioned in the *Rate* column, click the option box arrow that appears and then click *50* at the drop-down list.

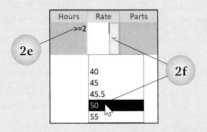

g. Click the Toggle Filter button (which displays the ScreenTip *Apply Filter*) in the Sort & Filter group on the Home tab. The records that meet the filter conditions display.

3. Review the seven filtered records in the datasheet.

4. Click the Toggle Filter button (which displays the ScreenTip *Remove Filter*) to redisplay all the records.

5. Click the Advanced button and then click *Filter By Form* at the drop-down list. Notice that the filter criteria in the *Hours* and *Rate* columns are intact.

6. Save the filter as a query so the criteria can be reused by completing the following steps:

a. Click the Advanced button and then click *Save As Query* at the drop-down list.

b. At the Save As Query dialog box, type WO2orMoreRate50 in the *Query Name* text box and then click OK.

c. Click the Advanced button and then click *Close* at the drop-down list to close the Filter By Form datasheet.

d. Close the WorkOrders table. Click No when prompted to save changes to the table design.

7. Expand the Navigation pane.

8. Double-click the query object *WO2orMoreRate50* to open the query and then review the results.

9. Hide the *Comments* column in the query results datasheet.

10. Print the datasheet in landscape orientation with the left and right margins set to 0.5 inch.

11. Switch to Design view. Notice that the query design grid for a query created from a filter includes columns only for those columns for which criteria have been defined.

12. Close the query. Click Yes when prompted to save changes to the query layout.

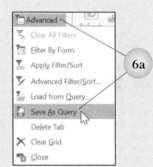

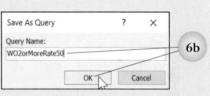

Check Your Work

Creating a Parameter Query

Creating a Parameter Query

In a parameter query, specific criteria for a field are not stored with the query design. Instead, the field(s) used to select records has a prompt message that displays when the query is run. The prompt message instructs the user to type the criteria to be used in selecting records.

Figure 3.1 shows the Enter Parameter Value dialog box, which displays when a parameter query to select by a technician's name is run. The message shown in the dialog box is created in the field for which the criterion will be applied. When the query is run, the user types the criterion at the Enter Parameter Value dialog box and Access selects the records based on the entry. If more than one field contains a parameter, Access prompts the user one field at a time.

A parameter query is useful if a query is run several times on the same field but different criteria are used each time. For example, suppose that a list of each technician's work orders is needed. Normally, a separate query would be made for each technician, resulting in several query objects being displayed in the Navigation pane. Creating a parameter query that prompts the user to enter the technician's name means that only one query is created.

To create a parameter query, start a new query in Design view and then add the desired tables and fields to the query design grid. Type a message enclosed in square brackets to prompt the user for the required criterion in the *Criteria* row of the field to be used to select records. Access does not allow punctuation at the end of the message. The text inside the square brackets is displayed in the Enter Parameter Value dialog box when the query is run. Figure 3.2 displays the entry in the *Criteria* row of the *FName* field that generated the Enter Parameter Value message shown in Figure 3.1.

Figure 3.1 Enter Parameter Value Dialog Box

Figure 3.2 Criterion to Prompt for the Name in the *FName* Field

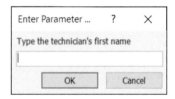

Type a message enclosed in square brackets to prompt the user for the criterion to use in selecting records.

Activity 1b Creating a Parameter Query to Prompt for Technician Names — Part 2 of 3

1. With **3-RSRCompServ** open, create a query in Design view to select records from the Technicians table and WorkOrders table by completing the following steps:
   a. Click the Create tab and then click the Query Design button in the Queries group.
   b. At the Show Table dialog box, add the Technicians table and WorkOrders table to the query.
   c. Close the Show Table dialog box.

d. At the top of the query, drag down the bottom borders of both table field list boxes until all the field names are visible. If necessary, resize the query design grid.

e. Double-click the following field names in the Technicians table field list box and WorkOrders table field list box to add the fields to the query design grid (click the field names in the order indicated): *WO, FName, LName, ServDate, Hours, Rate.*

These fields are added to the query design grid in Step 1e.

Field:	WO	FName	LName	ServDate	Hours	Rate
Table:	WorkOrders	Technicians	Technicians	WorkOrders	WorkOrders	WorkOrders
Sort:						
Show:	☑	☑	☑	☑	☑	☑

2. Click the Run button in the Results group on the Query Tools Design tab to run the query.

3. Add parameters to select records by a technician's first and last names by completing the following steps:

a. Switch to Design view.

b. Click in the *Criteria* row in the *FName* column in the query design grid, type [Type the technician's first name], and then press the Enter key.

c. Position the mouse pointer on the vertical line between *FName* and *LName* in the gray field selector bar above the field names until the pointer changes to a left-and-right-pointing arrow with a vertical line in the middle. Double-click to expand the width of the *FName* column so the entire criterion entry can be seen.

d. With the insertion point positioned in the *Criteria* row in the *LName* column, type [Type the technician's last name] and then press the Enter key.

e. Expand the width of the *LName* column so the entire criterion entry can be seen.

4. Click the Save button on the Quick Access Toolbar, type PromptedTechLabor in the *Query Name* text box at the Save As dialog box, and then click OK.

5. Close the query.

6. Run the parameter query and extract a list of work orders for the technician named *Pat Hynes* by completing the following steps:

a. Double-click the query named *PromptedTechLabor* in the Navigation pane.

b. Type pat at the first Enter Parameter Value dialog box, which displays the message *Type the technician's first name,* and then click OK. (Note that Access is not case-sensitive for text strings.)

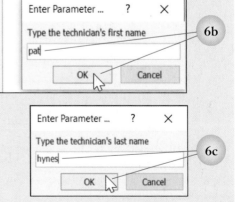

c. Type hynes at the second Enter Parameter Value dialog box, which displays the message *Type the technician's last name,* and then click OK.

7. Review the records in the query results datasheet.

8. Print the query results datasheet.

9. Close the query.

Check Your Work

1. With **3-RSRCompServ** open, create a parameter query in Design view to prompt the user for the starting and ending dates for selecting records in the WorkOrders table by completing the following steps:
   a. Click the Create tab and then click the Query Design button.
   b. At the Show Table dialog box, add the WorkOrders table to the query and then close the dialog box.
   c. Drag down the bottom border of the table field list box until all the field names are visible. If necessary, resize the query design grid.
   d. Add the following fields to the query design grid in this order: *WO*, *CustID*, *Descr*, *ServDate*, *Hours*, *Rate*, *Parts*.
   e. Click in the *Criteria* row in the *ServDate* column, type the entry between [Type starting date] and [Type ending date], and then press the Enter key. Access capitalizes *Between* and *And*.

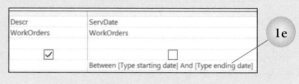

   f. Expand the width of the *ServDate* column until the entire criterion entry is visible.
2. Save the query, typing PromptedServiceDate as the query name, and then click OK.
3. Close the query.
4. Double-click *PromptedServiceDate* in the Navigation pane. At the first Enter Parameter Value dialog box with *Type starting date* displayed, type November 1, 2021 and then click OK. At the second Enter Parameter Value dialog box with *Type ending date* displayed, type November 13, 2021 and then click OK.

Work Order ▾	Customer ID ▾	Description ▾	Service Date ▾	Hours ▾	Rate ▾	Parts ▾
65012	1000	Biannual computer maintenance	Mon, Nov 01 2021	1.25	50.00	$10.15
65013	1000	Replace keyboard	Wed, Nov 03 2021	0.50	45.50	$42.75
65014	1005	Replace power supply	Wed, Nov 03 2021	1.75	50.00	$62.77
65015	1008	Restore operating system	Fri, Nov 05 2021	2.25	50.00	$0.00
65016	1010	Install upgraded video card	Fri, Nov 05 2021	1.00	50.00	$48.75
65017	1015	Replace hard drive	Tue, Nov 09 2021	2.50	50.00	$55.87
65018	1020	Upgrade Office Suite	Thu, Nov 11 2021	1.00	45.00	$0.00
65019	1025	Upgrade to new Windows	Thu, Nov 11 2021	1.50	55.00	$0.00

Records within the date range November 1, 2021 to November 13, 2021 are selected in Step 4.

5. Print the query results datasheet in landscape orientation.
6. Close the query.

Check Your Work

You will create and modify queries that obtain various results based on changing the join properties for related tables.

Modifying Join Properties in a Query

Hint Edit a relationship if the join type is always to be a left or right outer join. To do this, click the Join Type button in the Edit Relationships dialog box to open the Join Properties dialog box. Select the desired join type and then click OK.

Ŏuick Steps

Create Query with Left Outer Join
1. Create new query in Design view.
2. Add tables to query window.
3. Double-click join line between tables.
4. Select option 2.
5. Click OK.
6. Add fields to query design grid.
7. Save and run query.

Hint Changing the join type at a query window does not change the join type for other objects based on the relationship. The revised join property applies only to the query.

Modifying Join Properties in a Query

The term *join properties* refers to how Access matches the field values in the common field between the two tables in a relationship. The method used determines how many records are selected for inclusion in the query results datasheet. Access provides for three join types in a relationship: an inner join, a left outer join, and a right outer join. By default, Access uses an inner join between the tables; records are selected for display in a query only when a match on the joined field value exists in both tables. If a record exists in either table with no matching record in the other table, the record is not displayed in the query results datasheet. This means that in some cases, not all the records are seen from both tables when a query is run.

Specifying the Join Type

Double-click the join line between tables in a query to open the Join Properties dialog box, shown in Figure 3.3. By default, option *1*, referred to as an *inner join*, is selected. In an inner join, only those records are displayed for which the primary key field value in the primary table matches a foreign key field value in the related table.

Options *2* and *3* are referred to as *outer joins*. Option *2* is a left outer join. In this type of join, the primary table (referred to as the *left table*) displays all the rows, whereas the related table (referred to as the *right table*) displays only rows with matching values in the foreign key field. For example, examine the query results datasheet shown in Figure 3.4. This query was created with the Technicians table and TechSkills table. All the technician records are shown in the datasheet. However, notice that some technician records display with empty field values for the columns from the TechSkills table. These records reflect the technicians for which no information has yet been entered in the TechSkills table. In a left outer join, all the records from the primary table in the relationship are shown in the query results datasheet.

Figure 3.3 Join Properties Dialog Box

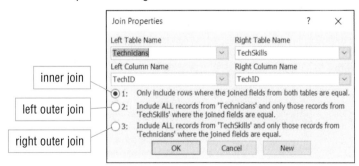

Figure 3.4 Left Outer Join Example

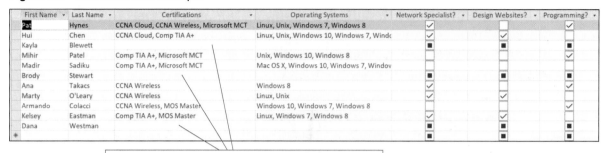

First Name	Last Name	Certifications	Operating Systems	Network Specialist?	Design Websites?	Programming?
Pat	Hynes	CCNA Cloud, CCNA Wireless, Microsoft MCT	Linux, Unix, Windows 7, Windows 8	✓		✓
Hui	Chen	CCNA Cloud, Comp TIA A+	Linux, Unix, Windows 10, Windows 7, Windc	✓	✓	
Kayla	Blewett			■	■	■
Mihir	Patel	Comp TIA A+, Microsoft MCT	Unix, Windows 10, Windows 8			✓
Madir	Sadiku	Comp TIA A+, Microsoft MCT	Mac OS X, Windows 10, Windows 7, Windov			
Brody	Stewart			■	■	■
Ana	Takacs	CCNA Wireless	Windows 8	✓		✓
Marty	O'Leary	CCNA Wireless	Linux, Unix	✓	✓	
Armando	Colacci	CCNA Wireless, MOS Master	Windows 10, Windows 7, Windows 8		✓	✓
Kelsey	Eastman	Comp TIA A+, MOS Master	Linux, Windows 7, Windows 8	✓	✓	
Dana	Westman			■	■	■

Left outer join query results show blank related *TechSkills* fields for those technicians for which records have not yet been entered in the TechSkills table.

Quick Steps

Create Query with Right Outer Join

1. Create new query in Design view.
2. Add tables to query window.
3. Double-click join line between tables.
4. Select option *3*.
5. Click OK.
6. Add fields to query design grid.
7. Save and run query.

Option *3* is a right outer join. In this type of join, the related table (right table) shows all the rows, whereas the primary table (left table) shows only rows with matching values in the common field. For example, examine the partial query results datasheet shown in Figure 3.5. This datasheet illustrates 15 of the 39 records in the query results datasheet from the Technicians table and WorkOrders table. Notice that the first four records have no technician first or last names. These are the work orders that have not yet been assigned to a technician. In a right outer join, all the records from the related table in the relationship are shown in the query results datasheet. In a left or right outer join, Access displays an arrow at the end of the join line pointing to the table that shows only matching rows. To display only the records that have not yet been assigned to a technician, type *null* in the criteria of either the first or last name fields.

To illustrate the difference in query results when no change is made to the join type, examine the query results datasheet shown in Figure 3.6. This is the datasheet created in Activity 2a. In this activity, a list of technician names and qualifications will be created. Compare the number of records shown in Figure 3.6 with the number of records shown in Figure 3.4. Notice that fewer records display in the datasheet in Figure 3.6. Since an inner join displays only those records for which matching entries exists in both tables, records from either table that do not have matching records in the other table are not displayed. Understanding that an inner join (the default join type) may not display all the records that exist in the tables when a query is run is important.

Figure 3.5 Right Outer Join Example

Right outer join query results show blank related technician fields for those work orders that have not yet been assigned to a technician.

First Name	Last Name	Work Order	WO Date	Description
		65047	Mon, Nov 29 2021	Set up automatic backup
		65048	Mon, Nov 29 2021	Replace LCD monitor
		65049	Tue, Nov 30 2021	Set up dual monitor system
		65050	Tue, Nov 30 2021	Reinstall Windows 10
Pat	Hynes	65020	Fri, Nov 12 2021	Troubleshoot noisy fan
Pat	Hynes	65033	Tue, Nov 23 2021	Install Windows 10 and Office 365
Pat	Hynes	65038	Fri, Nov 26 2021	Install latest version of Windows
Hui	Chen	65014	Wed, Nov 03 2021	Replace power supply
Hui	Chen	65019	Tue, Nov 09 2021	Upgrade to new Windows
Hui	Chen	65026	Wed, Nov 17 2021	Upgrade RAM
Hui	Chen	65032	Tue, Nov 23 2021	Install second storage drive
Hui	Chen	65035	Tue, Nov 23 2021	Office 365 training
Kayla	Blewett	65023	Mon, Nov 15 2021	Upgrade RAM
Kayla	Blewett	65036	Wed, Nov 24 2021	Set up home network
Kayla	Blewett	65041	Fri, Nov 26 2021	Biannual computer maintenance

Figure 3.6 Inner Join Example

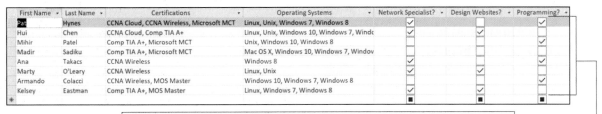

An inner join displays a record from either table only when a matching value in the joined field exists in the other table. No blank records appear in the query results. However, notice that not all the records in the Technicians table display.

Activity 2a Selecting Records in a Query Using an Inner Join
Part 1 of 4

1. With **3-RSRCompServ** open, create a query in Design view to display a list of technicians that notes each individual's skill specialties by completing the following steps:
   a. Create a new query in Design view. At the Show Table dialog box, add the Technicians table and TechSkills table. Close the Show Table dialog box and then drag down the bottom borders of both table field list boxes until all the field names are visible. If necessary, resize the query design grid.
   b. Double-click the join line between the two tables to open the Join Properties dialog box.

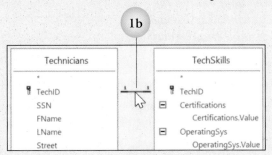

   c. At the Join Properties dialog box, notice that option *1* is selected by default. Option *1* selects records only when the joined fields from both tables are equal. This represents an inner join. Click OK.

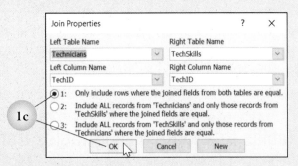

   d. Add the following fields to the query design grid in order: *FName, LName, Certifications, OperatingSys, NetworkSpc, WebDesign, Programming*.
   e. Run the query.
2. Save the query, typing TechSpecialties as the query name, and then click OK.
3. Close the query.

1. With **3-RSRCompServ** open, modify the TechSpecialties query to a left outer join to check whether information for any technicians has not yet been entered in the TechSkills table by completing the following steps:
   a. Right-click the *TechSpecialties* query and then click *Design View* at the shortcut menu.
   b. Right-click the join line between the two tables and then click *Join Properties* at the shortcut menu.

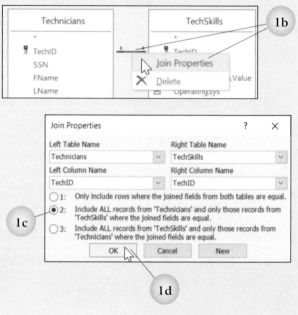

   c. At the Join Properties dialog box, click option *2*. Option *2* includes all the records from the Technicians table and only those records from the TechSkills table for which the joined fields are equal. The left table (Technicians) is the table that will show all the records. If there is not a matching record for a technician in the other table, the columns display empty fields next to the technician's name. *Note: Do not assume that a left join always occurs with the table that is the left table in the query window. Although Technicians is the left table in the query window, the term "left" refers to the table that represents the "one" side (primary table) in the relationship*.
   d. Click OK.
   e. Notice that the join line between the two tables now displays with an arrow pointing to the joined field in the TechSkills table.
   f. Run the query.
2. Minimize the Navigation pane and then compare your results with the query results datasheet displayed in Figure 3.4 (on page 60). Notice that 11 records display in this datasheet, whereas only 8 records display in the query results from Activity 2a.
3. Click the File tab and then click the *Save As* option. At the Save As backstage area, click *Save Object As* in the *File Types* section and then click the Save As button. At the Save As dialog box, type AllTechSkills in the *Save 'TechSpecialties' to* text box and then click OK.
4. Close the query.
5. Expand the Navigation pane.

> Check Your Work

Adding Tables to and Removing Tables from a Query

Show Table

Open a query in Design view to add a table to a query. Click the Show Table button in the Query Setup group on the Query Tools Design tab and then add the desired table using the Show Table dialog box. Close the Show Table dialog box when finished.

To remove a table from a query, click any field within the table field list box to activate the table in the query window and then press the Delete key. The table is removed from the window and all the fields associated with the table that were added to the query design grid are automatically removed. A table can also be removed by right-clicking the table in the query window and then clicking *Remove Table* at the shortcut menu.

Activity 2c Selecting Records in a Query Using a Right Outer Join

Part 3 of 4

1. With **3-RSRCompServ** open, modify an existing query to create a new query to check for work orders that have not been assigned to a technician by completing the following steps:
   a. Open the TechSpecialties query in Design view.
   b. Right-click the *TechSkills* table name in the table field list box in the query window and then click *Remove Table* at the shortcut menu. Notice that the last five columns are removed from the query design grid along with the table.

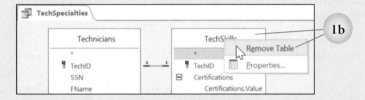

   c. Click the Show Table button in the Query Setup group on the Query Tools Design tab.
   d. At the Show Table dialog box, double-click *WorkOrders*, click the Close button, and then drag down the bottom border of the WorkOrders table field list box until all the fields are visible. If necessary, resize the query design grid.
   e. Double-click the join line between the two tables.
   f. At the Join Properties dialog box, click option *3*. Option *3* includes all the records from the WorkOrders table and only those records from the Technicians table for which the joined fields are equal. The right table (WorkOrders) is the table that will show all the records. If a work order does not have a matching record in the other table, the columns display empty fields for the technician names. ***Note: Do not assume that a right join always occurs with the table that is the right table in the query window. Although WorkOrders is the right table in the query window, the term "right" refers to the table that represents the "many" side (related table) in the relationship***.

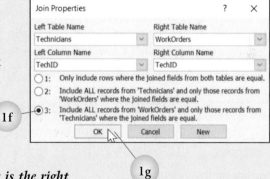

   g. Click OK.
   h. Notice that the join line between the two tables now displays with an arrow pointing to the joined field in the Technicians table.
   i. Add the following fields from the WorkOrders table to the query design grid in this order: *WO*, *WODate*, *Descr*.
2. Click the File tab and then click the *Save As* option. At the Save As backstage area, click *Save Object As* in the *File Types* section and then click the Save As button. At the Save As dialog box, type UnassignedWO in the *Save 'TechSpecialties' to* text box and then click OK.
3. Run the query.

4. Compare your results with the partial query results datasheet shown in Figure 3.5 (on page 60). Notice that the first four records in the query results datasheet have empty fields in the *First Name* column and *Last Name* column.
5. Double-click the right boundary of the *Description* column to adjust the width.
6. Print only the records that have not been assigned to a technician by completing the following steps.
   a. Switch to Design view.
   b. Click in the *FName Criteria* row and then type null.
   c. Run the query.
   d. Print the query results datasheet with the right margin set to 0.5 inch.
7. Close the query. Click Yes to save the changes.

 Check Your Work

 Tutorial

Creating a Self-Join Query and Assigning Aliases

💡 **Hint** In a self-join query, the two fields joined must have the same data type.

Quick Steps

Create Self-Join Query and Assign Aliases
1. Create query in Design view.
2. Add two copies of table to query.
3. Right-click second table name.
4. Click *Properties.*
5. Click in *Alias* property box and delete existing table name.
6. Type alias table name.
7. Close Property Sheet.
8. Drag field name from left table to field name with matching values in right table.
9. Add fields to query design grid.
10. Save and run query.

Creating a Self-Join Query

Assume that a table has two fields that contain similar field values. For example, look at the *Technician ID* column and *Tier 2 Supervisor* column in the Technicians table datasheet, shown in Figure 3.7. Notice that each column contains a technician's ID number. Tier 2 supervisors are senior technicians who are called in when a work order is too complex for a regular technician to solve. The ID number in the *Tier 2 Supervisor* column is the ID of the senior technician who is assigned to the technician.

Viewing the list of technicians with the Tier 2 supervisors' last names may be more informative than viewing the list with the supervisors' ID numbers. If a table has matching values in two separate fields, create a self-join query; this creates a relationship between fields in the same table. To create a self-join query, add two copies of the same table to the query window. The second occurrence of the table is named using the original table name with *_1* added to the end. Assign an alias to the second table to provide it with a more descriptive name in the query. To join the two tables, drag the field with matching values from one table field list to the other. Add the required fields to the query design grid and then run the query.

Creating an Alias for a Table

An alias is an additional name that can be used to reference a table in a query. The alias is temporary and applies only to the query. Generally, the reason for creating an alias is to assign a shorter name to a table (or a more descriptive name, in the case of a self-join query). For example, one of the tables used in the query in Activity 2d is named *Technicians_1*. Assign the table a more descriptive name, such as *Supervisors*, to more accurately describe its role in the query.

To assign an alias to a table, right-click the table name in the query window and then click *Properties* at the shortcut menu to open the Property Sheet task pane. Click in the *Alias* property box, delete the existing table name, and then type the reference name of the table. Access replaces all the occurrences of the table name in the query design grid with the alias.

Figure 3.7 Technicians Table Datasheet with the Fields Used in a Self-Join Query

Technician ID ⋅	SSN ⋅	First Name ⋅	Last Name ⋅	Street Address ⋅	City ⋅	State ⋅	ZIP Code ⋅	Home Phone ⋅	Cell Phone ⋅	🔗	Tier 2 Supervisor ⋅
⊞ 01	000-45-5368	Pat	Hynes	206-31 Woodland Street	Detroit	MI	48202-1138	313-555-6874	313-555-6412	🔗(1)	03
⊞ 02	000-47-3258	Hui	Chen	12905 Hickory Street	Detroit	MI	48205-3462	313-555-7468	313-555-5234	🔗(1)	06
⊞ 03	000-62-7468	Kayla	Blewett	1310 Jarvis Street	Detroit	MI	48220-2011	313-555-3265	313-555-6486	🔗(1)	
⊞ 04	000-33-1485	Mihir	Patel	8213 Elgin Street	Detroit	MI	48234-4092	313-555-7458	313-555-6385	🔗(1)	11
⊞ 05	000-48-7850	Madir	Sadiku	8190 Kenwood Street	Detroit	MI	48220-1132	313-555-6327	313-555-8569	🔗(1)	03
⊞ 06	000-75-8412	Brody	Stewart	3522 Moore Place	Detroit	MI	48208-1032	313-555-7499	313-555-3625	🔗(1)	
⊞ 07	000-55-1248	Ana	Takacs	14902 Hampton Court	Detroit	MI	48215-3616	313-555-6142	313-555-4586	🔗(0)	11
⊞ 08	000-63-1247	Marty	O'Leary	14000 Vernon Drive	Detroit	MI	48237-1320	313-555-9856	313-555-4125	🔗(0)	11
⊞ 09	000-84-1254	Armando	Colacci	17302 Windsor Avenue	Detroit	MI	48224-2257	313-555-9641	313-555-8796	🔗(0)	06
⊞ 10	000-43-5789	Kelsey	Eastman	550 Montclair Street	Detroit	MI	48214-3274	313-555-6315	313-555-7411	🔗(2)	06
⊞ 11	000-65-4185	Dana	Westman	18101 Keeler Streeet	Detroit	MI	48223-1322	313-555-5488	313-555-4158	🔗(0)	
✳										🔗(0)	

These two fields contain technician ID numbers. Tier 2 supervisors are senior-level technicians who are assigned to handle complex cases for regular technicians.

Activity 2d Creating a Self-Join Query

1. With **3-RSRCompServ** open, create a self-join query to display the last name of the Tier 2 supervisor instead of his or her ID number by completing the following steps:
   a. Create a new query in Design view.
   b. At the Show Table dialog box, double-click *Technicians* two times to add two copies of the Technicians table to the query and then close the Show Table dialog box. Notice that the second copy of the table is named *Technicians_1*.
   c. Drag down the bottom borders of both table field list boxes until all the field names are visible. If necessary, resize the query design grid.
   d. Create an alias for the second table by completing the following steps:
      1) Right-click the *Technicians_1* table name and then click *Properties* at the shortcut menu.

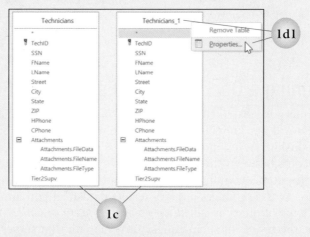

   2) Select the current name in the *Alias* property box in the Property Sheet task pane and then type Supervisors.
   3) Close the Property Sheet task pane.

e. Drag the field named *Tier2Supv* from the Technicians table field list box at the left to the field named *TechID* in the Supervisors table field list box at the right. This creates a join line between the two tables.

f. Add the *FName* and *LName* fields from the Technicians table to the query design grid.

g. Add the *LName* field from the Supervisors table to the query design grid.

2. Run the query. The last names displayed in the second *Last Name* column represent the *Tier2Supv* names.

3. Switch to Design view.

4. Right-click the second *LName* in the *Field* row (from the Supervisors table) in the query design grid and then click *Properties* at the shortcut menu. Click in the *Caption* property box in the Property Sheet task pane, type Tier 2 Supervisor, and then close the Property Sheet task pane.

5. Save the query, typing Tier2Supervisors as the query name, and then click OK.

6. Run the query.

7. Double-click the right boundary of the *Tier 2 Supervisor* column to adjust the width and then compare your results with the datasheet shown at the right.

8. Close the query. Click Yes to save changes.

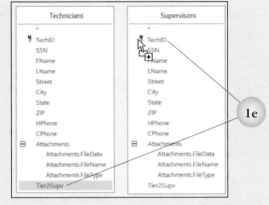

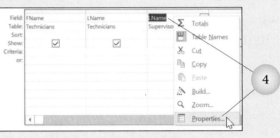

Check Your Work

Running a Query with No Established Relationship

If a query is created from two tables for which no join has been established, Access will not know how to relate the records in the tables. When there is no relationship, Access produces a datasheet representing every possible combination of records between the two tables. For example, if one table contains 20 records, the other table contains 10 records, and no join has been established between the tables, Access produces a query results datasheet containing 200 records (20 × 10). This type of query is called a *cross product query* or *Cartesian product query*. In most cases, the results of such a query provide data that serves no purpose.

If two tables are added to a query and no join line appears, create a join by dragging a field from one table field list box to a compatible field in the other table field list box. The two fields should contain the same data type and be logically related in some way. If no join can logically be established and a many-to-many relationship exists between the tables, add a junction table to the query.

You will use a subquery nested within another query to calculate the total amount earned from each work order. You will also use conditional logic to apply a discount if certain criteria are met.

 Tutorial

Creating and Using Subqueries

 Tutorial

Review: Performing Calcuations in a Query

 Quick Steps

Create Subquery
1. Start new query in Design view.
2. At Show Table dialog box, click Queries tab.
3. Double-click query to be used as subquery.
4. Add other queries or tables as required.
5. Close Show Table dialog box.
6. Add fields as required.
7. Save and run query.

Builder

Creating and Using Subqueries

When performing multiple calculations based on numeric fields, a user may decide to create a separate query for each individual calculation and then use subqueries to generate the final total. The term *subquery* is used to refer to a query nested inside another query. Using subqueries to break calculations into individual objects allows a calculated field to be reused in multiple queries.

For example, assume that the total amount for each work order needs to be calculated. The WorkOrders table contains fields with the number of hours for each service call, the labor rate, and the total value of the parts used. To find the total for each work order, the total labor needs to be calculated. This is done by multiplying the hours times the rate and then adding the parts value to the total labor value. However, the total labor value should be in a separate query so that other calculations can be performed, such as finding the average, maximum, and minimum labor on work orders. To be able to reuse the total labor value, create the calculated field in its own query.

Level 1, Chapter 3 demonstrated how to insert a calculated field in a query. Recall that the format for inserting an equation in a query is to type in a blank *Field* row the new field name followed by a colon and then the equation with the field names in square brackets—for example, *Total:[Sales]+[SalesTax]*. If the equation is typed in the Expression Builder, the square brackets can be omitted from the field names as Access will add them automatically. If an Enter Parameter dialog box displays when the query is run, check the spelling of the field names. Also check to make sure that the correct type of brackets have been used and that there are no extra spaces.

Activity 3a **Creating a Query to Calculate Total Labor** Part 1 of 3

1. With **3-RSRCompServ** open, create a query to calculate the total labor for each work order by completing the following steps:
   a. Create a new query in Design view. At the Show Table dialog box, add the WorkOrders table to the query window and then close the Show Table dialog box.
   b. Drag down the bottom border of the WorkOrders table field list box until all the fields are visible. If necessary, resize the query design grid.
   c. Add the following fields to the query design grid in this order: *WO, ServDate, Hours, Rate*.
   d. Click in the blank *Field* row next to *Rate* in the query design grid and then click the Builder button in the Query Setup group on the Query Tools Design tab.
   e. In the Expression Builder dialog box, type TotalLabor:Hours*Rate.
   f. Click OK.
   g. Expand the width of the calculated column until the entire formula in the *Field* row can be seen.

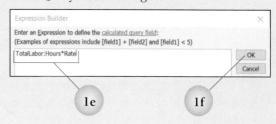

1e   1f

2. Run the query and view the query results. Notice that the *TotalLabor* column does not display a consistent number of decimal values.
3. Switch to Design view.
4. Format the *Total Labor* column by completing the following steps:
   a. Activate the field by clicking in the *TotalLabor* field row.
   b. Click the Property Sheet button in the Show/Hide group on the Query Tools Design tab.
   c. Click in the *Format* property box in the Property Sheet task pane, click the option box arrow that appears, and then click *Standard* at the drop-down list.
   d. Type Total Labor in the *Caption* property box.
   e. Close the Property Sheet task pane.

5. Save the query, typing TotalLabor as the query name, and then click OK.
6. Run the query. Notice that the last four rows contain no values because the service calls have not yet been completed.
7. Switch to Design view. Click in the *Criteria* row in the *Hours* column, type >0, and then press the Enter key.
8. Save the revised query and then run it. Adjust any column widths if necessary.
9. Close the query.

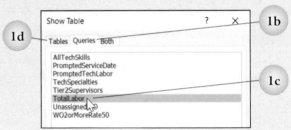

> Check Your Work

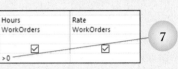

Hint Subqueries are not restricted to use in nested calculations. Use a subquery for any combination of fields that you want to reuse in multiple queries.

Once the query is created to calculate the total labor, nest the query inside another query to add the labor to the parts and then calculate the total for each work order. Creating subqueries provides flexibility in reusing calculations, thus avoiding duplication of effort and reducing the potential for calculation errors.

Small queries and subqueries are useful because they are easier to build and troubleshoot than large queries. Subqueries are also useful when creating a complex query. They can be created to build and test sections individually and then combined into the final larger query.

Activity 3b Creating a Subquery   Part 2 of 3

1. With **3-RSRCompServ** open, create a new query to calculate the total value of each work order using the TotalLabor query as a subquery by completing the following steps:
   a. Create a new query in Design view.
   b. At the Show Table dialog box, click the Queries tab.
   c. Double-click *TotalLabor* in the queries list box.
   d. Click the Tables tab.

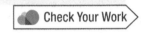

e. Double-click *WorkOrders* and then close the Show Table dialog box. Notice that Access automatically joins the two objects at the *WO* field.

subquery object

Access automatically joins the two objects at the *WO* field.

1e

TotalLabor
WO
ServDate
Hours
Rate
TotalLabor

WorkOrders
WO
CustID
WODate
TechID
ServDate

2. Add fields from the TotalLabor subquery and WorkOrders table by completing the following steps:

a. Double-click the asterisk (*) at the top of the TotalLabor table field list box. Access adds the entry *TotalLabor.** to the first column in the query design grid. This entry adds all the fields from the query. Individual columns do not display in the grid but when the query is run, the datasheet will show all the fields.

TotalLabor
WO
ServDate
Hours
Rate
TotalLabor

2a

b. Run the query. Notice that the query results datasheet shows all five columns from the TotalLabor query.

c. Switch to Design view. Apply the Currency format to the *Total Labor* column in this new query. To do this, add the column to the query design grid as an individual field and not as a group by completing the following steps:

   1) Right-click in the field selector bar (the gray bar above the *Field* row) for the *TotalLabor.** column and then click *Cut* at the shortcut menu to remove the column from the query design grid.

   2) Add the following fields from the TotalLabor query field list box to the query design grid in this order: *WO, ServDate, TotalLabor*.

   3) Format the *TotalLabor* column by applying the Currency format.

d. Drag down the bottom border of the WorkOrders table field list box until all the fields are visible and then double-click *Parts* to add the field to the query design grid. If necessary, resize the query design grid.

3. Create a calculated field to add the total labor and parts by completing the following steps:

a. Click in the blank *Field* row next to *Parts* in the query design grid, type TotalWorkOrder:[TotalLabor]+[Parts], and then press the Enter key.

b. Expand the width of the *TotalWorkOrder* column until the entire formula is visible.

c. Click in the *TotalWorkOrder* column, open the Property Sheet task pane and type Total Work Order in the *Caption* property box.

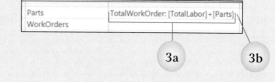

| Parts | TotalWorkOrder: [TotalLabor]+[Parts] |
| WorkOrders | |

3a   3b

e. Close the Property Sheet task pane.

4. Save the query, typing TotalWorkOrders as the query name, and then click OK. Run the query.

5. Double-click the right boundary for the *TotalWorkOrder* column to adjust the width.

TotalWorkOrders				
Work Order ▾	Service Date ▾	Total Labor ▾	Parts ▾	Total Work Order
65012	Mon, Nov 01 2021	$62.50	$10.15	$72.65
65013	Wed, Nov 03 2021	$22.75	$42.75	$65.50
65014	Wed, Nov 03 2021	$87.50	$62.77	$150.27

5

6. Close the query. Click Yes to save changes.

Check Your Work

Creating a Query Using Conditional Logic

Using conditional logic in an Access query is similar to using a logical formula in Excel. Using conditional logic requires Access to perform a calculation based on the outcome of a logical or conditional test. One calculation is performed if the test proves true and another calculation is performed if the test proves false. The structure of the IF function is =IIF(logical_test,value_if_true,value_if_false). Access uses an *Immediate IF* (IIF) function to differentiate between this function and the Visual Basic for Applications (VBA) IF function. VBA is discussed further in Chapter 7. Logical functions are also similar to calculated fields in Access in that the new field name is followed by a colon and field names are enclosed in square brackets.

In Activity 3c, a 2% discount will be processed on the total cost of the work order if the cost of the labor is greater than $80. The following IF function will be used to calculate this discount, where *Labor>80* in the field that will store the discounted total: *Labor>80:IIF([TotalLabor]>80,[TotalWorkOrder]-([TotalWorkOrder]*.02),[TotalWorkOrder])*.

To determine whether a discount should be provided, Access first tests the value of the record in the *TotalLabor* field to see if it is greater than $80. If the condition proves true, Access populates the *Labor>80* field with the value from the *TotalWorkOrder* field minus 2%. This is the discounted total that the customer owes. If the value in the *TotalLabor* field is not greater than $80, the condition proves false. Access populates the *Labor>80* field with the same value that appears in the *TotalWorkOrder* field since the work order did not qualify for a discount.

Figure 3.8 shows the TotalWorkOrders query after the IF function has been applied. For work order 65019, the value in the *TotalLabor* field is $82.50. Since this value is greater than $80, the test is true and Access returns the value_if_true value, or the value in the *TotalWorkOrder* field minus 2% of the value ($82.50 – $1.65 = $80.85). For work order 65012, the value in the *TotalLabor* field is $72.65. Since the test is false, Access returns the value_if_false value, or the value in the *TotalWorkOrder* field ($72.65).

Figure 3.8 Activity 3c TotalWorkOrders Query in Datasheet View

Work Order ▾	Service Date ▾	Total Labor ▾	Parts ▾	Total Work Order ▾	Work Order w/Disc ▾
65012	Mon, Nov 01 2021	$62.50	$10.15	$72.65	$72.65
65013	Wed, Nov 03 2021	$22.75	$42.75	$65.50	$65.50
65014	Wed, Nov 03 2021	$87.50	$62.77	$150.27	$147.26
65015	Fri, Nov 05 2021	$112.50	$0.00	$112.50	$110.25
65016	Fri, Nov 05 2021	$50.00	$48.75	$98.75	$98.75
65017	Tue, Nov 09 2021	$125.00	$55.87	$180.87	$177.25
65018	Thu, Nov 11 2021	$45.00	$0.00	$45.00	$45.00
65019	Thu, Nov 11 2021	$82.50	$0.00	$82.50	$80.85
65020	Mon, Nov 15 2021	$75.00	$72.50	$147.50	$147.50
65021	Mon, Nov 15 2021	$178.75	$0.00	$178.75	$175.18
65022	Tue, Nov 16 2021	$82.50	$400.00	$482.50	$472.85
65023	Mon, Nov 15 2021	$62.50	$100.00	$162.50	$162.50

The caption *Work Order w/Disc* was added to the *Labor>80* field.

Since the *Total Labor* value is less than $80, Access returns the value from the *Total Work Order* field in the *Work Order w/Disc* field.

Since the *Total Labor* value is greater than $80, Access subtracts 2% from the *Total Work Order* field and displays this value in the *Work Order w/Disc* field.

1. With **3-RSRCompServ** open, open the TotalWorkOrders query in Design view.
2. Modify the TotalWorkOrders query to calculate a 2% discount if the total labor cost is greater than $80 by completing the following steps:
   a. Right-click in the blank *Field* row next to *TotalWorkOrder* in the query design grid and then click *Zoom* at the drop-down list.

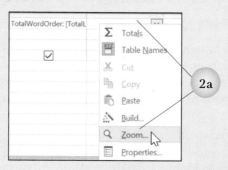

   b. Type Labor>80:IIF([TotalLabor]>80,[TotalWorkOrder]-([TotalWorkOrder]*.02), [TotalWorkOrder]).
   c. Click OK.

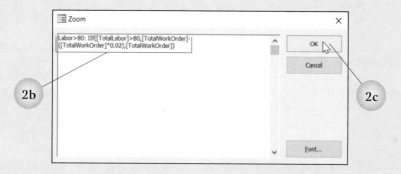

   d. Run the query. If an error message indicating that invalid syntax was entered or an Enter Parameter Value dialog box displays, close the message box and check to ensure that the formula was typed correctly.
3. Switch to Design view, open the Property Sheet task pane for the *Labor>80* column, and then make the following changes:
   a. Apply the Currency format.
   b. Type Work Order w/Disc in the *Caption* property box.
   c. Close the Property Sheet task pane.
4. Run the query.
5. Double-click the right boundary of the *Work Order w/Disc* column to adjust the width. Compare your results with the query results data displayed in Figure 3.8.
6. Print the query results datasheet with the right margin set to 0.5 inch.
7. Close the query. Click Yes to save changes.

 **Check Your Work**

You will select records using a multiple-value field in a query.

Tutorial

Using a Multiple-Value Field in a Query

Quick Steps

Show Multiple-Value Field in Separate Rows in Query
1. Open query in Design view.
2. Click in *Field* row of multiple-value field in design grid.
3. Move insertion point to end of field name.
4. Type period (.).
5. Press Enter key to accept *.Value.*
6. Save query.

Using a Multiple-Value Field in a Query

Recall from Chapter 2 that a multiple-value field displays a lookup list that allows for more than one check box to be checked. In a query, a multiple-value field can display as it does in a table datasheet, with the multiple field values in the same column separated by commas. Or each field value can be shown in a separate row.

To show each value in a separate row, add *.Value* at the end of the multiple-value field name in the *Field* row in the query design grid. Figure 3.9 displays the query design grid for the query used in Activity 4 that displays each entry in the *OperatingSys* field in a separate row in the datasheet.

To select records using criteria in a multiple-value field, type the criteria using the same procedures as for a single-value field. For example, in the TechSpecialties query, typing *Windows 10* in the *Criteria* row in the *OperatingSys* column causes Access to return the records of any technician with Windows 10 certification as one of the multiple field values.

Figure 3.9 Activity 4 Query Design Grid

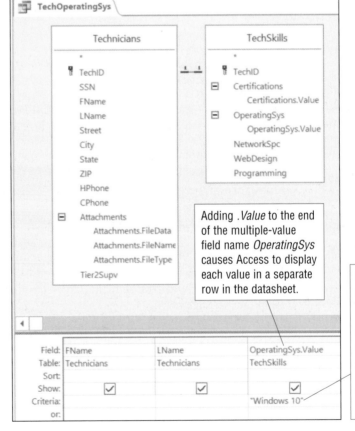

1. With **3-RSRCompServ** open, open the TechSpecialties query in Design view.
2. Right-click in the field selector bar above the *Certifications* field and then click *Cut* at the shortcut menu to remove the field from the query design grid.

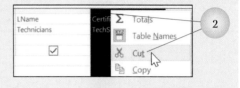

3. Delete the *NetworkSpc*, *WebDesign*, and *Programming* columns from the query design grid. Refer to Step 2.
4. Run the query. Notice that each record in the *Operating Systems* column displays the multiple values separated by commas.
5. Switch to Design view.
6. Click in the *OperatingSys* field row in the query design grid, move the insertion point to the end of the field name, and then type a period (.). Access automatically adds *.Value* to the end of the name in the *Field* row. Press the Enter key to accept *.Value* at the end of the field name. ***Note: When creating a query from scratch, drag the multiple-value field name with the .Value property already attached from the table field list box to the query design grid.***

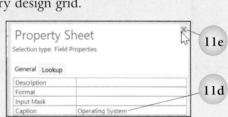

7. Save the query with a new name, typing TechOperatingSys as the query name, and then click OK.
8. Run the query. Notice that each entry in the multiple-value field now displays in a separate row.
9. Switch to Design view.
10. Click in the *Criteria* row in the *OperatingSys.Value* column in the query design grid, type Windows 10, and then press the Enter key. Access adds quotation marks around the text.

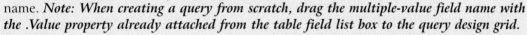

11. Run the query. Notice that the column title for the multiple-value field in the query results datasheet is now *TechSkills.OperatingSys. Value*. Change the column heading for the field by completing the following steps:
   a. Switch to Design view.
   b. Click in the *OperatingSys.Value* field row in the query design grid.
   c. Press the F4 function key to open the Property Sheet task pane.
   d. Click in the *Caption* property box, type Operating System, and then press the Enter key.
   e. Close the Property Sheet task pane.
12. Run the query.
13. Print the query results datasheet and then close the query. Click Yes to save changes.

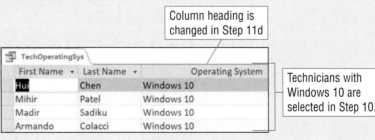

Column heading is changed in Step 11d

Technicians with Windows 10 are selected in Step 10.

Check Your Work

You will create a new table, add and delete records, and update field values using action queries.

Performing Operations Using Action Queries

An action query is used to perform an operation on a group of records. Building an action query is similar to building a select query but with the extra step of specifying the action to perform on the group of selected records. Four types of action queries are available, as described in Table 3.1.

Hint Create a backup copy of the database before running an action query.

To create an action query, first build a select query by adding tables, fields, and criteria to the query design grid. Run the select query to make sure the correct group of records is targeted for action. After verifying the results that the proper records will be modified, change the query type using the Make Table, Append, Update, or Delete button in the Query Type group on the Query Tools Design tab, shown in Figure 3.10. Once the query type has been changed to an action query, clicking the Run button causes Access to perform the make table, append, update, or delete operation. Once an action query has been run, the results cannot be reversed.

When you build queries, Access creates code behind the scenes using Structured Query Language (SQL). View the code by clicking the View Button arrow and then clicking *SQL View*. Three queries, the Union, Pass-Through, and Data Definition queries, are created directly in the SQL view and can be initiated using buttons on the Query Tools Design tab in the Query Type group as shown in Figure 3.10.

Table 3.1 Types of Action Queries

Query Type	Description
make table	A new table is created from selected records in an existing table—for example, a new table that combines fields from two other tables in the database.
append	Selected records are added to the end of an existing table. This action is similar to performing a copy and paste.
update	A global change is made to the selected group of records based on an update expression—for example, increasing the labor rate by 10% in one step.
delete	The selected group of records is deleted from a table.

Figure 3.10 Query Type Group on the Query Tools Design Tab

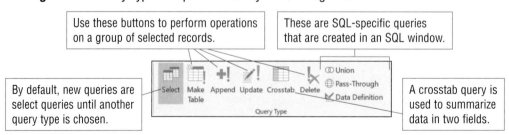

Use these buttons to perform operations on a group of selected records.

These are SQL-specific queries that are created in an SQL window.

By default, new queries are select queries until another query type is chosen.

A crosstab query is used to summarize data in two fields.

Creating a New Table Using a Query

 Make Table

A make table query creates a new table from selected records in the same database or another database. This type of query is useful for creating a history table before purging old records that are no longer required. The history table can be placed in the same database or in another database used as an archive copy.

Once a select query is created that will extract the records to be copied to a new table, click the Make Table button in the Query Type group on the Query Tools Design tab. This opens the Make Table dialog box, shown in Figure 3.11. Enter a table name, choose the destination database, and then click OK. Once a make table query has been run, do not double-click the query name in the Navigation pane; doing so instructs Access to run the query again. Open the query in Design view to make changes to the criteria and/or query type if the query needs to be run again.

Quick Steps

Create Make Table Query
1. Create query in Design view.
2. Add table(s) to query.
3. Add fields to query design grid.
4. If necessary, enter criteria to select records.
5. Run query.
6. Switch to Design view.
7. Click Make Table button.
8. Type table name.
9. If necessary, select destination database.
10. Click OK.
11. Run query.
12. Click Yes.
13. Save query.

Figure 3.11 Make Table Dialog Box

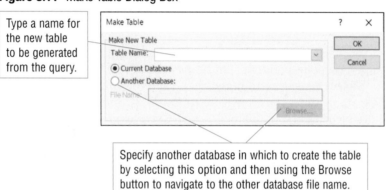

Type a name for the new table to be generated from the query.

Specify another database in which to create the table by selecting this option and then using the Browse button to navigate to the other database file name.

Activity 5a Creating a New Table Using a Query

Part 1 of 4

1. With **3-RSRCompServ** open, create a select query to select all the work order records for November 1, 2021 through November 7, 2021 by completing the following steps:
   a. Create a new query in Design view. Add the WorkOrders table to the query window and then close the Show Table dialog box. Drag down the bottom border of the table field list box until all the field names can be seen. If necessary, resize the query design grid.
   b. Double-click the WorkOrders table field list box title bar. This selects all the fields within the table.
   c. Position the mouse pointer within the selected field names in the table field list box and then drag the pointer to the first column in the query design grid. All the fields in the table are added to the query design grid.
   d. Click in the *Criteria* row in the *ServDate* column, type Between November 1, 2021 and November 7, 2021, and then press the Enter key.

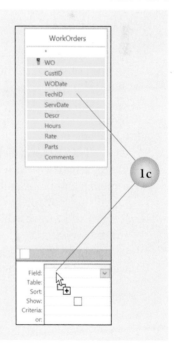

e. Run the query. The query results datasheet displays five records.

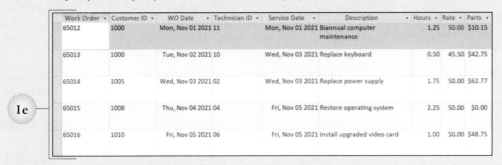

Work Order ▾	Customer ID ▾	WO Date ▾	Technician ID ▾	Service Date ▾	Description ▾	Hours ▾	Rate ▾	Parts ▾
65012	1000	Mon, Nov 01 2021	11	Mon, Nov 01 2021	Biannual computer maintenance	1.25	50.00	$10.15
65013	1000	Tue, Nov 02 2021	10	Wed, Nov 03 2021	Replace keyboard	0.50	45.50	$42.75
65014	1005	Wed, Nov 03 2021	02	Wed, Nov 03 2021	Replace power supply	1.75	50.00	$62.77
65015	1008	Thu, Nov 04 2021	04	Fri, Nov 05 2021	Restore operating system	2.25	50.00	$0.00
65016	1010	Fri, Nov 05 2021	06	Fri, Nov 05 2021	Install upgraded video card	1.00	50.00	$48.75

(callout: 1e)

f. It has been decided not to archive the *Comments* field data. Switch to Design view and then delete the *Comments* column from the query design grid.

2. Make a new table from the selected records and store the table in a history database to be used for archiving purposes by completing the following steps:
   a. If necessary, switch to Design view.
   b. Click the Make Table button in the Query Type group on the Query Tools Design tab.
   c. With the insertion point positioned in the *Table Name* text box, type Nov2021WO.
   d. Click the *Another Database* option and then click the Browse button.
   e. At the Make Table dialog box, navigate to the AL2C3 folder on your storage medium and then double-click the file named **3-RSRCompServHistory**.
   f. Click OK.

(callouts: 2c, 2d, 2e, 2f — Make Table dialog box; Make New Table; Table Name: Nov2021WO; Current Database; Another Database; File Name: E:\AL2C3\3-RSRCompServHistory.accdb; OK; Cancel; Browse....)

   g. Run the query.
   h. Click Yes at the Microsoft Access message box indicating that five rows are about to be pasted to a new table. You will verify that the query has run and created a table in later steps.
3. Save the query, typing Nov2021MakeTable as the query name, and then click OK.
4. Close the query.
5. Close **3-RSRCompServ**.
6. Open **3-RSRCompServHistory** and enable the content. Click OK to continue if a message appears stating that Access has to update object dependencies.
7. Open the Nov2021WO table in Datasheet view.
8. Review the records that were copied to the new table from the make table query that was run in Step 2g.
9. Close the table.
10. Close **3-RSRCompServHistory**.

Check Your Work

Creating a Delete Query

A delete query is used to delete in one step a group of records that meet specified criteria. Use this action query when the records to be deleted can be selected using a criteria statement. Using a query to remove the records is more efficient and reduces the chances of removing a record in error, which can happen when records are deleted manually.

The make table query used in Activity 5a created a duplicate copy of the records in the new table. The original records still exist in the WorkOrders table. The make table query used to archive the records can be changed to a delete query and then used to remove the records from the original table.

In the Navigation pane, an action query name displays with a black exclamation mark next to an icon. The icon indicates the type of action that will be performed when the query is run.

Activity 5b Deleting Records Using a Query

Part 2 of 4

1. Open **3-RSRCompServ** and enable the content if necessary.
2. Right-click *Nov2021MakeTable* in the Queries group in the Navigation pane and then click *Design View* at the shortcut menu. Do not double-click to open/run it or Access will try to make a new table.
3. Click the Delete button in the Query Type group on the Query Tools Design tab to change the query to a Delete query.
4. Save the query with a new name, typing Nov2021Delete as the query name, and then click OK.
5. With the Query Tools Design tab selected, run the query.
6. At the Microsoft Access message box indicating that five rows are about to be deleted from the table and that the action cannot be reversed, click Yes to delete the selected records.

7. Close the query.
8. Open the WorkOrders table. Notice that no records have a service date before November 8, 2021.
9. Close the table.

 Check Your Work

Creating an Append
Query

⊞ Append

Quick Steps

Create Append Query
1. Create query in
   Design view.
2. Add table to query.
3. Add fields to query
   design grid.
4. Enter criteria to
   select records.
5. Run query.
6. Switch to Design
   view.
7. Click Append
   button.
8. Type table name.
9. Select destination
   database.
10. Click OK.
11. Run query.
12. Click Yes.
13. Save query.

Creating an Append Query

An append query is used to copy a group of records from one or more tables to the end of an existing table. Consider using an append query when a duplicate copy of records needs to be made. For example, in Activity 5a, the make table query was used to create a new table to store archived records. Once a table exists, use append queries to copy subsequent archived records to the end of the existing history table.

Click the Append button in the Query Type group on the Query Tools Design tab to open the Append dialog box. This dialog box opens with the same options as the Make Table dialog box, as shown in Figure 3.12. Click the *Table Name* option box arrow to choose from a list of existing tables in the specified database. The receiving table should have the same structure as the query from which the records are selected.

Figure 3.12 Append Dialog Box

Activity 5c Adding Records to a Table Using a Query

Part 3 of 4

1. With **3-RSRCompServ** open, open the Nov2021MakeTable query in Design view.
2. Modify the criteria to select work order records for the second week of November 2021 by completing the following steps:
   a. Expand the width of the *ServDate* column until the entire criteria statement can be seen.
   b. Click in the *Criteria* row in the *ServDate* field, insert and delete text as necessary to modify the criteria statement to read *Between #11/8/2021# And #11/14/2021#*, and then press the Enter key.

3. Click the Append button in the Query Type group on the Query Tools Design tab.
4. Since the query is being changed from a make table query, Access inserts the same table name and database that were used to create the table in Activity 5a. Click OK to accept the table name *Nov2021WO* and the file name *3-RSRCompServHistory*.

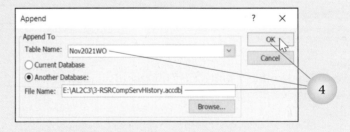

5. Save the query with a new name, typing Nov2021Append as the file name, and then click OK.
6. With the Query Tools Design tab selected, run the query.
7. Click Yes at the Microsoft Access message box indicating that three rows are about to be appended and that the action cannot be undone.
8. Close the query.
9. Close **3-RSRCompServ**.
10. Open **3-RSRCompServHistory**.
11. Open the Nov2021WO table and then print the datasheet in landscape orientation.
12. Close the table.
13. Close **3-RSRCompServHistory**.

 **Check Your Work**

 Tutorial

Creating an Update Query

 Update

Creating an Update Query

When the change needed to a group of records can be selected in a query and is the same for all the records, instruct Access to modify the data using an update query. Making a global change using an update query is efficient and reduces the potential for errors that can occur from manually editing multiple records. Increase or decrease fields such as quotas, rates, and selling prices by adding or subtracting specific amounts or by multiplying by desired percentages.

Clicking the Update button in the Query Type group on the Query Tools Design tab causes an *Update To* row to appear in the query design grid. Click in the *Update To* row in the column to be modified and then type the expression that will change the field values as needed. Run the query to make the global change.

Exercise caution when running an action query because the queries make changes to database tables. For example, in Activity 5d, if the update query is run a second time, the rates for the Plan A service plans will increase another 6%. Once the rates have changed, they cannot be undone. Reversing the update would require creating and running, a mathematical expression in a new update query to remove 6% from the prices.

Quick Steps

Create Update Query
1. Create query in Design view.
2. Add table to query.
3. Add fields to query design grid.
4. Enter criteria to select records.
5. Run query.
6. Switch to Design view.
7. Click Update button.
8. Click in *Update To* row in field to be changed.
9. Type update expression.
10. Run query.
11. Click Yes.
12. Save query.

Activity 5d Changing Service Plan Rates Using an Update Query   Part 4 of 4

1. Open **3-RSRCompServ** and enable the content.
2. Open the FeesSCPlans table and review the current values in the *Rate* column. For example, notice that the current rate for Plan A's six-month term for one computer is $58.00. Close the table after reviewing the current rates.
3. Create an update query to increase the Plan A service contract rates by 4% by completing the following steps:
    a. Create a new query in Design view.
    b. Add the FeesSCPlans table to the query and then close the Show Table dialog box.
    c. Add the *Plan* field and *Rate* field to the query design grid.

d. Click in the *Criteria* row in the *Plan* column, type A, and then click in the *Criteria* row in the next column in the query design grid. ***Note: The AutoComplete feature will show a list of functions as soon as A is typed. Ignore the AutoComplete drop-down list, since a mathematical expression is not being entered for the criteria. However, pressing the Enter key causes Access to add Abs to the Criteria row. Clicking in another box in the query design grid will remove the AutoComplete drop-down list.***

e. Run the query. Review the four records shown in the query results datasheet.

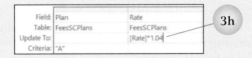

f. Switch to Design view.
g. Click the Update button in the Query Type group on the Query Tools Design tab. Access adds a row labeled *Update To* in the query design grid between the *Table* row and *Criteria* row.
h. Click in the *Update To* row in the *Rate* column, type [Rate]*1.04, and then press the Enter key.

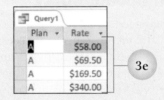

4. Save the query, typing RateUpdate as the query name, and then click OK.
5. Run the query. Click Yes at the Microsoft Access message stating that four rows are about to be updated.
6. Close the query.
7. Open the FeesSCPlans table. Notice that the rate values for the Plan A records have increased. For example, the value in the *Rate* column for Plan A's six-month term for one computer is now $60.32. Print the datasheet.
8. Close the table and then close **3-RSRCompServ**.

Chapter Summary

- Select queries extract records that meet specified criteria from a single table or multiple tables.

- A filter can be saved as a query by displaying the filter criteria in a Filter By Form window, clicking the Advanced button, and then clicking *Save As Query* at the drop-down list.

- Parameter queries prompt the user for the criteria by which to select records when the query is run. To create a parameter query, type a prompt message enclosed in square brackets in the *Criteria* row of the field to be used to select records.

- Changing the join property can alter the number of records that are displayed in the query results datasheet.

- An inner join displays records only if matching values are found in the joined field in both tables.

- A left outer join displays all the records from the left table and matching records from the related table. Empty fields display if no matching records exist in the related table.

- A right outer join displays all the records from the right table and matching records from the primary table. Empty fields display if no matching records exist in the primary table.

- Click the Show Table button in the Query Setup group on the Query Tools Design tab to add a table or query to the query window. Remove a table from a query by clicking a field within the table field list box and then pressing the Delete key or by right-clicking the table and clicking *Remove Table* at the shortcut menu.

- A self-join query is created by adding two copies of the same table to the query window and joining them by dragging the field with matching values from one table field list to the other.

- An alias is another name used to reference a table in a query. To assign an alias to a table, right-click the table name in the query window, click *Properties* at the shortcut menu, and then enter the alias for the table in the *Alias* property box in the Property Sheet task pane.

- A query that contains two tables that are not joined creates a cross product or Cartesian product query. This means that Access creates records for every possible combination from both tables—the results of which are generally not meaningful.

- A subquery is a query nested inside another query. Use subqueries to break down a complex query into manageable units. For example, a query with multiple calculations can be created by combining subqueries in which individual calculations are built individually. Another reason for using subqueries is to be able to reuse a smaller query in many other queries, avoiding the need to keep recreating the same structure.

- Using conditional logic requires Access to perform a calculation based on the outcome of a logical or conditional test.

- Create select queries on multiple-value fields using the same methods used for single-field criteria.

- Adding .*Value* at the end of a multiple-value field name in the *Field* row in the query design grid causes Access to place each field value in a separate row in the query results datasheet.

- A make table query creates a new table from selected records in the same database or another database using the structure defined in the query design grid.
- Delete a group of records in one step by creating and running a delete query. Add a group of records to the end of an existing table in the active database or another database by using an append query.
- An update query allows making a global change to records by entering an expression such as a mathematical formula in the query design grid.

Commands Review

FEATURE	RIBBON TAB, GROUP	BUTTON
advanced filter options	Home, Sort & Filter	
append query	Query Tools Design, Query Type	
create query in Design view	Create, Queries	
delete query	Query Tools Design, Query Type	
make table query	Query Tools Design, Query Type	
run query	Query Tools Design, Results	
show table	Query Tools Design, Query Setup	
update query	Query Tools Design, Query Type	

Microsoft®

Access®

Creating and Using Custom Forms

Performance Objectives

Upon successful completion of Chapter 4, you will be able to:

1 Create a custom form using Design view

2 Add fields to a form individually and as a group

3 Move, size, and format control objects

4 Change the tab order of fields

5 Add a tab control to a form and insert a subform

6 Add and format a calculation to a custom form

7 Adjust the alignment, sizing, and spacing of control objects

8 Add graphics to a form

9 Anchor a control object to a position in a form

10 Create a datasheet form and restrict form actions

11 Create a blank form

12 Add a list box and a combo box to a form

13 Locate a record using a wildcard character

A form provides an interface for data entry and maintenance that allows users to work more efficiently with data stored in the underlying tables. A form can also include fields from multiple tables, which allows data to be entered in one object and then used to update several tables. Generally, database designers provide forms for users to perform data maintenance and restrict access to tables, protecting the structure and integrity of the database. In this chapter, you will learn how to build custom forms.

Data Files

Before beginning chapter work, copy the AL2C4 folder to your storage medium and then make AL2C4 the active folder.

The online course includes additional training and assessment resources.

Tutorial

Creating Forms
Using Design View

Form Design

Quick Steps

**Start New Form Using
Design View**
1. Click Create tab.
2. Click Form Design
   button.

Creating Forms Using Design View

Access provides several tools that allow the user to build a form quickly, such as the Form tool, Split Form tool, and Form Wizard. A form generated using one of these tools can be modified in Layout view or Design view to customize the content, format, or layout. If several custom options are required in a form, begin in Design view and build the form from scratch. Click the Create tab and then click the Form Design button in the Forms group to begin a new form using the Design view window, shown in Figure 4.1.

In Design view, the form displays the *Detail* section, which is used to display fields from the table associated with the form. Objects are added to the form using buttons in the Controls, Header/Footer, and Tools groups on the Form Design Tools Design tab, shown in Figure 4.2.

Figure 4.1 Form Design View

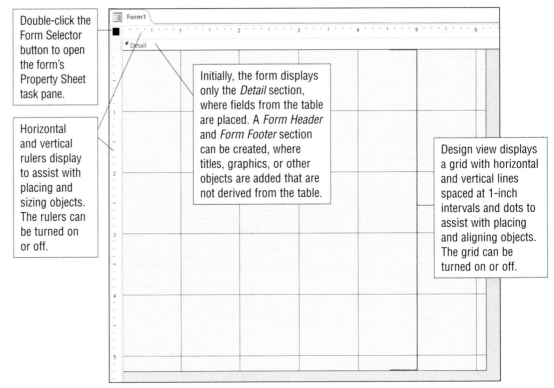

Double-click the Form Selector button to open the form's Property Sheet task pane.

Horizontal and vertical rulers display to assist with placing and sizing objects. The rulers can be turned on or off.

Initially, the form displays only the *Detail* section, where fields from the table are placed. A *Form Header* and *Form Footer* section can be created, where titles, graphics, or other objects are added that are not derived from the table.

Design view displays a grid with horizontal and vertical lines spaced at 1-inch intervals and dots to assist with placing and aligning objects. The grid can be turned on or off.

Figure 4.2 Controls, Header/Footer, and Tools Groups on the Form Design Tools Design Tab

Understanding Bound, Unbound, and Calculated Control Objects

Three types of control objects can be created in a form. A control object in a form may be bound, unbound, or calculated. A bound control object draws and displays data in the control object from the field in the table associated with the control object. In other words, the content that is displayed in the control object in Form view is drawn from a field in a record in a table. An unbound control object is used to display text or graphics and does not rely on a table for its content. For example, a control object that contains an image to enhance the visual appearance of a form or a control object that contains the hours of business for informational purposes are both unbound control objects. A calculated control object displays the result of a mathematical formula. Totals and percentages are examples of calculated control objects.

Creating Titles and Label Control Objects

 Title

Aa Label

Quick Steps

Add Form Title
1. Open form in Design view.
2. Click Title button.
3. Type title text.
4. Press Enter key.

Add Label Control Object
1. Open form in Design view.
2. Click Label button.
3. Drag to create control object of specified height and width.
4. Type label text.
5. Press Enter key.

Click the Title button in the Header/Footer group on the Form Design Tools Design tab to display the *Form Header* and *Form Footer* sections and automatically insert a label control object with the name of the form inside the *Form Header* section. Select the text inside the title control object to type new text, delete existing text, or otherwise modify the default title text. Click the Label button in the Controls group to draw a label control object within any section in the form and then type descriptive or explanatory text inside the control object.

After creating a title or label control object, format the text using buttons in the Font group on the Form Design Tools Format tab. Move and resize the control object to reposition it on the form as needed.

Use the *Form Header* section to create control objects that are to be displayed at the top of the form when scrolling through records in Form view. This section is printed at the top of the page when a record or group of records is printed from Form view. Titles and company logos are generally placed in the *Form Header* section. Use the *Form Footer* section to create control objects that are to be displayed at the bottom of the form when scrolling through records in Form view. This section is printed at the end of a printout when a record or group of records is printed from Form view. Consider adding a creation date and/or revision number in the *Form Footer* section.

1. Open **4-RSRCompServ** and enable the content.
2. Click the Create tab and then click the Form Design button in the Forms group.

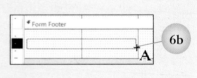

3. Add a title in the *Form Header* section of the form and then center the text within the control object by completing the following steps:
   a. With the Form Design Tools Design tab active, click the Title button in the Header/Footer group. Access displays the *Form Header* section above the *Detail* section and inserts a title control object with the text *Form1* selected.

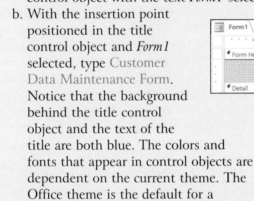

   b. With the insertion point positioned in the title control object and *Form1* selected, type Customer Data Maintenance Form. Notice that the background behind the title control object and the text of the title are both blue. The colors and fonts that appear in control objects are dependent on the current theme. The Office theme is the default for a database.

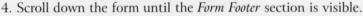

   c. Press the Enter key.
   d. With the title control object selected (as indicated by the orange border around the title text), click the Form Design Tools Format tab and then click the Center button in the Font group.
4. Scroll down the form until the *Form Footer* section is visible.
5. Position the mouse pointer at the bottom border of the form's grid until the pointer changes to an up-and-down-pointing arrow with a horizontal line in the middle and then drag down the bottom of the form to the 0.5-inch position on the vertical ruler.

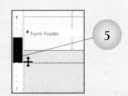

6. Add one label control object at the left edge of the form footer that contains a revision number and another at the right edge of the form footer that contains your name by completing the following steps:
   a. Click the Form Design Tools Design tab and then click the Label button in the Controls group (third button from the left).
   b. Position the crosshairs with the label icon attached at the left side of the *Form Footer* section and then drag to draw a label control object of the approximate height and width shown at the right. When the mouse is released, the insertion point appears inside the label control object.

   c. Type Revision number 1.0 and then press the Enter key.

d. Create another label control object at the right side of the *Form Footer* section that is similar in height and width to the one shown below. Type your first and last names inside the label control object and then press the Enter key. Refer to Steps 6a–6c for help with this step.

e. Click in a blank area of the form to deselect the label control object.

7. Click the Save button on the Quick Access Toolbar, type CustMaintenance in the *Form Name* text box at the Save As dialog box, and then click OK.

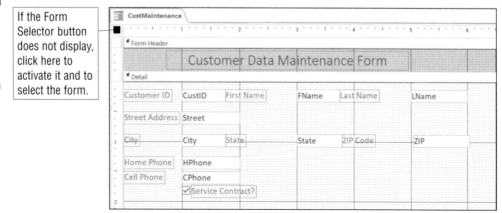

Check Your Work

Adding Fields to a Form

 Tutorial

Adding Fields to a Form

 Form Selector

 Add Existing Fields

Quick Steps

Connect Table to Form
1. Open form in Design view.
2. Double-click Form Selector button.
3. Click Data tab in Property Sheet task pane.
4. Click arrow in *Record Source* property box.
5. Click table.
6. Close Property Sheet task pane.

Add Fields to Form
1. Click Add Existing Fields button.
2. Drag field name from Field List task pane to *Detail* section.
OR
1. Click first field name in Field List task pane.
2. Press and hold down Shift key.
3. Click last field name in Field List task pane.
4. Release Shift key.
5. Drag selected fields from Field List task pane to *Detail* section.

Before fields can be added to the *Detail* section, a table must be connected to the form. Specify the table that will be connected with the form using the *Record Source* property box on the Data tab of the form's Property Sheet task pane. To do this, press the F4 function key or double-click the Form Selector button above the vertical ruler and left of the horizontal ruler; this opens the Property Sheet task pane at the right side of the work area. Click the Data tab, click the option box arrow in the *Record Source* property box, click the table at the drop-down list, and then close the Property Sheet task pane.

To add fields to the form, click the Add Existing Fields button in the Tools group on the Form Design Tools Design tab to open the Field List task pane. Select and drag individual or groups of field names from the Field List task pane to the *Detail* section of the form. Release the mouse button when the mouse pointer is near the location in the *Detail* section where the data will display.

For each field added to the form, Access inserts two control objects. A label control object is placed at the left of where the mouse button is released and a text box control object is placed at the right. The label control object contains the field name or a caption, if a caption was entered in the *Caption* property box. The text box control object is bound to the field and displays table data from the record in Form view. Figure 4.3 displays the fields from the Customers table that will be added to the CustMaintenance form in Activity 1b.

Figure 4.3 Fields from the Customers Table Added to the CustMaintenance Form in Activity 1b

If the Form Selector button does not display, click here to activate it and to select the form.

1. With **4-RSRCompServ** open and the CustMaintenance form open in Design view, scroll to the top of the form in the work area.
2. Connect the Customers table to the form by completing the following steps:
    a. Double-click the Form Selector button (which displays as a black square) at the top of the vertical ruler and left of the horizontal ruler to open the form's Property Sheet task pane. *Note: If the **Form Selector button does not display, click at the top of the vertical ruler and left of the horizontal ruler to display it.***

    b. Click the Data tab if necessary in the Property Sheet task pane.
    c. Click the option box arrow in the *Record Source* property box and then click *Customers* at the drop-down list.
    d. Close the Property Sheet task pane.
3. Add fields individually from the Customers table to the *Detail* section of the form by completing the following steps:
    a. Click the Add Existing Fields button in the Tools group on the Form Design Tools Design tab. The Field List task pane opens at the right side of the work area.
    b. Position the mouse pointer on the right border of the form's grid until the pointer changes to a left-and-right-pointing arrow with a vertical line in the middle and then drag the right edge of the form to the 6.5-inch position on the horizontal ruler.

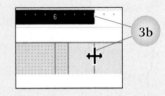

    c. If necessary, click *CustID* in the Field List task pane to select the field and then drag the field name to the *Detail* section. Release the mouse button when the pointer is near the top of the section at the 1-inch position on the horizontal ruler.

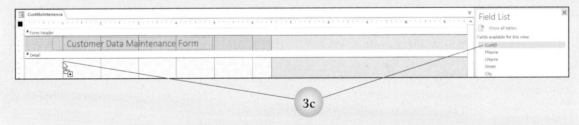

    d. Click to select *FName* in the Field List task pane and then drag the field to the same vertical position as *CustID* in the *Detail* section. Release the mouse button when the pointer is at the 3-inch position on the horizontal ruler. Some of the control objects may overlap. They will be adjusted in the following steps.
    e. Click to select *LName* in the Field List task pane and then drag the field to the same vertical position as *CustID* in the *Detail* section. Release the mouse button when the pointer is at the 5-inch position on the horizontal ruler.

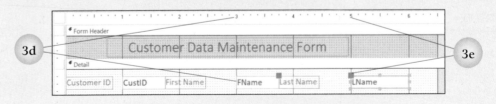

f. Drag the *Street* field from the Field List task pane to the *Detail* section below *CustID*. Release the mouse button when the pointer is at the 1-inch position on the horizontal ruler and approximately three rows of grid dots below *CustID*.

g. Drag the *City* field from the Field List task pane to the *Detail* section below *Street*. Release the mouse button when the pointer is at the 1-inch position on the horizontal ruler and approximately three rows of grid dots below *Street*.

h. Drag the *State* field from the Field List task pane to the *Detail* section at the same horizontal position as *City*. Release the mouse button when the pointer is at the 3-inch position on the horizontal ruler.

i. Drag the *ZIP* field from the Field List task pane to the *Detail* section at the same horizontal position as *City*. Release the mouse button when the pointer is at the 5-inch position on the horizontal ruler.

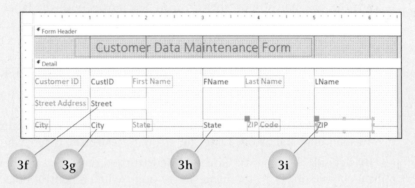

4. Add a group of fields from the Customers table to the *Detail* section of the form by completing the following steps:

a. Click the *HPhone* field name in the Field List task pane.

b. Press and hold down the Shift key, click the *ServCont* field name in the Field List task pane, and then release the Shift key. When the Shift key is held down, Access selects all the fields from the first field name clicked to the last field name clicked.

c. Position the mouse pointer within the selected group of fields in the Field List task pane and then drag the group to the *Detail* section below *City*. Release the mouse button when the pointer is at the 1-inch position on the horizontal ruler and approximately three rows of grid dots below *City*.

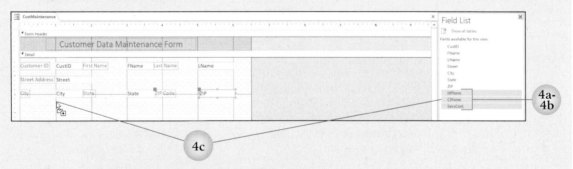

5. Click in any blank area of the form to deselect the group of fields.
6. Compare your form with the one shown in Figure 4.3 (on page 87).
7. Click the Save button on the Quick Access Toolbar.
8. Close the Field List task pane.

Tutorial

Moving, Resizing,
and Formatting
Control Objects

Quick Steps

**Move Control Objects
in Design View**
1. Select control object.
2. Drag using orange
   border or move
   handle to new
   location.

**Resize Control Objects
in Design View**
1. Select control object.
2. Drag middle top,
   bottom, left, or
   right sizing handle
   to resize height or
   width.
OR
1. Select control object.
2. Drag corner sizing
   handle to resize
   height and width at
   the same time.

**Toggle Snap to Grid
On or Off**
1. Click Form Design
   Tools Arrange tab.
2. Click Size/Space
   button.
3. Click *Snap to Grid*.

 Size/Space

💡 **Hint** Do not be
overly concerned with
exact placement and
alignment when you
begin to add fields
to a form. Use the
alignment and spacing
tools in the Form
Design Tools Arrange
tab to assist with
layout.

Moving and Resizing Control Objects

Once fields in a form have been placed, control objects can be moved or resized to change the layout. As shown in Activity 1b, Access places two control objects for each field in the form. A label control object, which contains the caption or field name, is placed left of a text box control object, which displays the field value from the record (or a blank entry box when a new record is added). Click the label control object or text box control object for the field to be moved or resized and the control object is surrounded by an orange border with eight handles. Access displays a large dark-gray square (called the *move handle*) at the top left of the label control object or text box control object for the selected field.

Point to the orange border around the selected control object until the mouse pointer displays with the four-headed arrow move icon attached and then drag the field to the new position on the form. Access moves both the label control and text box control objects to the new location. If a label control or text box control object is to be moved independently of its connected control object, point to the large dark-gray move handle at the top left of either control object and use it to drag the control object to the new position, as shown in Figure 4.4.

To resize a selected control object, point to one of the sizing handles (the small orange squares) on the border of the selected control object until the pointer displays with an up-and-down-pointing arrow, a left-and-right-pointing arrow, or a two-headed diagonal arrow. Drag the arrow to resize the height and/or width.

By default, the Snap to Grid feature is turned on in Design view. This feature pulls a control object to the nearest grid point when a control object is moved or resized. To move or resize a control object in small increments, turn off this feature. To do this, click the Form Design Tools Arrange tab, click the Size/Space button in the Sizing & Ordering group, and then click *Snap to Grid* at the drop-down list. Snap to Grid is a toggle feature, which means it is turned on or off by clicking the button.

Figure 4.4 Moving Control Objects

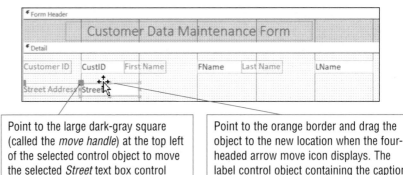

Point to the large dark-gray square (called the *move handle*) at the top left of the selected control object to move the selected *Street* text box control object independently of the *Street Address* label control object.

Point to the orange border and drag the object to the new location when the four-headed arrow move icon displays. The label control object containing the caption *Street Address* will move with the selected *Street* text box control object.

1. With **4-RSRCompServ** open and the CustMaintenance form open in Design view, preview the form to determine what control objects need to be moved or resized by clicking the View button in the Views group on the Form Design Tools Design tab. (Make sure to click the button and not the button arrow.)
2. The form is displayed in Form view and data from the first record is displayed in the text box control objects. Notice that some label control objects overlap text box control objects and that the street address in the first record is not entirely displayed.
3. Click the Design View button in the view area at the right side of the Status bar.
4. Move the control objects that overlap other control objects by completing the following steps:
    a. Click the *First Name* label control object.
    b. Point to the large dark-gray move handle at the top left of the selected label control object until the pointer displays with the four-headed arrow move icon attached and then drag right to the 2-inch position on the horizontal ruler. Notice that the connected *FName* text box control object does not move because the move handle is being used while dragging.

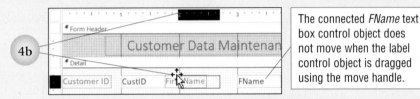

The connected *FName* text box control object does not move when the label control object is dragged using the move handle.

    c. Click the *Last Name* label control object and then use the move handle to drag right to the 4-inch position on the horizontal ruler.
    d. Move the *State* label control object right to the 2-inch position on the horizontal ruler.
    e. Move the *ZIP Code* label control object right to the 4-inch position on the horizontal ruler.

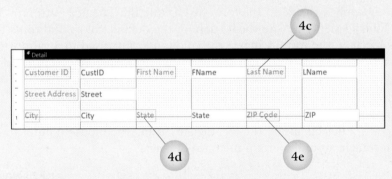

    f. Click in a blank area to deselect the *ZIP Code* label control.
5. Click the *Street* text box control object and then drag the right middle sizing handle right to the 3-inch position on the horizontal ruler.

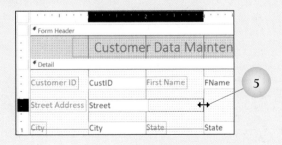

6. Click the *State* text box control object and then drag the right middle sizing handle left to the 3.5-inch position on the horizontal ruler. (Since the *State* field displays only two characters, this control object can be made smaller.)

7. Resize the *CustID* text box control object so the right edge of the object is at the 1.5-inch position on the horizontal ruler.

8. Click the *Service Contract?* label control object. Point to the selected control object's orange border (not on a sizing handle) and then drag the control object until the left edge is at the 3-inch position on the horizontal ruler, adjacent to the *HPhone* field. Notice that both the label control object and text box control object moved because the border was dragged rather than changed with the move handle.

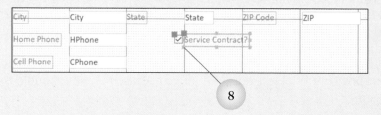

9. Deselect the *Service Contract?* control object.

10. Select the Cell Phone label control object and press the down arrow 3 times.

11. Save the form.

Formatting Control Objects

Quick Steps

Format Multiple Control Objects Using Shift Key
1. In Design view, click to select first control object.
2. Shift + click remaining control objects.
3. Click formatting options.
4. Deselect controls.

Format Multiple Control Objects Using Selection Rectangle
1. In Design view, position pointer above top left control to be formatted.
2. Drag down and right to draw rectangle around controls.
3. Release mouse.
4. Click formatting options.
5. Deselect controls.

 Themes

Use the buttons in the Font group on the Form Design Tools Format tab to change the font, font size, font color, background color, or alignment and to apply bold, italic, or underline formatting to the selected control object.

Apply a theme to the form by clicking the Themes button on the Form Design Tools Design tab. The theme defines the default colors and fonts for control objects in the form. Change the color and fonts used in a form with the Colors and Fonts buttons in the Themes group. A theme applied to one form is applied to all the forms and reports in the database.

Format multiple control objects at the same time by pressing and holding down the Shift key while clicking individual control objects. Select multiple control objects inside the rectangle by using the mouse pointer to draw a selection rectangle around a group of control objects.

1. With **4-RSRCompServ** open and the CustMaintenance form open in Design view, click the View button to preview the form in Form view.
2. Scroll through a few records in the form and then switch back to Design view.
3. Click the Design tab, click the Themes button in the Themes group and then click the Facet Theme.
4. Format multiple control objects using a selection rectangle by completing the following steps:
   a. Position the mouse pointer in the top left corner of the *Detail* section (above the *Customer ID* label control object), drag down and to the right until a rectangle has been drawn around all the control objects in the section, and then release the mouse button.

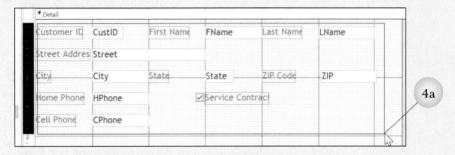

   b. Notice that all the control objects contained within the rectangle are selected.

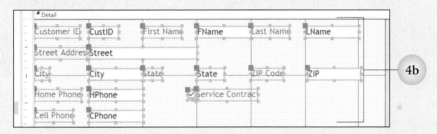

   c. Use the Font button in the Font group on the Form Design Tools Format tab to change the font to Candara.
   d. Use the Font Size button to change the font size to 10 points.
   e. Click in a blank area to deselect the control objects.
5. Apply formatting to multiple controls at once by completing the following steps:
   a. Click the *CustID* text box control object.
   b. Press and hold down the Shift key, click each of the other text box control objects in the *Detail* section, and then release the Shift key.

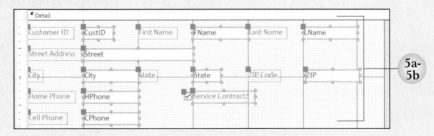

c. Click the Font Color button arrow in the Font group on the Form Design Tools Format tab and then click *Dark Red* at the color palette (first option in last row in the *Standard Colors* section).

d. Click the Bold button.

e. Click in a blank area to deselect the controls.

6. Click the Form Design Tools Design tab and then switch to Form view to view the formatting changes applied to the form. The zip code data remains red and does not change to dark red because of the formatting changes applied to the Format property in Chapter 1.

7. Switch to Design view and then save the form.

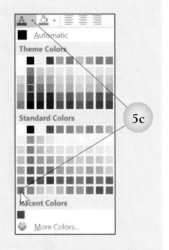

5c

Check Your Work

Tutorial

Changing the Tab Order of Fields

Tab Order

Quick Steps

Change Tab Order of Fields
1. Open form in Design view.
2. Click Tab Order button.
3. Click in field selector bar next to field name.
4. Drag field to new location.
5. Repeat Steps 3–4 as required.
6. Click OK.

Changing the Tab Order of Fields

The term *tab order* refers to the order that fields are selected when the Tab key is pressed while entering data in Form view. Data does not need to be entered into a record in the order the fields are presented. Click the Tab Order button in the Tools group on the Form Design Tools Design tab to open the Tab Order dialog box, shown in Figure 4.5.

The order of the fields in the *Custom Order* list box in the Tab Order dialog box is the order in which the fields will be selected when the Tab key is pressed in a record in Form view. To change the tab order of the fields, position the pointer in the field selector bar next to the field name to be moved until the pointer displays as a black right-pointing arrow and then click to select the field. Drag the selected field up or down to the appropriate position. Click OK to accept the changes and close the Tab Order dialog box. To quickly set the tab order as left-to-right and top-to-bottom, click the Auto Order button at the bottom of the Tab Order dialog box.

Figure 4.5 Tab Order Dialog Box

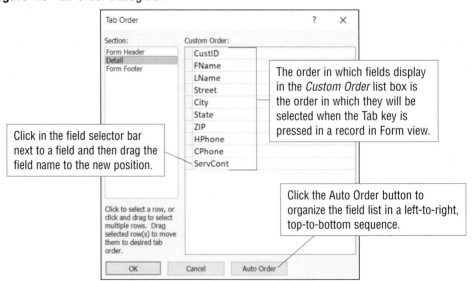

Click in the field selector bar next to a field and then drag the field name to the new position.

The order in which fields display in the *Custom Order* list box is the order in which they will be selected when the Tab key is pressed in a record in Form view.

Click the Auto Order button to organize the field list in a left-to-right, top-to-bottom sequence.

1. With **4-RSRCompServ** open and the CustMaintenance form open in Design view, click the View button to display the form in Form view.
2. With the insertion point positioned in the *CustID* field in the first record in the table, press the Tab key seven times. As the Tab key is pressed, notice that the fields are selected in a left-to-right, top-to-bottom sequence.
3. With the insertion point positioned in the *HPhone* field, press the Tab key. Notice that the selected field moves down to the *CPhone* field instead of right to the *ServCont* field.
4. With the insertion point in the *CPhone* field, press the Tab key to move to the *ServCont* field.
5. Switch to Design view.
6. Change the tab order of the fields so the *ServCont* field is selected after the *HPhone* field by completing the following steps:
   a. Click the Tab Order button in the Tools group on the Form Design Tools Design tab.
   b. At the Tab Order dialog box, hover the mouse pointer over the field selector bar next to *ServCont* until the pointer displays as a black right-pointing arrow.
   c. Click to select the field.
   d. With the pointer now displayed as a white arrow in the field selector bar, drag up *ServCont* until the black horizontal line indicating where the field will be moved is positioned between *HPhone* and *CPhone* in the *Custom Order* list and then release the mouse button.
   e. Click OK. Note that since the tabs follow the left-to-right, top-to-bottom sequence, this tab order could have been created by clicking the Auto Order button in the Tab Order dialog box.
7. Switch to Form view.
8. Press the Tab key nine times to move through the fields in the first record. Notice that when the *HPhone* field is reached and the Tab key is pressed, the *ServCont* field becomes active instead of the *CPhone* field.
9. Switch to Design view.
10. Save the form.

Adding a Tab Control Object to a Form

Adding a Tab Control Object to a Form

A tab control object is an object used to add pages to a form. Each page displays with a tab at the top. When viewing the form, click the page tab to display the contents of the page within the tab control object. Add a tab control object to a form to organize fields in a large table into smaller, related groups or to insert multiple subforms that display on separate pages within the tab control object.

Examine the tab control object that will be created in Activities 1f and 1g, as shown in Figure 4.6. The tab control object contains three pages. The tabs across the top display the captions assigned to the individual pages.

In Activities 1f and 1g, a subform will be created on each page within the tab control object to display fields from a related table. When completed, the CustMaintenance form will be used to enter or view customer-related data that includes fields from four tables.

Quick Steps

Add Tab Control Object to Form

1. Open form in Design view.
2. Click Tab Control button.
3. Position crosshairs in *Detail* section at specified location.
4. Drag down and right to draw object.
5. Release mouse.

Change Page Caption

1. Click tab in tab control object.
2. Click Property Sheet button.
3. Click in *Caption* property box.
4. Type text.
5. Close Property Sheet task pane.

Add Page to Tab Control

1. Right-click existing tab in tab control object.
2. Click *Insert Page*.

Delete Page from Tab Control

1. Right-click tab of page to delete.
2. Click *Delete Page*.

 Tab Control

Figure 4.6 Tab Control Object with Three Pages Created in Activities 1f and 1g

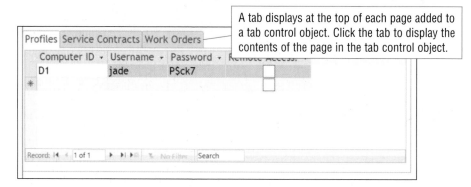

A tab displays at the top of each page added to a tab control object. Click the tab to display the contents of the page in the tab control object.

Figure 4.7 New Tab Control Object with Two Pages

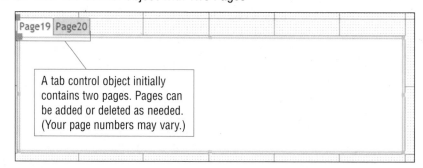

A tab control object initially contains two pages. Pages can be added or deleted as needed. (Your page numbers may vary.)

To add a tab control object to a form, click the Tab Control button in the Controls group on the Form Design Tools Design tab. Position the crosshairs with the tab control icon attached in the top left of the area of the *Detail* section where the tab control is to begin and then drag down and to the right to draw the control. When the mouse button is released, the tab control object initially displays with two pages, as shown in Figure 4.7.

Change the text displayed in the tab at the top of the page by editing the text in the *Caption* property box in the page's Property Sheet task pane. Add fields or create subforms on each page as needed. Add a page to the tab control object by right-clicking an existing tab in the tab control object and then clicking *Insert Page* at the shortcut menu. Remove a page from the tab control object by right-clicking the tab to be deleted and then clicking *Delete Page* at the shortcut menu.

 Tutorial

Adding a Subform

Adding a Subform

Use the Subform/Subreport button in the Controls group on the Form Design Tools Design tab to add a subform to a form. Create a subform to display fields from another related table within the existing form. The form that a subform is created in is called the *main form*.

Adding a related table as a subform creates a control object within the main form that can be moved, formatted, and resized independently of other objects. The subform displays as a datasheet within the main form in Form view. Data can be entered or updated in the subform while the main form is being viewed. Before clicking the Subform/Subreport button, make sure the Use Control Wizards button in the Controls group is toggled on so the subform can be added using the Subform Wizard; the first Subform Wizard dialog box is shown in

 Subform/ Subreport

 Use Control Wizards

Figure 4.8. The Use Control Wizards button displays with a gray background when the feature is active.

A subform is stored as a separate object outside the main form. An additional form name (with *subform* at the end of it) will appear in the Navigation pane when the Subform Wizard is finished. Do not delete a subform object from the Navigation pane. If the subform object is deleted, the main form will no longer display the fields from the related table in the tab control object.

Quick Steps

Create Subform
1. Click page tab in tab control object.
2. Make sure Use Control Wizards is toggled on.
3. Click More button in Controls group.
4. Click Subform/ Subreport button.
5. Click crosshairs inside selected page.
6. Click Next.
7. Choose table and fields.
8. Click Next.
9. Click Next.
10. Click Finish.
11. Delete subform label control object.
12. Move and resize subform object as required.

Figure 4.8 First Subform Wizard Dialog Box

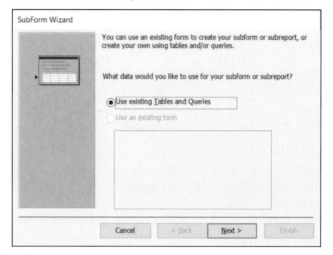

Activity 1f Adding a Tab Control Object and a Subform

Part 6 of 7

1. With **4-RSRCompServ** open and the CustMaintenance form open in Design view, add a tab control object to the form by completing the following steps:
   a. Click the Tab Control button in the Controls group on the Form Design Tools Design tab. The Tab Control button is the fifth button from the left in the Controls group.
   b. Position the crosshairs with the tab control icon attached at the left edge of the grid in the *Detail* section at the 2-inch position on the vertical ruler. Drag down to the 4-inch position on the vertical ruler and right to the 6-inch position on the horizontal ruler and then release the mouse button.

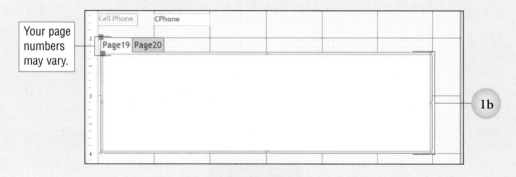

2. Change the page caption and add a subform to the first page within the tab control object by completing the following steps:
   a. Click the first tab in the tab control object that displays *Pagexx*, where *xx* is the page number to select the page. (For example, click *Page19* in the image shown above.)
   b. Click the Property Sheet button in the Tools group.
   c. Click the Format tab in the Property Sheet task pane, click in the *Caption* property box, type Profiles, and then close the Property Sheet task pane. The tab displays the caption text in place of *Pagexx*.

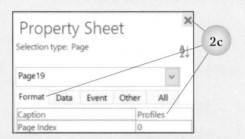

   d. By default, the Use Control Wizards feature is toggled on in the Controls group. Click the More button in the Controls group to expand the Controls group and view two rows of buttons and the Controls drop-down list. View the current status of the Use Control Wizards button. If the button displays with a gray background, the feature is active. If the button is gray, click in a blank area to remove the expanded Controls list. If the feature is not active (displays with a white background), click the Use Control Wizards button to turn it on.
   e. Click the More button in the Controls group and then click the Subform/Subreport button at the expanded Controls list.

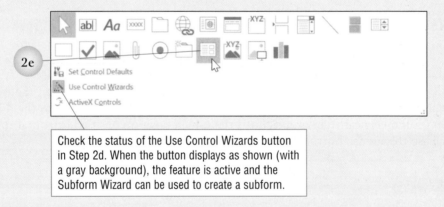

Check the status of the Use Control Wizards button in Step 2d. When the button displays as shown (with a gray background), the feature is active and the Subform Wizard can be used to create a subform.

   f. Move the crosshairs with the subform icon attached to the Profiles page in the tab control object. The background of the page turns black. Click the mouse to start the SubForm Wizard.

   g. At the first SubForm Wizard dialog box with *Use existing Tables and Queries* already selected, click the Next button.

h. At the second SubForm Wizard dialog box, select the table and fields to be displayed in the subform by completing the following steps:

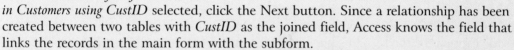

1) Click the *Tables/Queries* option box arrow and then click *Table: Profiles* at the drop-down list.
2) Move all the fields except *CustID* from the *Available Fields* list box to the *Selected Fields* list box.
3) Click the Next button.

i. At the third SubForm Wizard dialog box with *Show Profiles for each record in Customers using CustID* selected, click the Next button. Since a relationship has been created between two tables with *CustID* as the joined field, Access knows the field that links the records in the main form with the subform.

j. Click the Finish button at the last SubForm Wizard dialog box to accept the default subform name *Profiles subform*.

3. Access creates the subform within the active page in the tab control with a label control object above it. Click the label control object displaying the text *Profiles subform* to select the object and then press the Delete key.

4. Click the edge of the subform control object to display the orange border and sizing handles and then use the techniques learned in Activity 1c to move and resize the object so that the subform fills the tab control object as shown at the right.

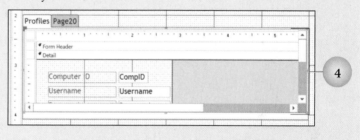

5. Click in a blank area outside the grid to deselect the subform control object and then switch to Form view. Notice that the subform displays as a datasheet within the tab control object in the CustMaintenance form.

6. In the field names row in the datasheet, position the mouse pointer on each boundary line that separates the columns and then double-click to adjust each column's width to best fit.

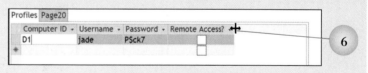

7. Notice that two sets of navigation buttons are displayed: one at the bottom of the main form (just above the Status bar) and another at the bottom of the datasheet in the subform. Use the navigation buttons at the bottom of the main form to scroll through a few records and watch the fields update in both the main form and subform upon moving to the next customer record.

8. Switch to Design view.

9. Save the form.

10. In the Navigation pane, notice that a form object exists with the name *Profiles subform*. Subforms are separate objects within the database. If the main form is closed, the subform can still be opened to edit data.

Check Your Work

In Design view, the control objects within the subform display one below another, but in Form view, the subform displays using a datasheet layout. Change the setting of the *Default View* property box in the subform's Property Sheet task pane to *Single Form* to match the layout of the controls in Design view to the layout of the fields in Form view. The fields display one below another in a single column in Form view.

To do this, open the subform's Property Sheet task pane by double-clicking the Form Selector button at the top of the vertical ruler and left of the horizontal ruler in the subform control object in Design view. Click in the *Default View* property box on the Format tab, click the option box arrow, and then click *Single Form* at the drop-down list.

Activity 1g Adding More Subforms and Adding a New Page to the Tab Control Object   Part 7 of 7

1. With **4-RSRCompServ** open and the CustMaintenance form open in Design view, change the caption for the second page in the tab control object to *Service Contracts* by completing steps similar to those in Steps 2a–c of Activity 1f.
2. With the Service Contracts page selected in the tab control object, add a subform to display the fields from the ServiceContracts table on the page by completing the following steps:
    a. Click the More button in the Controls group and then click the Subform/Subreport button in the expanded Controls group.
    b. Click inside the selected Service Contracts page in the tab control object.
    c. Click the Next button at the first SubForm Wizard dialog box.
    d. At the second SubForm Wizard dialog box, change the table displayed in the *Tables/Queries* option box to *Table: ServiceContracts*.
    e. Move all the fields from the table except the *CustID* field to the *Selected Fields* list box.
    f. Click the Next button.
    g. At the third SubForm Wizard dialog box with *Show ServiceContracts for each record in Customers using CustID* selected, click the Next button.
    h. Click the Finish button at the last SubForm Wizard dialog box to accept the default subform name *ServiceContracts subform*.
3. Select and then delete the label control object above the subform (which displays the text *ServiceContracts subform*).
4. Click the subform control object to display the orange border and sizing handles and then move and resize the form as shown below.

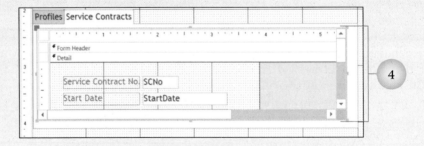

5. Deselect the subform control object and then switch to Form view.

6. Click the Service Contracts
tab and then adjust the
width of each column in
the datasheet to best fit. If
necessary, use the scroll bar to
access the right border for the *PlanID* field.

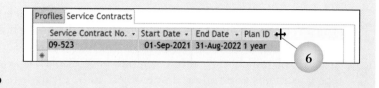

7. Switch to Design view and then save the form.
8. Add a new page to the tab control object and add a subform to display selected fields from
the WorkOrders table in the new page by completing the following steps:
   a. Right-click the Service Contracts tab and then click *Insert Page* at the shortcut menu.
   b. With the new page already selected, display the Property Sheet task pane, type Work
   Orders as the page caption, and then close the Property Sheet task pane.
   c. Click the More button in the Controls group and then click the Subform/Subreport
   button. Click inside the selected Work Orders page. Create a subform to display selected
   fields from the WorkOrders table by completing the following steps:
      1) Click the Next button at the first SubForm Wizard dialog box.
      2) At the second SubForm Wizard dialog box, change the table displayed in the
      Tables/Queries option box to *Table: WorkOrders*.
      3) Move the following fields from the *Available Fields* list box to the *Selected Fields* list box:
         WO
         WODate
         Descr
      4) Click the Next button.
      5) Click the Next button at the third SubForm Wizard dialog box.
      6) Click the Finish button at the last SubForm Wizard dialog box to accept the default
      subform name.
9. Select and then delete the label control object above the subform (which displays the text
WorkOrders subform).
10. Click the subform control object to display the orange border and sizing handles and then
move and resize the form as shown below.
11. Access automatically extends the width of the form and widens the tab control object if a
table with many fields is added to a subform. If necessary, select the tab control object
and decrease the width so the right edge of the tab control is at the 6-inch position on the
horizontal ruler. If necessary, decrease the width of the form so the right edge of the grid is
at the 6.5-inch position on the horizontal ruler. **Hint: If Access resizes the tab control object
to the edge of the form, the grid may have to be temporarily widened to see the middle sizing
handle at the right edge of the tab control object.**

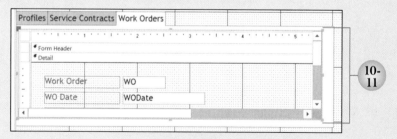

12. Deselect the subform control object and then switch to Form view.
13. While viewing the form, you decide the title would look better if it was not centered. Switch
to Design view, click the Title control object in the *Form Header* section, click the Form
Design Tools Format tab, and then click the Align Left button in the Font group.

14. Click the Form Design Tools Design tab and then switch to Form view.
15. Click the Work Orders tab and adjust the width of each column in the datasheet to best fit. Compare the CustMaintenance form with the one shown in Figure 4.9. (Notice that adding the tab control with a separate page displaying a subform for each table related to the Customers table allowed one object to be created that can be used to view and update fields in multiple tables.)
16. Print only the selected record. To do this, open the Print dialog box, click *Selected Record(s)* in the *Print Range* section, and then click OK.
17. Save and then close the CustMaintenance form.

Figure 4.9 Completed CustMaintenance Form

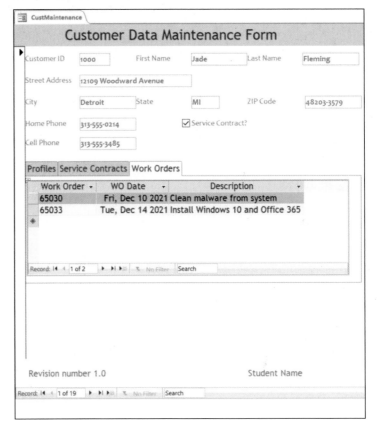

Activity 2 **Create a New Form with Calculations and Graphics** **5 Parts**

You will create a new form using the Form Wizard, add two calculations to the form, use features that assist with alignment and spacing of multiple control objects, and add graphics to the form.

Tutorial

Adding Calculations to a Form

  Text Box

Quick Steps

Add Calculated Control Object to Form

1. Open form in Design view.
2. Click Text Box button.
3. Position crosshairs in *Detail* section at specified location.
4. Drag to create control object of required height and width.
5. Release mouse.
6. Click in text box control object.
7. Type formula.
8. Press Enter key.
9. Delete text in label control object.
10. Type label text and press Enter key.

Adding Calculations to a Form

To display a calculated value in a form, create a query that includes a calculated column and then create a new form based on that query. Alternatively, create a calculated control object in an existing form using Design view.

To do this, click the Text Box button in the Controls group on the Form Design Tools Design tab and then drag the crosshairs in the *Detail* section to create a control object the approximate height and width required to show the calculation. Access displays a text box control object with *Unbound* inside the object and a label control object to the left displaying *Textxx* (where *xx* is the text box control object number). A calculated control object is considered as unbound because the data displayed in the control is not drawn from a stored field value in a record. Click inside the text box control object (*Unbound* disappears) and then type the formula, beginning with an equals sign. For example, the formula =[Hours]*[Rate] multiplies the value in the *Hours* field times the value in the *Rate* field. The field names in a formula are enclosed in square brackets.

Edit the label control object next to the calculated control to add a descriptive name that identifies the calculated value. Open the Property Sheet task pane for the calculated control object and then apply the Fixed, Standard, or Currency format, as appropriate for the calculated value. Since the data displayed in a calculated control object is based on a formula, there is no need to tab to this control object when entering data or moving through a record. Change the *Tab Stop* property box on the Other tab of the Property Sheet task pane from *Yes* to *No* to avoid stopping at any text box control object.

Activity 2a Adding and Formatting Calculated Control Objects

Part 1 of 5

1. With **4-RSRCompServ** open, use the Form Wizard to create a new form based on the WorkOrders table by completing the following steps:
   a. Click *WorkOrders* in the Tables group in the Navigation pane and then click the Create tab.
   b. Click the Form Wizard button in the Forms group.
   c. With *Table: WorkOrders* selected in the *Table/Queries* option box, complete the steps in the Form Wizard as follows:
      1) Move the *WO*, *Descr*, *ServDate*, *Hours*, *Rate*, and *Parts* fields from the *Available Fields* list box to the *Selected Fields* list box and then click the Next button.
      2) With *Columnar* selected as the layout option, click the Next button.
      3) With *WorkOrders* the default text in the *What title do you want for your form?* text box, click *Modify the form's design* at the last dialog box and then click the Finish button.
2. With the WorkOrders form displayed in Design view, add a calculated control object to display the total labor for the work order by completing the following steps:

a. Position the pointer on the top border of the gray *Form Footer* section bar until the pointer displays as an up-and-down-pointing arrow with a horizontal line in the middle and then drag down just below the 3-inch position on the vertical ruler. (The 3-inch mark will not appear until the border is dragged below it and the mouse button is released.) Doing this creates more grid space in the *Detail* section so that controls can be added.

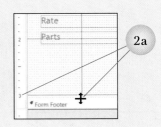

b. Click the Text Box button in the Controls group.

c. Position the crosshairs with the text box icon attached below the *Parts* text box control, drag to create an object of the approximate height and width shown at the right, and then release the mouse button.

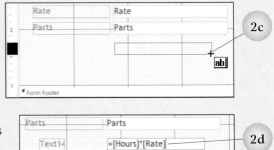

d. Click in the text box control (which displays *Unbound*) and then type =[Hours]*[Rate].

e. Press the Enter key.

f. With the calculated control object selected, click the Property Sheet button in the Tools group. With the Format tab in the Property Sheet task pane active, click the option box arrow in the *Format* property box, click *Standard* at the drop-down list, and then close the Property Sheet task pane. By default, calculated values display right-aligned in Form view.

g. Click to select the label control object to the left of the calculated control object (which displays *Textxx* [where *xx* is the text box label number]). Click in the selected label control object a second time to place the insertion point. Delete *Textxx*, type Total Labor, and then press the Enter key. Notice that the label control automatically expands to accommodate the width of the typed text.

3. With the Form Design Tools Design tab selected, click the View button to display the form in Form view and then scroll through a few records to view the calculated field. Do not be concerned with the size, position, alignment, and/or spacing of the controls; the format will be fixed in a later activity.

4. Switch to Design view and then save the form.

5. A calculated control object can be used as a field in another formula. To do this, reference the calculated object in the formula by its Name property enclosed in square brackets. Change the name for the calculated object created in Step 2 to a more descriptive name by completing the following steps:

a. Click the calculated control object (which displays the formula =[Hours]*[Rate]).

b. Click the Property Sheet button in the Tools group.

c. Click the Other tab in the Property Sheet task pane.

d. Select and delete the existing text (which displays *Textxx* [where *xx* is the text box number]) in the *Name* property box.

e. Type LaborCalc, press the Enter key, and then close the Property Sheet task pane.

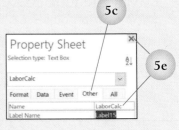

6. Add another calculated control object to the form to include labor and parts and determine the total value for the work order by completing the following steps:

a. Click the Text Box button in the Controls group.

b. Position the crosshairs with the text box icon attached below the calculated control created in Step 2, drag to create an object the approximate height and width as the first calculated control, and then release the mouse button.

c. Click in the text box control (which displays *Unbound*), type =[LaborCalc]+[Parts], and then press the Enter key.

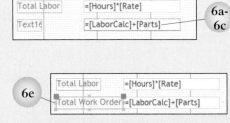

d. Apply the Currency format to the calculated control. Refer to Step 2g for assistance with this step.

e. Type Total Work Order in the label control object. Refer to Step 2h for assistance with this step.

7. Remove the tab stops from the two new calculated control objects by completing the following steps:

a. Press and hold the Shift key, use the left mouse button to click the calculated control objects that display the formulas *=[Hours]*[Rate]* and *=[LaborCalc]+[Parts]*, and then release the Shift key.

b. Click the Property Sheet button in the Tools group.

c. Click the Other tab in the Property Sheet task pane.

d. Double-click *Yes* in the *Tab Stop* property box to change it to *No*.

e. Close the Property Sheet task pane.

8. Save the form. Display the form in Form view and then scroll through a few records to view the calculations. Tab through a record and notice that when the Tab key is pressed after the *Parts* field, the insertion point does not stop at *Total Labor* or *Total Work Order* but instead moves to the next record.

9. Switch to Design view.

> Check Your Work

> Tutorial

Sizing, Aligning, and Spacing Multiple Control Objects

 Align

Adjusting Control Objects for Consistency in Appearance

When working in Design view, use the tools that Access provides for positioning, aligning, sizing, and spacing multiple controls to create forms with a consistent and professional appearance. Locate these tools using the Size/Space button and Align button in the Sizing & Ordering group on the Form Design Tools Arrange tab.

Aligning Multiple Control Objects

Quick Steps

Align Multiple Control Objects
1. In Design view, select control objects.
2. Click Form Design Tools Arrange tab.
3. Click Align button.
4. Click option at drop-down list.
5. Deselect control objects.

The options in the Align button drop-down list in the Sizing & Ordering group on the Form Design Tools Arrange tab are shown in Figure 4.10. Use these options to align multiple selected control objects at the same horizontal or vertical position and avoid having to adjust each control object individually.

Figure 4.10 Options in the Align Button Drop-Down List

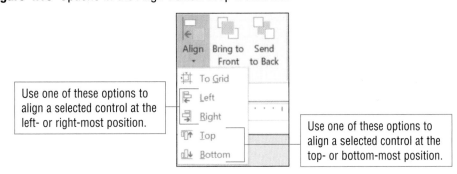

Use one of these options to align a selected control at the left- or right-most position.

Use one of these options to align a selected control at the top- or bottom-most position.

Adjusting the Sizing and Spacing between Control Objects

 Size/Space

Quick Steps

Adjust Spacing between Control Objects
1. In Design view, select control objects.
2. Click Form Design Tools Arrange tab.
3. Click Size/Space button.
4. Click option in *Spacing* section.
5. Deselect controls.

The options provided by the Size/Space button in the Sizing & Ordering group on the Form Design Tools Arrange tab are shown in Figure 4.11. Use these options to assist with the consistent sizing of and spacing between control objects. Use options in the *Size* section of the drop-down list to adjust the height or width to the tallest, shortest, widest, or narrowest of the selected control objects. Use options in the *Spacing* section to adjust the horizontal and vertical spacing between control objects, increase or decrease the space, or provide equal spaces between all the selected control objects.

Using these tools is helpful when creating a new form by adding control objects manually to the grid or after editing an existing form because the space between control objects can be changed easily after objects are added or deleted. To adjust the spacing by moving individual control objects would be too time consuming.

Figure 4.11 Size and Spacing Options in the Size/Space Button Drop-Down List

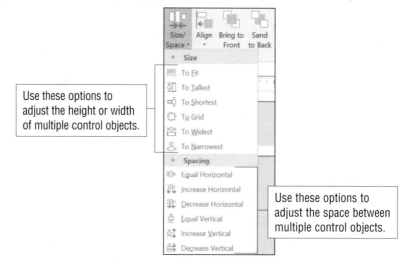

Use these options to adjust the height or width of multiple control objects.

Use these options to adjust the space between multiple control objects.

Activity 2b Sizing, Aligning, and Spacing Multiple Control Objects

Part 2 of 5

1. With **4-RSRCompServ** open and the WorkOrders form open in Design view, change the font of the text in the *Title* control object to 16 pts and then change it to read *Work Orders with Calculations*. Widen the *Title* control object to fit the title on one line.
2. With the title control object still selected, position the mouse pointer on the orange border until the pointer changes to the four-headed arrow move icon and then drag the control object to the approximate center of the *Form Header* section.

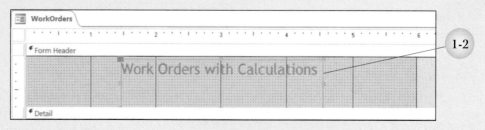

3. Point to the bottom gray border in the *Form Footer* section bar until the pointer displays as an up-and-down-pointing arrow with a horizontal line in the middle and then drag down approximately 0.5 inch to create space in the *Form Footer* section. Create a label control object with your name in the center of the *Form Footer* section.

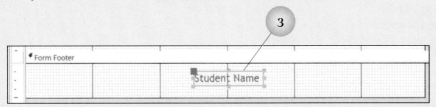

4. Click to select the *Descr* text box control object and then drag the right middle sizing handle left until the control object is resized to approximately the 4.5-inch position on the horizontal ruler.
5. Press and hold down the Shift key, use the left mouse button to click the six text box control objects for the fields above the two calculated control objects, and then release the Shift key. Click the Form Design Tools Arrange tab, click the Size/Space button in the Sizing & Ordering group, and then click *To Widest* at the drop-down list. The six text box control objects are now all the same width. (The width is set to fit the widest selected control object).

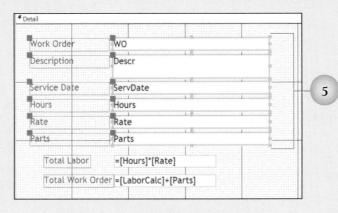

6. Click in a blank area to deselect the control objects.
7. Use the Align button to align multiple control objects by completing the following steps:

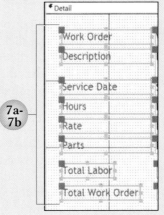

   a. Draw a selection rectangle around all the label control objects at the left side of the form. This selects all eight label control objects.
   b. Click the Align button in the Sizing & Ordering group and then click *Left* at the drop-down list. All the label control objects align at the left edge of the left-most control object.
   c. Deselect the control objects.
   d. Draw a selection rectangle around all the text box control objects at the right of the form to select all eight text box control objects.

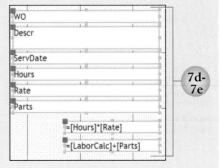

   e. Click the Align button in the Sizing & Ordering group and then click *Right* at the drop-down list. All the control objects align at the right edge of the right-most control object.
   f. Deselect the control objects.

8. Adjust the vertical spaces between control objects to make all the control objects in the *Detail* section equally spaced by completing the following steps:

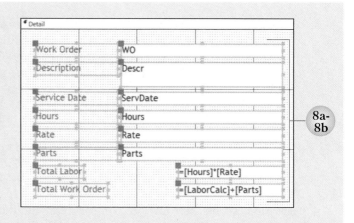

a. Draw a selection rectangle around all the control objects in the *Detail* section.

b. Click the Size/Space button and then click *Equal Vertical* in the *Spacing* section of the drop-down list. All the control objects are now separated by the same amount of vertical space.

c. Deselect the control objects.

9. Save the form.

10. Display the form in Form view and then scroll through a few records to view the revised alignment and spacing. The numeric fields will be formatted in Activity 2e to align the numbers correctly.

11. Switch to Design view.

Check Your Work

 Tutorial

Adding Graphics to a Form

Adding and Anchoring Graphics to a Form

 Logo

 Insert Image

 Line

Quick Steps

Add Image to Form
1. Open form in Design view.
2. Click Insert Image button.
3. Click *Browse* and navigate to folder.
4. Double-click image.
5. Click and drag crosshairs.
6. Move and resize.
7. If necessary, display Property Sheet task pane and change Size Mode property.

A picture that is saved in a graphic file format can be added to a form using the Logo button or Insert Image button in the Controls group on the Form Design Tools Design tab. Click the Logo button and the Insert Picture dialog box opens. Navigate to the drive and/or folder in which the graphic file is stored and then double-click the image file name. Access automatically adds the image to the left side of the *Form Header* section. Move and/or resize the image as needed. Access supports the BMP, GIF, JPEG, JPG, and PNG graphic file formats for a logo control object.

Use the Insert Image button to place the picture in another section or draw a larger control object to hold the picture. Click the Insert Image button and then click *Browse* at the drop-down list to open the Insert Picture dialog box. Navigate to the drive and/or folder in which the graphic file is stored and then double-click the image file name. Position the crosshairs with the image icon attached at the location in the form where the image is to be placed and then drag the crosshairs to draw a control object of the approximate height and width desired. Access supports the GIF, JPEG, JPG, and PNG file formats for an image control object.

Use the Line button in the Controls group to draw horizontal or vertical lines in a form. Press and hold down the Shift key while dragging to draw a straight line. Once the line has been drawn, use the Shape Outline button in the Control Formatting group on the Form Design Tools Format tab to modify the line thickness, type, and color.

Online images can be added to a form. Access does not provide an online pictures button in the Controls group. However, an online picture can be inserted in a Microsoft Word document and then saved to a folder using standard Windows commands. Use the Insert Image button to insert the saved image.

1. With **4-RSRCompServ** open and the WorkOrders form open in Design view add an image to the WorkOrders form by completing the following steps:
   a. Click the Insert Image button in the Controls group, click the *Browse* option, navigate to your AL2C4 folder, and then double click *CompRepair*.

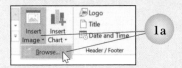

   b. Position the crosshairs with the picture icon attached to the right of the text box control objects, drag to create an object of the approximate height and width shown, and then release the mouse button. Access inserts the image and displays the orange border with selection handles.

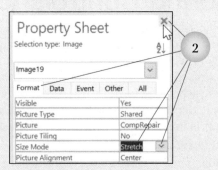

2. Click the Property Sheet button in the Tools group. If necessary, make Format the active tab in the Property Sheet task pane. Click in the *Size Mode* property box (displays *Zoom*), click the option box arrow that appears, and then click *Stretch* at the drop-down list. Close the Property Sheet task pane. Changing the *Size Mode* property box to display *Stretch* instructs Access to resize the image to fit the height and width of the control object. Note that using *Stretch* may skew the appearance of the image.

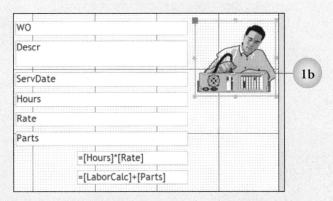

3. Deselect the control object containing the image and then display the form in Form view. Add a line below the title to improve the appearance of the form. Draw and modify the line by completing the following steps:
   a. Switch to Design view and then click the Line button in the Controls group.

b. Position the crosshairs with the line icon attached below the title in the *Form Header* section beginning a few rows of grid dots below the first letter in the title. Press and hold down the Shift key, drag to the right, release the mouse button below the last letter in the title, and then release the Shift key.

c. Click the Form Design Tools Format tab, click the Shape Outline button in the Control Formatting group, point to *Line Thickness* at the drop-down list, and then click *3 pt* (fourth option from the top).

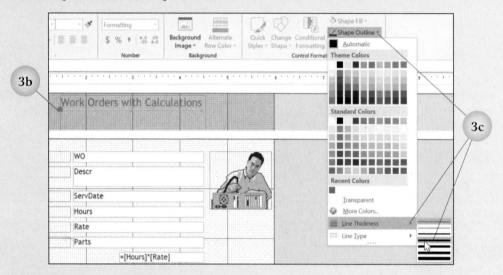

d. With the line control object still selected, click the Shape Outline button and then click the *Dark Red* color (first option in last row of *Standard Colors* section).

e. Deselect the line control object.

4. Display the form in Form view to see the line under the title.

5. Switch to Design view. If desired, adjust the length and/or position of the line.

6. Save the form.

Check Your Work

Anchoring Controls to a Form

A control object in a form can be anchored to a section or another control object using the Anchoring button in the Position group on the Form Design Tools Arrange tab. When a control object is anchored, its position is maintained when the form is resized. For example, if an image is anchored to the top right of the *Detail* section, when the form is resized in Form view, the image automatically moves so that the relative distance between the image and the top right of the *Detail* section is maintained. If the image is not anchored and the form is resized, the position of the image relative to the edges of the form can change.

By default, *Top Left* is selected as the anchor position for each control object in a form. To change the anchor position, select the object(s), click the Form Design Tools Arrange tab, click the Anchoring button, and then click one of these options: *Stretch Down, Bottom Left, Stretch Across Top, Stretch Down and Across, Stretch Across Bottom, Top Right, Stretch Down and Right,* or *Bottom Right*. Click the option that represents how the control object is to move as the form is resized. Note that some options will cause a control object to resize as well as move when the form is changed.

1. With **4-RSRCompServ** open and the WorkOrders form open in Design view, anchor the image to the top of the *Detail* section of the form by completing the following steps:
   a. Click to select the image.
   b. Click the Form Design Tools Arrange tab.
   c. Click the Anchoring button in the Position group and then click *Top Right* at the drop-down list.

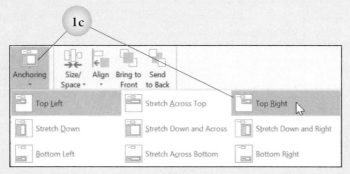

   d. Take note of the distance between the top border of the selected image and the top of the *Detail* section. ***Note: If you have a wide monitor, you may want to change the anchoring back to* Top Left**.
2. Display the form in Form view. Notice that the image has shifted to the right of the *Detail* section. However, it has maintained the distance between the top of the control object boundary and the top of the *Detail* section.

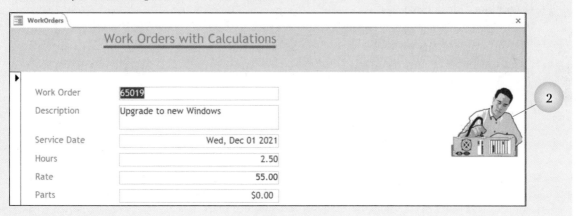

3. Save the form and then switch to Design view.

Check Your Work

1. With **4-RSRCompServ** open and the WorkOrders form open in Design view, adjust the alignment and formatting of numeric fields by completing the following steps:
   a. Press and hold down the Shift key; click the left mouse button to select the *Rate* and *Parts* text box control objects and both calculated text box control objects; and then release the mouse button.
   b. Click the Form Design Tools Format tab, click the Align Right button in the Font group, and then click the Currency button in the Number group.
   c. Deselect the control objects.
2. Switch to Form view. Notice that the decimal point in the data in the *Hours* text box control object does not align with decimal points in the other numbers in the form. Switch back to Design view. To align the decimal point, complete the following steps:
   a. Right click the *Hours* text box control object and then click *Properties* at the shortcut menu.
   b. Click the Format tab, scroll down if necessary, and then change the Right Margin property from *0* to *0.07*.
   c. Close the Property Sheet task pane.
3. Click to select the title control object box in the *Form Header* section and then drag the bottom middle sizing handle up to decrease the height of the control object to approximately 0.4 inch on the vertical ruler. If necessary, drag down the bottom border of the *Form Header* section in order to select the sizing handle

4. Position the pointer on the top of the gray *Detail* section bar until the pointer displays as an up-and-down-pointing arrow with a horizontal line in the middle and then drag up to decrease the height of the *Form Header* section to approximately 0.5 inch on the vertical ruler.
5. Save the form.
6. Display the form in Form view and compare it with the one shown in Figure 4.12.
7. Print only the selected record with the left and right margins both set to 0.5 inch.
8. Close the form.

Check Your Work

Figure 4.12 Completed WorkOrders Form

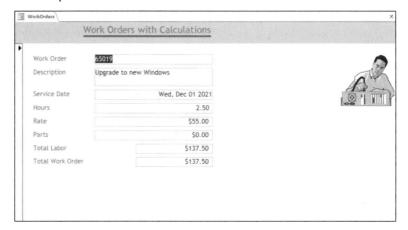

> You will create a datasheet form for use in entering information into a table and set the form's properties to prevent records from being deleted.

Tutorial

Creating a Datasheet Form and Restricing Actions

Quick Steps

Create Datasheet Form

1. Select table in Navigation pane.
2. Click Create tab.
3. Click More Forms button.
4. Click *Datasheet*.
5. Save form.

 More

 datasheet

Quick Steps

Restrict Actions in Form

1. Open form in Design view.
2. Double-click Form Selector button.
3. Click Data tab.
4. Change *Allow Additions*, *Allow Deletions*, *Allow Edits*, or *Allow Filters* to *No*.
5. Close Property Sheet task pane.
6. Save form.

Creating a Datasheet Form and Restricting Actions

A form can be created that looks just like the datasheet of a table. With the appropriate table selected in the Navigation pane, click the Create tab, click the More Forms button in the Forms group, and then click *Datasheet* at the drop-down list. Access creates a form including all the fields from the selected table presented in a datasheet layout. The datasheet form is similar in appearance and purpose of a table datasheet, and providing the form to users instead of the underlying table prevents them from accessing and modifying the structure of the table.

To restrict what actions users can perform, display the form in Form view and use options on the Data tab of the form's Property Sheet, as shown in Figure 4.13. For example, prevent new records from being added and/or existing records from being deleted, edited, and/or filtered. Setting the *Data Entry* property to *Yes* means the user will see only a blank form when the form is opened. A data entry form is intended to be used only to add new records. The user is prevented from scrolling through existing records in the form.

Figure 4.13 Form Property Sheet Task Pane with the Data Tab Selected

Use these form properties to restrict use of the form.

1. With **4-RSRCompServ** open, click *Technicians* in the Tables group in the Navigation pane and then click the Create tab.

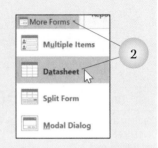

2. Click the More Forms button in the Forms group and then click *Datasheet* at the drop-down list.
3. Review the Technicians form in the work area. Notice that the form resembles a table datasheet.
4. Switch to Design view.
5. Modify the properties of the Technicians form to prevent users from deleting records by completing the following steps:

    a. Click in a blank area to deselect the control objects.

    b. Double-click the Form Selector button at the top of the vertical ruler and to the left of the horizontal ruler to open the form's Property Sheet task pane.

    c. Click the Data tab.

    d. Double-click in the *Allow Deletions* property box to change *Yes* to *No*.

    e. Close the Property Sheet task pane.

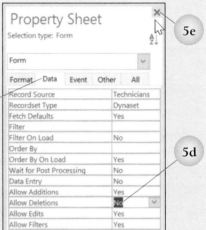

6. Click the Save button and then click OK to accept *Technicians* as the form name.
7. Click the View button arrow in the Views group on the Form Design Tools Design tab. Notice that *Datasheet View* and *Design View* are the only views available. The *Form View* option is not available at the drop-down list or in the View buttons at the right side of the Status bar.

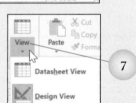

8. Click in a blank area to close the drop-down list.
9. Close the form.
10. Double-click *Technicians* in the Forms group in the Navigation pane. (Be careful to open the form object and not the table object.)
11. Click the record selector bar next to the first row in the datasheet (for technician ID 01) to select the record.
12. Click the Home tab and then look at the Delete button in the Records group. Notice that the Delete button is dimmed. This feature is unavailable because the *Allow Deletions* property box is set to *No*.

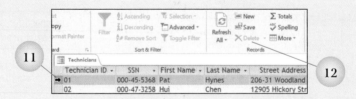

13. Best fit the width of each column, print the first page of the Technicians form in landscape orientation, and then close the form.

14. Right-click *Technicians* in the Forms group in the Navigation pane and then click *Layout View* at the shortcut menu. Notice that the datasheet form displays in a columnar layout in Layout view. The Technicians form includes a field named *Attachments*. In this field in the first record, a picture of the technician has been attached to the record. In Layout view, Access automatically opens the image file and displays the contents.

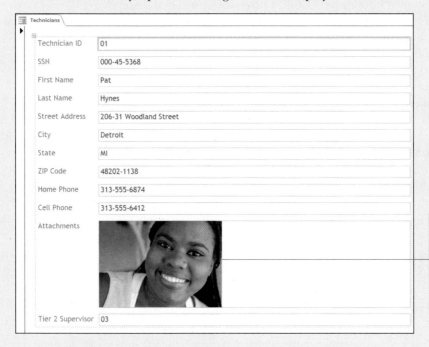

The *Attachments* field automatically displays an attached image file if one has been provided.

15. Close the form.

 Check Your Work

---

Activity 4 Create a Blank Form with Lists   3 Parts

You will use the Blank Form tool to create a new form for maintaining the FeesSCPlans table, which is used to track service plan fees. In the form, you will create list boxes to provide an easy way to enter data for new service contract plans.

 Tutorial

Creating a Blank
Form

 Blank Form

Creating a Blank Form

To quickly build a form that contains a small number of fields, use the Blank Form tool. A blank form begins with no controls or formatting and displays as a blank white page in Layout view.

To begin a new form, click the Create tab and then click the Blank Form button in the Forms group. Access opens the Field List task pane at the right side of the work area. Expand the list for the appropriate table and then add fields to the form as needed. If the Field List task pane displays with no table names, click the Show all tables hyperlink at the top of the pane.

1. With **4-RSRCompServ** open, click the Create tab and then click the Blank Form button in the Forms group.
2. If the Field List task pane at the right side of the work area does not display the table names, click the Show all tables hyperlink. Otherwise, proceed to Step 3.
3. Add fields from the FeesSCPlans table to the form by completing the following steps:
   a. Click the plus symbol (+) next to the FeesSCPlans table to expand the field list.
   b. Click the first field (named *ID*) in the Field List task pane and then drag the field to the top left of the form.
   c. Click the second field (named *Term*) in the Field List task pane, press and hold down the Shift key, click the last field, (named *Rate*) in the Field List task pane to select the remaining fields in the FeesSCPlans table, and then release the Shift key.
   d. Position the mouse pointer within the selected field names and then drag the group of fields to the form below the *ID* field. Release the mouse when the pink bar displays below *ID*.
   e. With the four fields added to the table still selected, press and hold down the Shift key, click the *ID* field, and then release the Shift key.
   f. Position the mouse pointer on the orange border at the right of any of the selected label control objects until the pointer changes to a left-and-right-pointing arrow and then drag the right edge of the label control objects to the right until all the label text is visible, as shown below.

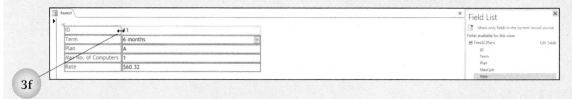

4. Close the Field List task pane.
5. Save the form with the name *SCPlans*.

Check Your Work

Adding a List Box to a Form

A list box displays a list of values for a field within the control object. In Form view, the user can easily see the entire list for the field. Create the list of values when the control object is created or instruct Access to populate the list using values from a table or query. When a list box control object is added to the form, the List Box Wizard begins (as long as the Use Control Wizards button is toggled on). Within the List Box Wizard, specify the values to be shown in the list box.

1. With **4-RSRCompServ** open and the SCPlans form open in Layout view, add a list box control object to show the plan letters in a list by completing the following steps:
   a. Click the List Box button in the Controls group on the Form Layout Tools Design tab.
   b. Position the pointer with the list box icon attached below the *Plan* field text box control object in the form. Click the mouse when the pink bar displays between *A* and *1* in the right column. The List Box Wizard starts when the mouse is released. ***Note: If the Wizard does not start and the Use Control Wizards button is active, undo the addition, change to Design view and then back to Layout view and repeat Steps 1a and 1b.***

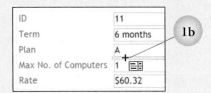

   c. At the first List Box Wizard dialog box, click *I will type in the values that I want* and then click the Next button.

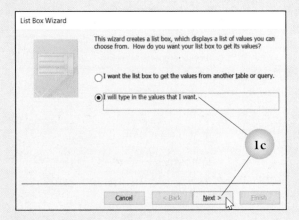

   d. At the second List Box Wizard dialog box, click in the first cell below *Col1*, type A, and then press the Tab key.
   e. Type B, press the Tab key, type C, press the Tab key, type D, and then click the Next button.

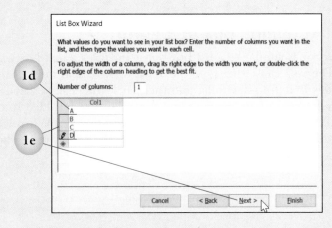

f. At the third List Box Wizard dialog box, click *Store that value in this field* option, click the option box arrow, and then click *Plan* at the drop-down list.

g. Click the Next button.

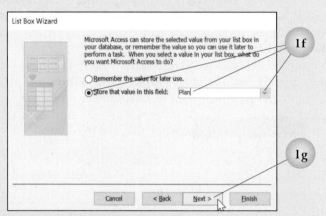

h. At the last List Box Wizard dialog box with the current text already selected in the *What label would you like for your list box?* text box, type PlanList and then click the Finish button. Access adds the list box to the form, displaying all the values entered in the list.

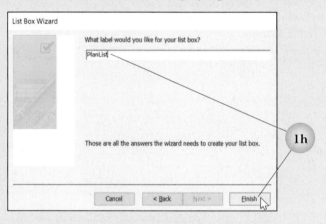

2. Save the form.

Adding a Combo Box to a Form

Adding a Combo Box

  Combo Box

Quick Steps

Add Combo Box
1. Open form in Layout or Design view.
2. Click Combo Box button in Controls group.
3. Click within form.
4. Create values within Combo Box Wizard.
5. Save form.

A combo box is similar to a list box but it includes a text box within the control object. The user can either type the value for the field or click the arrow to display field values in a drop-down list and then click the desired value. As happens when adding a list box, when a combo box control object is added to the form, the Combo Box Wizard begins (as long as the Use Control Wizards button is toggled on). Within the Combo Box Wizard, specify the values to be shown in the drop-down list.

1. With **4-RSRCompServ** open and the SCPlans form open in Layout view, add a combo box control object to enter the maximum number of computers in a plan by completing the following steps:

   a. Click the Combo Box button in the Controls group on the Form Layout Tools Design tab.

   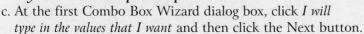

   b. Position the pointer with the combo box icon attached below the *Max No. of Computers* text box control object in the form. Click the mouse when the pink bar displays between *1* and *$60.32* in the right column. The Combo Box Wizard starts when the mouse button is released. ***Note: If the Wizard does not start and the Use Control Wizards button is active, undo the addition, change to Design view and then back to Layout view and repeat Step 1a and 1b.***

   c. At the first Combo Box Wizard dialog box, click *I will type in the values that I want* and then click the Next button.

   d. At the second Combo Box Wizard dialog box, click in the first cell below *Col1*, type 1, and then press the Tab key.

   e. Type 2, press the Tab key, type 3, press the Tab key, type 4, press the Tab key, type 5, and then click the Next button.

   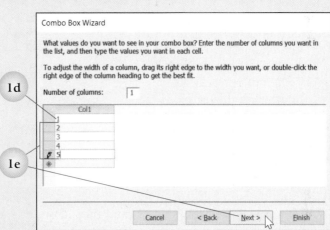

   f. At the third Combo Box Wizard dialog box, click *Store that value in this field* option, click the option box arrow, click *MaxCptr* at the drop-down list, and then click the Next button.

   g. At the last Combo Box Wizard dialog box with the current text already selected in the *What label would you like for your combo box?* text box, type CptrList and then click the Finish button. Access adds the combo box to the form, displaying a value and an option box arrow at the right side of the text box.

   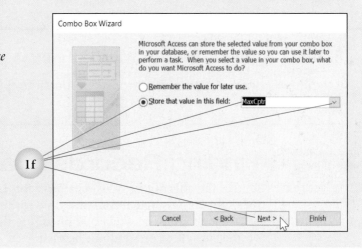

2. Double-click the label for the combo box added in Step 1g (which currently reads *CptrList*), select the text if it is not already selected, and then type Maximum Computers. Edit the label for the list box added in Activity 4b (which currently reads *PlanList*) to add a space between *Plan* and *List*.

3. Right-click the label control object above the combo box (which displays the text *Max No. of Computers*) and then click *Select Entire Row* at the shortcut menu. Press the Delete key to remove the selected row from the form.

4. Click the Title button in the Header/Footer group on the Form Layout Tools Design tab and then type the text Service Contract Plans.

5. Switch to Form view and then scroll through the records in the form.

6. Add a new record to the table by completing the following steps:
   a. Click the New button in the Records group on the Home tab.
   b. Press the Tab key to move past the *ID* field since this field is an AutoNumber data type field.
   c. Click the *Term* option box arrow and then click *2 years* at the drop-down list.
   d. Click *B* in the *Plan List* list box. Notice that *B* is entered in the *Plan* field text box control object when the letter *B* is clicked in the list box.
   e. Click the *Maximum Computers* option box arrow and then click *3* at the drop-down list.
   f. Click in the *Rate* text box and then type 360.50.

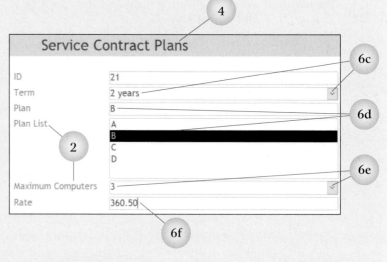

7. Print the selected record, save, and then close the form.

Check Your Work

Activity 5 Finding Records within a Form   1 Part

You will open a custom-built form and use it to find records by using a wildcard character.

 Tutorial

Finding Records in a Form

Finding Records in a Form

One of the advantages of using a form for data entry and maintenance is that the form displays a single record at a time within the work area. Seeing one record at a time reduces the likelihood of editing the wrong record; the user can focus on the current record and not be distracted by other records. In a table with many records, quickly finding the specific record to be maintained or viewed is important. Use the Find feature to move to records quickly.

Find

The Find feature allows searching for records without specifying the entire field value. To do this, substitute wildcard characters in the positions for which the exact text is not specified. Two commonly used wildcard characters are the asterisk (*) and the question mark (?).

For example, suppose that a search is needed for a record by a person's last name but the correct spelling is not known. Use the asterisk wildcard character in a position for which one or more characters vary. Use the question mark wildcard character in a fixed-width word for which the records with the same number of characters in the field are to be viewed. In this case, substitute one question mark for each character not specified. Table 4.1 provides examples of how to use the asterisk and question mark wildcard characters. In Activity 5, the asterisk wildcard character will be used to locate customer records for a specified street.

Table 4.1 Examples of Using Wildcard Characters with the Find Feature

Find What Entry	In This Field	Will Find
104?	*Customer ID*	customer records with customer IDs that begin with *104* and have one more character, such as *1041, 1042, 1043,* and so on up to *1049*
4820?	*ZIP Code*	customer records with zip codes that begin with *4820* and have one more character, such as *48201, 48202,* and so on
650??	*Work Order*	work order records with work order numbers that begin with *650* and have two more characters, such as *65023, 65035, 65055,* and so on
313*	*Home Phone*	customer records with telephone numbers that begin with the *313* area code
Peter*	*Last Name*	customer records with last names that begin with *Peter* and have any number of characters following, such as *Peters, Peterson, Petersen, Peterovski,* and so on
4820*	*ZIP Code*	customer records with zip codes that begin with *4820* and have any number of characters following, such as *48201* and *48203-4841*
oak	*Street Address*	customer records with street addresses that have *oak* within them, such as *1755 Oak Drive, 12-234 Oak Street,* and *9 Oak Boulevard*

1. With **4-RSRCompServ** open, open the CustMaintenance form in Form view.
2. To locate the name of the customer who resides on Roselawn Street when neither the house number nor the customer's name is known, complete the following steps to find the record using a wildcard character in the criterion:
   a. Click in the *Street Address* field.
   b. Click the Find button in the Find group on the Home tab.
   c. With the insertion point positioned in the *Find What* text box, type *roselawn* and then click the Find Next button. The entry *roselawn* means "Find any record in which any number of characters before *roselawn* and any number of characters after *roselawn* exist in the active field." Access displays the first record in the form in which a match was found.

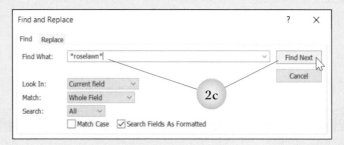

   d. Click the Find Next button to see if any other records exist for customers on Roselawn Street.
   e. At the Microsoft Access message box indicating that Access has finished searching the records, click OK.
   f. Close the Find and Replace dialog box.
3. Close the CustMaintenance form.
4. Close **4-RSRCompServ**.

Chapter Summary

- In Design view initially, a new form displays the *Detail* section, which is used to display records from the table associated with the form.

- A *Form Header* section and *Form Footer* section can be added to a form. Control objects placed in the *Form Header* section display at the top of the form or print at the beginning of a printout of records from Form view. Control objects placed in the *Form Footer* section display at the bottom of the form or print at the end of a printout of records from Form view.

- A form can contain three types of control objects: bound, unbound, and calculated.

- Click the Title button to display the *Form Header* and *Form Footer* sections and add a label control object in the *Form Header* section that contains the form name.

- Click the Label button in the Controls group to add a label control object within any section of the form.

- To specify the table to be connected to the form, use the *Record Source* property box on the Data tab of the form's Property Sheet task pane. Open this task pane by double-clicking the Form Selector button.

- Once a table has been connected to a form, click the Add Existing Fields button to open the Field List task pane.
- To add fields to a form, drag selected field names individually or in groups from the Field List task pane to the *Detail* section of the form.
- Use the move handle (the large dark-gray square at the top left of the selected control object) to move a selected control object independently of its connected label control or text box control object.
- Use the buttons in the Font group on the Form Design Tools Format tab to apply various types of formatting to selected control objects.
- Format multiple control objects at the same time by drawing a selection rectangle around a group of adjacent control objects or pressing and holding down the Shift key while clicking individual control objects.
- Open the Tab Order dialog box to change the order that fields are selected in when using the Tab key to move from field to field in Form view.
- Add a tab control object to a form to organize groups of related fields or insert subforms that display on separate pages.
- Click the Subform/Subreport button in the Controls group to add a subform to a page.
- Add a calculated control object to a form using the Text Box button in the Controls group.
- Type a formula in a text box control object (which displays *Unbound*) beginning with an equals sign. Enclose field names within the formula in square brackets.
- Use the Size/Space button and Align button in the Sizing & Ordering group on the Form Design Tools Arrange tab to position, align, size, or adjust spacing between multiple selected control objects.
- Add graphics to a form using the Logo button in the Header/Footer group or the Insert Image button in the Controls group.
- Draw a horizontal or vertical line in a form using the Line button in the Controls group. To draw a straight line, press and hold down the Shift key while dragging. Use the Shape Outline button on the Form Design Tools Format tab to adjust a line's thickness, type, or color.
- Online images can be added to a form after being inserted in a Microsoft Word document and then saved to a folder using standard Windows commands. Insert the saved image in the form using the Insert Image button.
- Change a control object's Size Mode property to resize an image within a control object while maintaining the original proportions of height and width.
- Use the Anchoring button in the Position group to anchor a control object to a section or another control object so that its position relative to the edges of the form is maintained when the form is resized.
- A datasheet form looks like a table datasheet.
- Use options on the Data tab of the form's Property Sheet task pane to restrict what actions a user can perform when viewing records in Form view.
- Use the Blank Form tool in the Forms group on the Create tab to create a new form with no controls or formatting applied. The form displays as a blank white page in Layout view with the Field List task pane open at the right of the work area.

- A list box displays a list of values for a field within the control object. The values to be shown in the list box can be specified within the List Box Wizard.
- A combo box includes a text box within the control object so the user can type the field value into the text box or click the option box arrow to pick the field value from a drop-down list. The values to be shown in the drop-down list can be specified within the Combo Box Wizard.
- Find specific records quickly using the Find feature.
- The Find feature allows the use of wildcard characters, such as the asterisk and question mark, to search records without specifying the entire field value.
- To find a record in a form, click in a field, click the Find button, and then enter the search criterion in the *Find What* text box.

Commands Review

FEATURE	RIBBON TAB, GROUP	BUTTON, OPTION	KEYBOARD SHORTCUT
add existing fields	Form Design Tools Design, Tools		
adjust size of multiple controls	Form Design Tools Arrange, Sizing & Ordering		
align multiple controls at same position	Form Design Tools Arrange, Sizing & Ordering		
anchor controls to form	Form Design Tools Arrange, Position		
blank form	Create, Forms		
change tab order of fields	Form Design Tools Design, Tools		
combo box	Form Layout Tools Design, Controls		
create form in Design view	Create, Forms		
datasheet form	Create, Forms		
Design view	Home, Views OR Form Design Tools Design, Views		
equal spacing between control objects	Form Design Tools Arrange, Sizing & Ordering		
Find	Home, Find		Ctrl + F
Form view	Home, Views OR Form Design Tools Design, Views		
insert image	Form Design Tools Design, Controls		
label control object	Form Design Tools Design, Controls	*Aa*	
line	Form Design Tools Design, Controls		
list box	Form Design Tools Design, Controls		

FEATURE	RIBBON TAB, GROUP	BUTTON, OPTION	KEYBOARD SHORTCUT
Property Sheet task pane	Form Design Tools Design, Tools		F4
subform	Form Design Tools Design, Controls		
tab control object	Form Design Tools Design, Controls		
text box control object	Form Design Tools Design, Controls	abl	
title control object	Form Design Tools Design, Header/Footer		

Microsoft
Access® Level 2

Unit 2
Advanced Reports, Access Tools, and Customizing Access

Microsoft®

Access®

Creating and Using Custom Reports

Performance Objectives

Upon successful completion of Chapter 5, you will be able to:

1. Create a custom report in Design view using all five report sections
2. Connect a table or query to a report and add fields
3. Move, size, format, and align control objects
4. Insert a subreport into a report
5. Add page numbers and date and time control objects to a report
6. Add graphics to a report
7. Group records and add functions and totals
8. Add a calculated field to a report
9. Modify section or group properties to control print options
10. Insert and edit a chart in a report
11. Create a blank report
12. Add tab control objects, list boxes, combo boxes, and hyperlinks to a report
13. Change the shape of a control object
14. Change the tab order of fields

Reports are used to generate printouts from the tables in a database. Although the Print feature can be used to print data from a table datasheet, query results datasheet, or form, the formatting of the data in this printout cannot be changed or customized. The Report feature provides tools and options that can be used to control the content and formatting to produce professional-quality reports that serve particular purposes. In this chapter, you will learn how to build custom reports.

 Data Files

Before beginning chapter work, copy the AL2C5 folder to your storage medium and then make AL2C5 the active folder.

The online course includes additional training and assessment resources.

You will create a custom report in Design view with fields from two tables and insert a subreport.

Tutorial

Creating Custom
Reports Using
Design View

Report
Design

Quick Steps

**Start New Report
in Design View**
1. Click Create tab.
2. Click Report Design button.

Add Report Title
1. Open report in Design view.
2. Click Title button.
3. Type title text.
4. Press Enter key.

**Add Label
Control Object**
1. Open report in Design view.
2. Click Label button.
3. Drag to create control object.
4. Type label text.
5. Press Enter key.

Creating Custom Reports Using Design View

Access provides the Report tool and Report Wizard to help quickly create reports that can be modified later. Customize the content, format, and layout of a report in Layout view or Design view. In most cases, use one of the report tools to generate the report structure and then customize the report using a different view. However, if a report requires several custom options, begin in Design view and build the report from scratch. Click the Create tab and then click the Report Design button in the Reports group to create a new report using Design view, as shown in Figure 5.1.

Creating a custom report in Design view involves using the same techniques taught in Chapter 4 for designing and building a custom form. Adding a title; connecting a table or query to the report; adding fields; and aligning, moving, resizing, and formatting control objects is done the same way as customizing a form.

A report can contain up to five sections, each of which is described in Table 5.1. *Group Header* and *Group Footer* sections can be added to group records that contain repeating values in a field, such as a department or city. How to use these additional sections will be demonstrated in Activity 3. A report that is grouped by more than one field can have multiple *Group Header* and *Group Footer* sections.

Figure 5.1 Report Design View

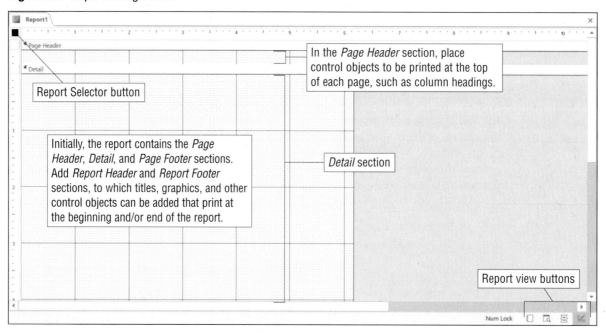

In the *Page Header* section, place control objects to be printed at the top of each page, such as column headings.

Report Selector button

Initially, the report contains the *Page Header*, *Detail*, and *Page Footer* sections. Add *Report Header* and *Report Footer* sections, to which titles, graphics, and other control objects can be added that print at the beginning and/or end of the report.

Detail section

Report view buttons

Table 5.1 Report Sections

Report Section	Description
Report Header	Content prints at the beginning of the report. Add control objects to this section for the report title and company logo or another image.
Page Header	Content prints at the top of each page in the report. Add control objects to this section for column headings in a tabular report format.
Detail	Add control objects to this section for the fields from the table or query that make up the body of the report.
Page Footer	Content prints at the bottom of each page in the report. Add a control object to this section to print a page number at the bottom of each page.
Report Footer	Content prints at the end of the report. Add a control object in this section to print a grand total or perform another function, such as calculating the average, maximum, minimum, or count.

Activity 1a Starting a New Report Using Design View and Adding a Title and Label Control Object

Part 1 of 5

1. Open **5-RSRCompServ** and enable the content.
2. Click the Create tab and then click the Report Design button in the Reports group.
3. Add a title in the *Report Header* section of the report by completing the following steps:
   a. With the Report Design Tools Design tab active, click the Title button in the Header / Footer group. Access adds the *Report Header* section above the *Page Header* section and places a title object containing the selected text *Report1*.
   b. Type RSR Computer Service Work Orders and then press the Enter key.

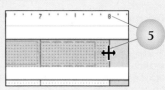

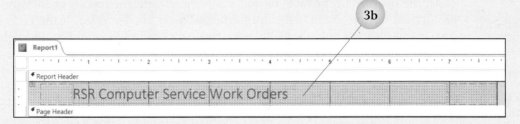

4. With the title control object still selected, click the Report Design Tools Format tab and then click the Center button in the Font group.
5. Drag the right edge of the report grid to the right until it is aligned at the 8-inch position on the horizontal ruler.
6. Scroll down the report until the *Page Footer* and *Report Footer* sections are visible. The *Report Footer* section was added to the design grid at the same time the *Report Header* section was added when the title was created in Step 3.

7. Drag down the bottom edge of the *Report Footer* section until the bottom of the report is aligned at the 0.5-inch position on the vertical ruler.
8. Click the Report Design Tools Design tab and then click the Label button in the Controls group. Add a label control object containing your first and last names at the left side of the *Report Footer* section.
9. Click in a blank area of the report to deselect the label control object.
10. Save the report and name it *WorkOrders*.

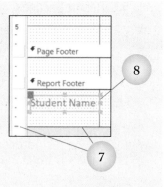

Check Your Work

 Tutorial

Connecting a Table to a Report and Adding Fields

Quick Steps

Connect Table or Query to Report
1. Open report in Design view.
2. Double-click Report Selector button.
3. Click Data tab in Property Sheet task pane.
4. Click arrow in *Record Source* property box.
5. Click table or query.
6. Close Property Sheet task pane.

Add Fields to Report
1. Click Add Existing Fields button.
2. Drag field name(s) from Field List task pane to *Detail* section.

Add Field from Related Table
1. Open Field List task pane.
2. Click Show all tables hyperlink.
3. Click expand button next to table name in *Fields available in related tables* section.
4. Drag field name from related table list to *Detail* section.

 Report Selector

 Add Existing Fields

Connecting a Table or Query to a Report and Adding Fields

A new report that is started in Design view does not have a table or query associated with it. To display data in the report, Access needs to know what source to draw the data from. Connect a table or query to the report using the Record Source property in the report's Property Sheet task pane. Make sure to complete this step before adding fields to the *Detail* section.

The steps for connecting a table or query to a report are the same as the steps for connecting a table to a form. Double-click the Report Selector button above the vertical ruler and left of the horizontal ruler to open the report's Property Sheet task pane. Click the Data tab and then select the table or query name in the drop-down list at the *Record Source* property box. Display the Field List task pane and then drag individual fields or a group of fields from the table or query to the *Detail* section. After adding the fields, move and resize the control objects as needed.

The Field List task pane displays in one of two ways: showing one section only, with the fields from the table or query associated with the report, or showing two additional sections, with fields from other tables in the database. If the Field List task pane contains only the fields from the associated table or query, add fields from other tables by displaying other table names in the database within the Field List task pane.

At the top of the Field List task pane, Access displays a Show all tables hyperlink. Click the hyperlink to display the *Fields available in related tables* and *Fields available in other tables* sections. Next to each table name is an expand button, which displays as a plus symbol. Click the expand button next to the table name to display the fields stored within the table and then drag the field name(s) to the *Detail* section of the report. You will perform these steps in Activity 1b.

1. With **5-RSRCompServ** open and the WorkOrders report open in Design view, scroll to the top of the report in the work area.
2. Connect the WorkOrders table to the report so that Access knows which fields to display in the Field List task pane by completing the following steps:
   a. Double-click the Report Selector button at the top of the vertical ruler and left of the horizontal ruler to open the report's Property Sheet task pane.
   b. Click the Data tab in the Property Sheet task pane, click the *Record Source* property option box arrow, and then click *WorkOrders* at the drop-down list.
   c. Close the Property Sheet task pane.

3. Click the Add Existing Fields button in the Tools group on the Report Design Tools Design tab to open the Field List task pane.
4. Add fields from the WorkOrders table and related fields from the Customers table by completing the following steps:
   a. Click the Show all tables hyperlink at the top of the Field List task pane. Access adds two sections to the pane: one containing related tables and the other containing tables in the database that do not have established relationships with the report's table. Next to each table name is an expand button (plus symbol), which is used to display field names for the table. *Note: Skip this step if the **Show only fields in the current record source** hyperlink displays at the top of the Field List task pane. This means that the other sections have already been added to the pane.*

   b. Click the expand button next to *Customers* in the *Fields available in related tables* section of the Field List task pane. Access expands the list to display the field names in the Customers table.

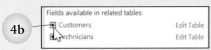

   c. Drag the *WO*, *CustID*, *WODate*, and *Descr* fields from the WorkOrders table to the design grid, as shown below.
   d. Drag the *FName* and *LName* fields from the Customers table to the design grid, as shown at the right. Notice that the Customers table and field names move to the *Fields available for this view* section in the Field List task pane after the first field has been added from the Customers table to the *Detail* section.
5. Close the Field List task pane.
6. Save the report.

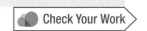

Tutorial

Moving Control
Objects to Another
Section

Moving Control Objects to Another Section

When a field is added to the *Detail* section of a report, a label control object containing the caption or field name is placed to the left of a text box control object that displays the field value from the record when the report is viewed or printed. Recall from Chapter 4 that the same thing happens when a form is customized.

In the WorkOrders report, the label control object for each field needs to be moved to the *Page Header* section so that the field names or captions print at the top of each page as column headings. In Activity 1c, the control objects will be cut from the *Detail* section and pasted into the *Page Header* section.

Quick Steps

**Move Control Objects
to Another Section**

1. Open report in Design view.
2. Select control objects to be moved.
3. Click Home tab.
4. Click Cut button.
5. Click bar of section to move control objects to.
6. Click Paste button.
7. Deselect controls.

Activity 1c Moving Control Objects to Another Section

1. With **5-RSRCompServ** open and the WorkOrders report open in Design view, move the label control objects from the *Detail* section to the *Page Header* section by completing the following steps:
   a. Click to select the *Work Order* label control object.
   b. Press and hold down the Shift key, click to select each of the other label control objects, and then release the Shift key.
   c. Click the Home tab and then click the Cut button in the Clipboard group.
   d. Click the *Page Header* section bar.
   e. Click the Paste button in the Clipboard group. (Do not click the button arrow.) Access pastes the label control objects and expands the *Page Header* section.
   f. Deselect the control objects.

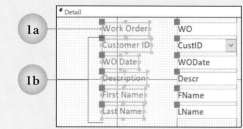

2. Click to select the *Customer ID* label control object and then move the control object to the top of the *Page Header* section next to the *Work Order* label control object by hovering the mouse pointer over the *Customer ID* label control object until the four-headed arrow move icon displays and then dragging the control object to the location shown below.

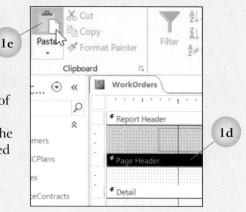

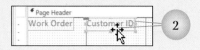

3. Move the remaining four label control objects to the top of the *Page Header* section in the order shown in the image below by completing a step similar to Step 2.
4. Drag the top of the *Detail* section bar up until the top of the bar is aligned at the bottom edge of the label control objects in the *Page Header* section, as shown below.

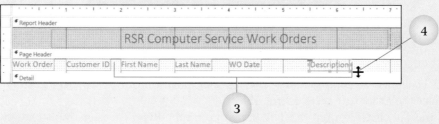

5. Save the report.

Check Your Work

Tutorial

Formatting a Report

Themes

Applying a Theme

Apply a theme to a report using the Themes button on the Report Design Tools Design tab. The theme sets the default colors and fonts for the report. The themes available in Access align with the themes available in Word, Excel, and PowerPoint. This allows having the same look in Access reports as in other documents, worksheets, and presentations. Note that changing a theme for one report in a database automatically changes the theme for all the reports and forms in the database.

Activity 1d Moving Controls, Resizing Controls, and Applying a Theme   Part 4 of 5

1. With **5-RSRCompServ** open and the WorkOrders report open in Design view, move each text box control object in the *Detail* section below its associated label control object in the *Page Header* section so that the field values align below the correct column headings in the report, as shown below.

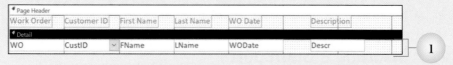

2. Click the Report Design Tools Design tab, click the View button arrow in the Views group, and then click *Print Preview* at the drop-down list. (Note that there is also a Print Preview button in the view area at the right side of the Status bar.) Notice that the field value in the *WO Date* column displays pound symbols (#); this indicates that the field's text box control object needs to be widened. **Note: If you receive an error message with the text The section width is greater than the page width, click OK.**

3. Return to Design view by clicking the Design View button in the view area at the bottom right side of the Status bar, next to the Zoom slider.

4. Resize the *WODate* text box control object in the *Detail* section so that the right edge of the control meets the left edge of the *Descr* text box control object.
5. Resize the *Descr* text box control in the *Detail* section so that the right edge of the control object is aligned at approximately the 7.5-inch position on the horizontal ruler.
6. Deselect the *Descr* text box control object.
7. Click the Themes button in the Themes group on the Report Design Tools Design tab and then click *Organic* at the drop-down gallery (third column, second row in the *Office* section).

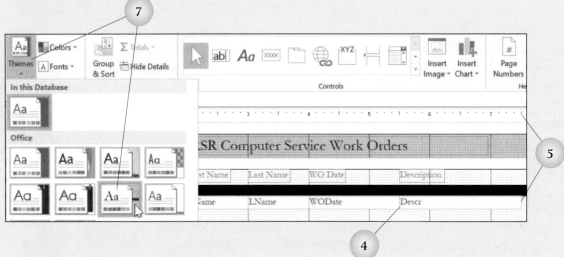

8. Save the report.
9. Display the report in Print Preview to review the changes made in this activity. Switch back to Design view when finished.

Check Your Work >

Tutorial >

Inserting a Subreport

Inserting a
Subreport

A report that is inserted inside another report is called a *subreport*. Similar to a nested query, a subreport allows for reusing a group of fields, formats, and calculations in more than one report without having to recreate the setup each time.

Subform/
Subreport

Use the Subform/Subreport button in the Controls group on the Report Design Tools Design tab to insert a subreport into a report. The report into which the subreport is inserted is called the *main report*. Adding a related table or query as a subreport creates a control object within the main report that can be moved, formatted, and resized independently of the other control objects. Make sure the Use Control Wizards button is toggled on in the expanded Controls group before clicking the Subform/Subreport button. This will enable the subreport to be added using the SubReport Wizard, shown in Figure 5.2.

Insert Subreport
1. Open report in Design view.
2. Make sure *Use Control Wizards* is active.
3. Click Subform/ Subreport button.
4. Drag crosshairs to appropriate height and width in *Detail* section.
5. Click Next.
6. Choose table or query and fields.
7. Click Next.
8. Choose field by which to link main report with subreport.
9. Click Next.
10. Click Finish.

Figure 5.2 First Dialog Box in the SubReport Wizard

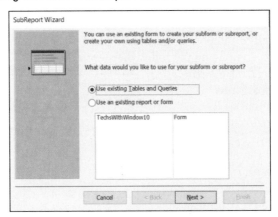

A subreport is stored as a separate object outside the main report. An additional report name will display in the Navigation pane with *subreport* at the end. Do not delete a subreport object in the Navigation pane. If the subreport object is deleted, the main report will no longer display the fields from the related table or query in the report.

Activity 1e Inserting a Subreport

Part 5 of 5

1. With **5-RSRCompServ** open and the WorkOrders report open in Design view, insert a subreport into the WorkOrders report with fields from a query for the service date, labor, and parts for each work order by completing the following steps:
   a. By default, the Use Control Wizards button is toggled on in the Controls group. Click the More button in the Controls group to expand the Controls group and view the buttons and the Controls drop-down list. View the current status of the Use Control Wizards button; if it displays with a gray background, the feature is active. If the button is gray, click in a blank area to close the expanded Controls list. If the feature is not active (displays with a white background), click the Use Control Wizards button to turn on the feature.
   b. Click the More button in the Controls group and then click the Subform/Subreport button.
   c. Move the crosshairs with the subreport icon attached to the *Detail* section below the *WO* text box control object and then drag down and to the right to create a subreport object of the approximate height and width shown below. Release the mouse and the SubReport Wizard begins.

> A gray background means the Use Control Wizards button is toggled on. Check the status and click the button to turn on the feature if necessary in Step 1a.

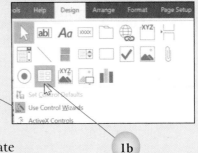

1b

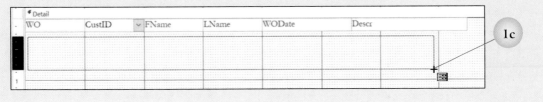

1c

d. With *Use existing Tables and Queries* already selected at the first SubReport Wizard dialog box, click the Next button.

e. At the second SubReport Wizard dialog box, select the query fields to be displayed in the subreport by completing the following steps:

1) Click the *Tables/Queries* option box arrow and then click *Query: TotalWorkOrders* at the drop-down list.

2) Move all the fields from the *Available Fields* list box to the *Selected Fields* list box.

3) Click the Next button.

f. At the third SubReport Wizard dialog box, choose the field by which to link the main report with the subreport by completing the following steps:

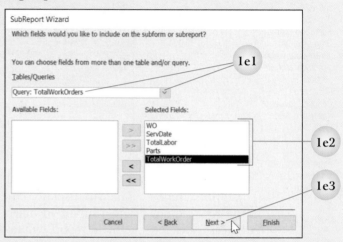

1) With *Choose from a list* and the first option in the list box selected, read the description of the linked field below the list box. The text indicates that the main report will be linked to the subreport using the *CustID* field. This is not the correct field; your report is to show the service date, labor, and parts based on the work order number.

2) Click the second option in the list box and then read the text below the list box.

3) Since the second option indicates that the two reports will be linked using the *WO* field, click the Next button.

g. Click the Finish button at the last SubReport Wizard dialog box to accept the default subreport name *TotalWorkOrders subreport*.

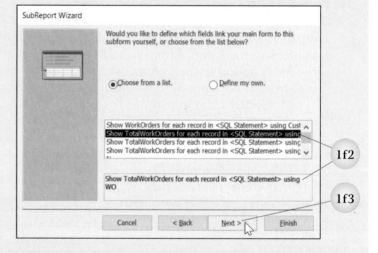

2. Access inserts the subreport with a label control object above it that contains the name of the subreport. Click the label control displaying the text *TotalWorkOrders subreport* to select the object and then press the Delete key.

3. Click the Report View button in the Views group on the Report Design Tools Design tab. (Note that a Report View button is also included in the View group, the first button on the right side of the Status bar.) Report view is not the same as Print Preview. Report view displays the report with data in the fields and is useful for viewing reports within the database. However, this view does not show how the report will look when printed. For printing purposes, always use Print Preview to resize and adjust control objects.
4. Notice that the work order number in the subreport is the same work order number that is displayed in the first record in the main report.

WorkOrders

RSR Computer Service Work Orders

Work Order	Customer ID	First Name	Last Name	WO Date	Description
65030	1000	Jade	Fleming	Fri, Dec 10 2021	Clean malware from system

Work Order	Service Date	Total Labor	Parts	Total Work Order
65030	Mon, Dec 13 2021	$40.00	$0.00	$40.00

4

5. Switch back to Design view.
6. Now that you know the subreport is linked correctly to the main report, the work order number does not need to be displayed in the subreport. Delete the work order number control objects in the subreport by completing the following steps:
   a. Click to select the subreport control object and then drag down the bottom middle sizing handle to increase the height of the subreport until all the controls are visible in the *Report Header* and *Detail* sections.
   b. Click to select the *Work Order* label control object in the *Report Header* section, press and hold down the Shift key, click the *WO* text box control object in the *Detail* section in the subreport, and then release the Shift key.
   c. Press the Delete key.
   d. Select all the label control objects in the *Report Header* section and all the text box control objects in the *Detail* section. Move the control objects to the left so the left edge of the *ServDate* control objects are at the left margin. Click outside the subreport to deselect the control objects.

6b

7. Click to select the subreport control object and then drag up the bottom middle sizing handle of the control until the height of the subreport is approximately 0.5 inch.
8. Scroll down the report until the *Page Footer* section bar is visible.
9. Drag up the top of the *Page Footer* section bar until the section bar is just below the subreport control object at the 1-inch mark in the *Detail* section, as shown below.

9 **7**

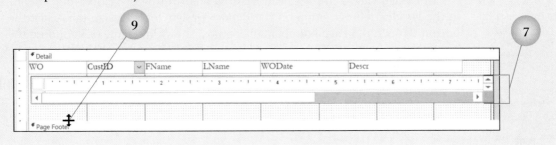

10. Save the report and then switch to Report view to view the revised report. Resizing the *Detail* section in Step 9 reduces the space between sections and allows more records and related subreport records to display on the page.

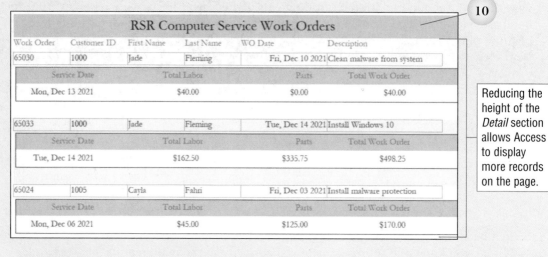

RSR Computer Service Work Orders

Work Order	Customer ID	First Name	Last Name	WO Date	Description
65030	1000	Jade	Fleming	Fri, Dec 10 2021	Clean malware from system

Service Date	Total Labor	Parts	Total Work Order
Mon, Dec 13 2021	$40.00	$0.00	$40.00

Work Order	Customer ID	First Name	Last Name	WO Date	Description
65033	1000	Jade	Fleming	Tue, Dec 14 2021	Install Windows 10

Service Date	Total Labor	Parts	Total Work Order
Tue, Dec 14 2021	$162.50	$335.75	$498.25

Work Order	Customer ID	First Name	Last Name	WO Date	Description
65024	1005	Cayla	Fahri	Fri, Dec 03 2021	Install malware protection

Service Date	Total Labor	Parts	Total Work Order
Mon, Dec 06 2021	$45.00	$125.00	$170.00

Reducing the height of the *Detail* section allows Access to display more records on the page.

11. Close the report.

Check Your Work

Activity 2 Add Features and Enhance a Report   2 Parts

You will modify the WorkOrders report to add page numbers, date and time control objects, and graphics.

 Tutorial

Adding Page Numbers and the Date and Time Control Objects to a Report

Adding Page Numbers and the Date and Time Control Objects to a Report

When creating a report using the Report tool, page numbers and the current date and time are automatically added to the top right of the report. The Report Wizard automatically inserts the current date at the bottom left and page numbers at the bottom right of the report.

Quick Steps

Add Page Numbers
1. Open Report in Design view.
2. Click Page Numbers button.
3. Select format, position, and alignment options.
4. Click OK.

 Page
Numbers

In Design view, add page numbers to a report using the Page Numbers button in the Header / Footer group on the Report Design Tools Design tab. Click the button to open the Page Numbers dialog box, shown in Figure 5.3. Choose the appropriate format, position, and alignment for the page number and then click OK. Access inserts a control object in the *Page Header* or *Page Footer* section, depending on the *Position* option selected in the dialog box. Including page numbers in a report allows referring to specific pages and putting pages back in order if they get rearranged.

Add the current date and/or time in the *Report Header* section by clicking the Date and Time button in the Header / Footer group to open the Date and Time dialog box, shown in Figure 5.4. By default, both the *Include Date* and *Include Time* check boxes contain check marks. Access creates one control object for the date format and another for the time format. Access places the date control object above the time control object and aligns both at the right edge of the *Report Header* section. Once the control objects have been inserted, they can be moved to another section in the report. Adding a date and/or time control object means that the current date and/or time the report is printed will be included on the printout. At a minimum, always include a date control. Depending on users' needs, the time control may or may not be necessary.

Figure 5.3 Page Numbers Dialog Box

Figure 5.4 Date and Time Dialog Box

Activity 2a Adding Page Numbers and the Date and Time to a Report   Part 1 of 2

1. With **5-RSRCompServ** open, right-click *WorkOrders* in the Report group in the Navigation pane and then click *Design View* at the shortcut menu.
2. When the subreport was inserted in Activity 1e, the width of the report may have been automatically extended beyond the page width. Look at the Report Selector button.
   If a green diagonal triangle displays in the upper left corner, correct the page width by completing the following steps. (Skip this step if the Report Selector button does not display with a green diagonal triangle.)
   a. Click the subreport control object to display the orange border and sizing handles. Point to the orange border and then drag the subreport left until the left edge is at the left edge of the *Detail* section.
   b. Drag the right middle sizing handle left to decrease the subreport width until the right edge of the subreport is aligned with the right edge of the *Descr* text box control object above it.

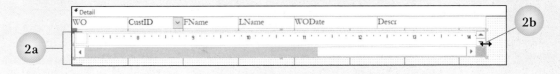

c. Click the green triangle to display the Error-Checking Options button and then click the button to display the drop-down list of options.

d. Click *Remove Extra Report Space* at the drop-down list to automatically decrease the width of the report. Notice that the green diagonal triangle is removed from the Report Selector button once the report width has been corrected.

3. Add a page number at the bottom center of each page by completing the following steps:

a. Click the Page Numbers button in the Header / Footer group on the Report Design Tools Design tab.

b. Click *Page N of M* in the *Format* section of the Page Numbers dialog box.

c. Click *Bottom of Page [Footer]* in the *Position* section.

d. With the *Alignment* option box set to *Center* and a check mark in the *Show Number on First Page* check box, click OK. Access adds a control object in the center of the *Page Footer* section that contains the codes required to print the page numbers centered at the bottom of all pages, including the first page.

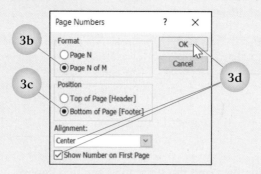

4. Add the current date and time to the end of the report along with a label control object that contains the text *Date and Time Printed:* by completing the following steps:

a. Click the Date and Time button in the Header / Footer group on the Report Design Tools Design tab.

b. Click the second option in the *Include Date* section in the Date and Time dialog box, which displays the date in the format *dd-mmm-yy*—for example, *17-Jan-21*.

c. Click the second option in the *Include Time* section, which displays the time in the format *hh:mm AM/PM*—for example, *2:05 PM*.

d. Click OK. Access adds two control objects, one above the other, at the right side of the *Report Header* section with the date code *=Date()* and time code *=Time()*.

e. Select the date and time control objects added to the *Report Header* section. Click the Home tab and then click the Cut button.

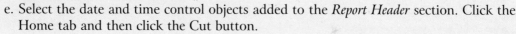

f. Click the *Report Footer* section bar and then click the Paste button. Access pastes the two control objects at the left side of the *Report Footer* section. With the date and time control objects still selected, position the mouse pointer on the orange border until the pointer displays with the four-headed arrow move icon and then drag the control objects to the right side of the *Report Footer* section, aligning the right edge of the control objects near the right edge of the report grid.

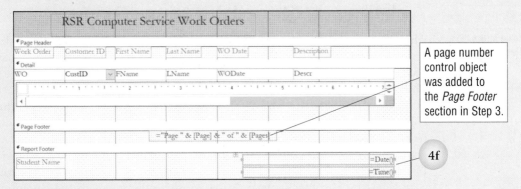

A page number control object was added to the *Page Footer* section in Step 3.

4f

g. Resize the date and time controls to be approximately 0.75-inch wide as shown at the right.
h. Create the two label control objects, type the text Date Printed: and Time Printed:, and then position the label control objects left of the date control object and time control object, as shown at the right.

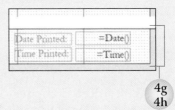

4g
4h

5. Save the report and then display it in Print Preview.
6. Scroll to the bottom of the first page to view the page number.
7. Click the Last Page button in the Page Navigation bar to scroll to the last page in the report and view the date and time.
8. Notice that in Print Preview, the subreport data is cut off at the right edge of the report, which means the total work order value is not visible. Exit Print Preview to switch back to Design view.
9. Adjust the size and placement of the subreport control objects by completing the following steps:
   a. Since the subreport control objects are not visible within the WorkOrders report, it is easier to work within the separate TotalWorkOrders subreport to make changes to its content. Close the WorkOrders report.
   b. Right-click *TotalWorkOrders subreport* in the Navigation pane and then click *Design View* at the shortcut menu.
   c. Using the diagram below as a guide, adjust the widths of the Total Labor and/or Parts label control objects and text box control objects. Remove any space between the control objects.
   d. Drag the right edge of the grid left to approximately the 7-inch position on the horizontal ruler.

9c

9d

10. Save and close the TotalWorkOrders subreport.
11. Open the WorkOrders report. Review the data.
12. Display the report in Design view.

Adding a Graphic to
a Report

Adding a Graphic to a Report

The same techniques learned in Chapter 4 for adding online images or drawing lines in a form in Design view can be applied to a report. An image in a standard picture file format (such as .gif, .jpg, or .png) can be inserted in an image control object. Click the Insert Image button in the Controls group on the Report Design Tools Design tab, browse to the drive and folder containing the image, double-click the image file name, and then drag to create an image control object of the required height and width within the report. Recall from Chapter 4 that when a control object containing an image is resized, parts of the image can be cut off. Display the Property Sheet task pane for the control object and then change the Size Mode property to *Zoom* or *Stretch* to resize the image to the height and width of the control object.

Activity 2b Adding Graphics and Formatting Control Objects   Part 2 of 2

1. With **5-RSRCompServ** open and the WorkOrders report open in Design view, insert a company logo in the report by completing the following steps:
   a. Position the mouse pointer on the top of the *Page Header* section bar until the pointer displays as an up-and-down-pointing arrow with a horizontal line in the middle and then drag down approximately 0.25 inch to increase the height of the *Report Header* section.
   b. Click the Insert Image button in the Controls group on the Report Design Tools Design tab and then click *Browse* at the drop-down list.
   c. At the Insert Picture dialog box, navigate to the drive and/or folder for the AL2C5 data files on your storage medium and then double-click the file named **RSRlogo**.
   d. Position the crosshairs with the image icon attached at the top of the *Report Header* section near the 6-inch position on the horizontal ruler and then drag to create an image control object of the approximate height and width shown.

2. Select the title control object in the *Report Header* section and then click the Bold button in the Font group on the Report Design Tools Format tab.

3. Draw and format a horizontal line below the title by completing the following steps:
   a. Click the More button in the Controls group on the Report Design Tools Design tab and then click the Line button in the expanded Controls group.
   b. Position the crosshairs with the line icon attached below the first letter in the title in the *Report Header* section, press and hold down the Shift key, click the left mouse button and drag to the right, release the mouse button below the last letter in the title, and then release the Shift key.
   c. Click the Report Design Tools Format tab and then click the Shape Outline button in the Control Formatting group.
   d. Point to *Line Thickness* and then click *3 pt* (fourth option).
   e. With the line still selected, click the Shape Outline button, click *Blue-Gray, Accent 3* (seventh column, first row in the *Theme Colors* section), and then deselect the line.

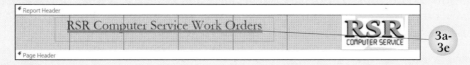

4. Draw and format a horizontal line below the column headings by completing the following steps:
   a. Drag down the top of the *Detail* section bar approximately 0.25 inch to add grid space in the *Page Header* section.
   b. Click the Report Design Tools Design tab, click the More button in the Controls group, and then click the Line button in the expanded Controls group.
   c. Draw a straight horizontal line that extends the width of the report along the bottom of the label control objects in the *Page Header* section.
   d. Click the Shape Outline button on the Report Design Tools Format tab, point to *Line Thickness*, and then click *2 pt*. Change the line color to the same Blue-Gray applied to the line below the report title.
   e. Deselect the line.

5. Format, move, and resize the control objects as follows:
   a. Select all the label control objects in the *Page Header* section, apply bold formatting, change the font size to 12 points, and then apply the Blue-Gray, Accent 3, Darker 50% font color (seventh column, last row in the *Theme Colors* section).
   b. Resize the control objects as needed to show all the label text after increasing the font size.
   c. Move the *WO Date* label control object in the *Page Header* section until the left edge of the control object is aligned at the 4.5-inch position on the horizontal ruler.
   d. Click the *Report Header* section bar, click the Shape Fill button in the Control Formatting group on the Report Design Tools Format tab, and then click *White, Background 1* (first column, first row in the *Theme Colors* section).
   e. Select all the text box control objects in the *Detail* section. Open the Property Sheet task pane, click the Format tab, click in the *Border Style* property box, click the arrow that appears, and then click *Transparent* at the drop-down list. This removes the borders around the data in the fields.

f. With all the text box control objects still selected, click the *Fore Color* property box on the Format tab of the Property Sheet task pane. Click the Build button to open the color palette and then click *Black, Text 1* (second column, first row in the *Theme Colors* section).

g. Deselect the text box control objects and then select the *WODate* text box control object. Click in the *Width* property box in the Property Sheet task pane, change *1.5* to *1.4*.

h. Select the *Descr* text box control object, change the number in the *Width* property box to *2.25*, and then close the Property Sheet task pane.

6. Display the report in Report view.

7. Compare the report with the partial report shown in Figure 5.5. If necessary, return to Design view; adjust the formats, alignments, or positions of the control objects; and then redisplay the report in Report view.

8. Save the report.

9. Print the report. *Note: This report is four pages long. Check with your instructor before printing it.*

10. Close the WorkOrders report.

 Check Your Work

Figure 5.5 Partial View of the Completed WorkOrders Report

Work Order	Customer ID	First Name	Last Name	WO Date	Description
65030	1000	Jade	Fleming	Fri, Dec 10 2021	Clean malware from system

Service Date	Total Labor	Parts	Total Work Order
Mon, Dec 13 2021	$40.00	$0.00	$40.00

Work Order	Customer ID	First Name	Last Name	WO Date	Description
65033	1000	Jade	Fleming	Tue, Dec 14 2021	Install Windows 10

Service Date	Total Labor	Parts	Total Work Order
Tue, Dec 14 2021	$162.50	$335.75	$498.25

Work Order	Customer ID	First Name	Last Name	WO Date	Description
65024	1005	Cayla	Fahri	Fri, Dec 03 2021	Install malware protection

Service Date	Total Labor	Parts	Total Work Order
Mon, Dec 06 2021	$45.00	$125.00	$170.00

Activity 3 Group Records and Add Functions to Count and Sum   **3 Parts**

You will create a new report using the Report Wizard and then modify the report in Design view to add count and sum functions.

Tutorial

Review: Creating a Report Using the Report Wizard

Tutorial

Grouping Records in a Report

Grouping Records and Adding Functions and a Calculated Field in a Report

A field that contains repeated values—such as a department, city, or name—is an appropriate field to group records on in a report. For example, organize a report to show all the records for the same department or city. By summarizing the records by a common field value, totals can be produced using functions to calculate the

Figure 5.6 Example Report with the Work Order Records Grouped by Customer

RSR COMPUTER SERVICE

Work Orders by Customer

23-Jan-22
2:49 PM

Customer ID	First Name	Last Name	Work Order	Description	Service Date	Total Work Order
1000	Jade	Fleming	65030	Clean malware from system	Mon, Dec 13 2021	$40.00
			65033	Install Windows 10	Tue, Dec 14 2021	$498.25
					Customer Total	**$538.25**
1005	Cayla	Fahri	65024	Install malware protection	Mon, Dec 06 2021	$170.00
					Customer Total	**$170.00**
1008	Leslie	Carmichael	65032	Install second storage drive	Mon, Dec 13 2021	$175.00
			65036	Set up home network	Fri, Dec 17 2021	$220.22
			65044	Noisy fan	Wed, Dec 22 2021	$105.40
					Customer Total	**$500.62**
1010	Randall	Lemaire	65025	Troubleshoot hard drive noise	Mon, Dec 06 2021	$75.00
			65027	Replace hard drive with SSD	Tue, Dec 07 2021	$185.00
			65038	Install latest version of Windows	Mon, Dec 20 2021	$125.00
			65046	Windows 10 training	Tue, Dec 28 2021	$100.00
					Customer Total	**$485.00**

The report is grouped on the *CustID* field (displays with column heading *Customer ID*), allowing the owners to see how many work orders and how much revenue each customer generated.

Quick Steps
Group Records Using Report Wizard
1. Click Create tab.
2. Click Report Wizard button.
3. Choose table or query and fields.
4. Click Next.
5. If necessary, remove default grouped field name.
6. Double-click field name by which to group records.
7. Click Next.
8. Choose field(s) by which to sort.
9. Click Next.
10. Choose layout options.
11. Click Next.
12. Enter title for report.
13. Click Finish.

 Group & Sort

sum, average, maximum, minimum, or count for each group. For example, a report similar to the partial report shown in Figure 5.6, which organizes work orders by customer, allows the owners of RSR Computer Service to see which customer has provided the most revenue to their service business. In this report, the *CustID* field (column heading *Customer ID*) is used to group the records and a Sum function has been added to each group.

Recall from Level 1, Chapter 6 that either the Group & Sort button or the Report Wizard can be used to group records in a report. At the Report Wizard dialog box, shown in Figure 5.7, double-click a field name in the list box to add a grouping level. The preview window updates to display the grouped field in blue. More than one grouping level can be added to a report. To remove a grouping level, use the Remove Field button (the button with the left-pointing arrow) to remove the grouped level. Use the Priority buttons (the buttons with up and down arrows) to change the grouping order when there are multiple grouped fields.

If a report was not grouped when using the Report Wizard, group records can be added after the report has been generated using Layout view or Design view. In Layout view, click the Group & Sort button in the Grouping & Totals group on the Report Layout Tools Design tab. In Design view, click the Group & Sort button in the Grouping & Totals group on the Report Design Tools Design tab.

Figure 5.7 Grouping by a Field Using the Report Wizard

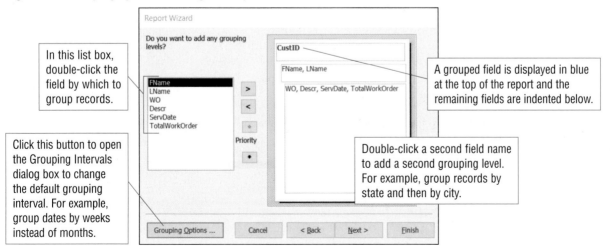

In this list box, double-click the field by which to group records.

Click this button to open the Grouping Intervals dialog box to change the default grouping interval. For example, group dates by weeks instead of months.

A grouped field is displayed in blue at the top of the report and the remaining fields are indented below.

Double-click a second field name to add a second grouping level. For example, group records by state and then by city.

Figure 5.8 Group, Sort, and Total Pane

Clicking the button in either view opens the Group, Sort, and Total pane, shown in Figure 5.8, at the bottom of the work area. Click the Add a group button and then click the field name by which to group records in the pop-up list.

Activity 3a Creating a Report with a Grouping Level Using the Report Wizard   Part 1 of 3

1. With **5-RSRCompServ**
   open, modify the TotalWorkOrders query
   to add two fields to include in a report by
   completing the following steps:
   a. Open the TotalWorkOrders query in
      Design view.
   b. Drag the *CustID* field from the *WorkOrders*
      table field list box to the *Field* row in the
      second column in the query design grid.
      ServDate and other fields will shift right to
      accommodate the new field.
   c. Drag the *Descr* field from the *WorkOrders*
      table field list box to the *Field* row in the
      third column in the query design grid.
   d. Save the revised query.
   e. Run the query.
   f. Close the query.

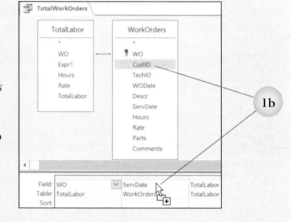

In Step 1e, the revised query is run with the new fields added.

Work Order ▾	CustID ▾	Descr ▾	Service Date ▾	Total Labor ▾	Parts ▾	Total Work Order ▾
65019	1025	Upgrade to nev	Wed, Dec 01 2021	$137.50	$0.00	$137.50
65020	1030	Troubleshoot n	Thu, Dec 02 2021	$75.00	$72.50	$147.50

2. Create a report based on the TotalWorkOrders query that is grouped by service date by
   week using the Report Wizard by completing the following steps:
   a. Click the Create tab and then click the Report Wizard button in the Reports group.
   b. At the first Report Wizard dialog box with *Query: TotalWorkOrders* already selected in
      the *Tables/Queries* option box, move all the fields from the *Available Fields* list box to the
      Selected Fields list box and then click the Next button.
   c. At the second Report Wizard dialog box, specify that the report is to be grouped by the
      ServDate field by completing the following steps:
      1) With *CustID* displayed in blue in the
         preview section, indicating that the
         report will be grouped by customer
         number, click the Remove field
         button (the left-pointing arrow) to
         remove the grouping level.

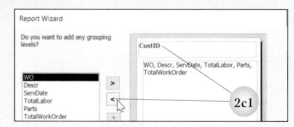

2) Double-click *ServDate* in the field
list box to add a grouping level by
the service date field. By default,
Access groups a date field by month.
3) Click the Grouping Options button.
4) At the Grouping Intervals dialog box
click the Grouping intervals option
box arrow, click *Week*, and then click
OK.
5) With the preview section now
displaying that the report will
be grouped by *ServDate by
Week*, click the Next button.

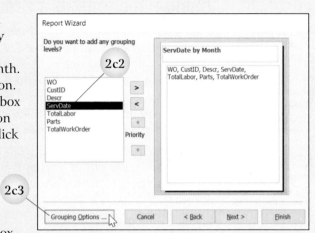

d. At the third Report Wizard dialog box,
click the arrow at the right of the first
sort option box, click *WO* at the drop-
down list to sort each group by work
order number in ascending order, and
then click the Next button.

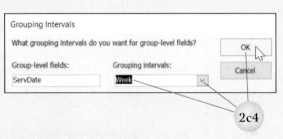

e. At the fourth Report Wizard dialog
box, click *Landscape* in the *Orientation*
section and then click the Next button.
f. At the last Report Wizard dialog box, select the existing text in
the *What title do you want for your report?* text box, type WorkOrdersbyWeek, and then
click the Finish button.
3. Switch to Design view.
4. Add an alternate row color by completing the following steps:
a. Click the *Detail* section bar.
b. Click the Report Design Tools Format tab.
c. Click the Alternate Row Color button in the Background group.
d. Click *Blue-Gray, Accent 3, Lighter 80%* (seventh column, second row in the Theme Colors
section).
5. Switch to Report view.
6. Minimize the Navigation pane.
7. Preview the report and then switch to Layout view or Design view. Edit the text in the
report title and column heading labels, adjust column widths and the *Report Header* section
height as necessary until the report looks similar to the one shown below. The Organic
theme was applied in Activity 1d. Modify the colors for the *Report Header* and *Page Header*
sections to most closely match the theme colors shown below.

Work Orders by Week

Serv Date by Week	Work Order	CustID	Descr	Service Date	Total Labor	Parts	Total Work Order
49							
	65019	1025	Upgrade to new Windows	Wed, Dec 01 2021	$137.50	$0.00	$137.50
	65020	1030	Troubleshoot noisy fan	Thu, Dec 02 2021	$75.00	$72.50	$147.50
	65021	1035	Customer has blue screen upon boot	Thu, Dec 02 2021	$178.75	$0.00	$178.75
	65022	1040	Customer reports screen is fuzzy	Fri, Dec 03 2021	$82.50	$400.00	$482.50
	65023	1045	Upgrade RAM	Sat, Dec 04 2021	$62.50	$100.00	$162.50
50							
	65024	1005	Install malware protection	Mon, Dec 06 2021	$45.00	$125.00	$170.00
	65025	1010	Troubleshoot hard drive noise	Mon, Dec 06 2021	$75.00	$0.00	$75.00

8. Save the report.

Check Your Work

Adding Functions to a Group

When a report is grouped, the Group, Sort, and Total pane can be used to add a calculation below a numeric field at the end of each group. Functions can also be added to more than one field within the group. For example, calculate a Sum function on a sales field and a Count function on an invoice field. The following functions are available for numeric fields: Sum, Average, Count Records, Count Values, Maximum, Minimum, Standard Deviation, and Variance. A non-numeric field can have a Count Records or Count Values function added.

The Group, Sort, and Total pane for a report with an existing grouping level looks similar to the one shown in Figure 5.9. Click the More Options button next to the group level to which a total is to be added to expand the available group options.

Click the option box arrow next to *with no totals* to open a *Totals* option box, similar to the one shown in Figure 5.10. Use the option boxes within this box to select the field a function should be added to and the type of aggregate function to calculate. Use the check boxes to choose to add a grand total to the end of the report, calculate group subtotals as percentages of the grand total, and decide whether to add the subtotal function to the *Group Header* or *Group Footer* section. Continue adding functions to other fields as needed and then click outside the *Totals* option box to close it.

Figure 5.9 Group, Sort, and Total Pane with a Grouping Level Added

Figure 5.10 *Totals* Option Box in the Group, Sort, and Total Pane

1. With **5-RSRCompServ** open, display the WorkOrdersbyWeek report in Design view.
2. Add two functions at the end of each week to show the number of work orders and the total value of the work orders by completing the following steps:

    a. In the Grouping & Totals group on the Report Design Tools Design tab, click the Group & Sort button.

    b. At the Group, Sort, and Total pane at the bottom of the work area, click the More Options button next to *from oldest to newest* in the *Group on ServDate* group options.

    c. Click the option box arrow next to *with no totals* in the expanded group options.

    d. At the *Totals* option box with *WO* selected in the *Total On* option box, specify the type of function and the placement of the result by completing the following steps:

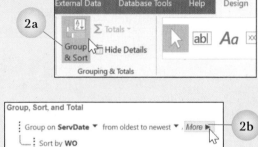

1) Click the *Type* option box arrow and then click *Count Records* at the drop-down list.
2) Click the *Show Grand Total* check box to insert a check mark. Access adds a Count function in a control object in the *Report Footer* section below the *WO* column.
3) Click the *Show subtotal in group footer* check box to insert a check mark. Access displays a new section with the title *ServDate Footer* in the section bar below the *Detail* section and inserts a Count function in a control object below the *WO* column.

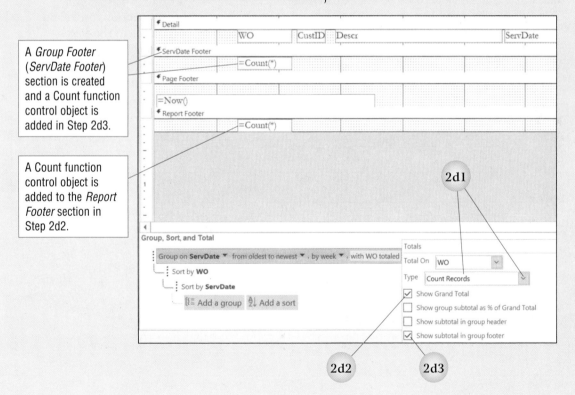

A *Group Footer* (*ServDate Footer*) section is created and a Count function control object is added in Step 2d3.

A Count function control object is added to the *Report Footer* section in Step 2d2.

e. With the *Totals* option box still open, click the *Total On* option box arrow and then click *TotalWorkOrder* at the drop-down list. The *Type* option defaults to *Sum* for a numeric field.

f. Click the *Show Grand Total* check box to insert a check mark. Access adds a Sum function in a control object in the *Report Footer* section.

g. Click the *Show subtotal in group footer* check box to insert a check mark. Access adds a Sum function in a control object in the *ServDate Footer* section.

h. Click outside the *Totals* option box to close it.

3. Click the Group & Sort button to close the Group, Sort, and Total pane.

4. Review the two Count functions and two Sum functions added to the report in Design view.

5. Display the report in Print Preview to view the calculated results. Notice that the printout requires two pages and that the report's grand totals print on page 2. If any totals display with pound signs (#) switch to Design view, adjust the width of the control object, and then switch back to Print Preview. Also notice that Access added the Sum function below the Count function rather than at the bottom of the *Total Work Order* column.

6. Close Print Preview to switch back to Design view.

7. Click to select the Sum function control object in the *ServDate Footer* section, press and hold down the Shift key, click to select the Sum function control object in the *Report Footer* section, and then release the Shift key. Position the mouse pointer on the orange border of one of the selected control objects and then drag to move the two control objects simultaneously to the right below the TotalWorkOrder control object in the *Detail* section.

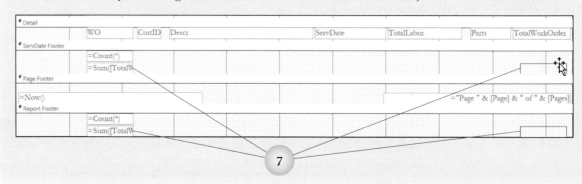

8. Add a label control object left of the Count function in the *ServDate Footer* section that displays the text *Work Order Count:* and another label control object left of the Sum function that displays the text *Weekly Total:*. Apply bold formatting and the red font color and right-align the text in the two label control objects. Resize and align the two label control objects as necessary. ***Note: Access displays an error flag on the two label control objects, indicating that these control objects are not associated with another control object. Ignore these error flags because the label control objects have been added for descriptive text only.***

9. Move up the Sum function control object and *Weekly Total:* label control object until they are at the same horizontal position as the Count function and then decrease the height of the *ServDate Footer* section as shown below.

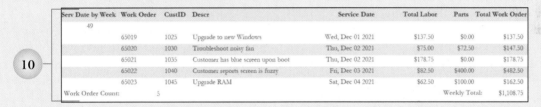

10. Display the report in Print Preview to view the labels. If necessary, return to Design view to further adjust the size and alignment.

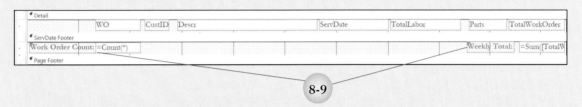

Serv Date by Week	Work Order	CustID	Descr	Service Date	Total Labor	Parts	Total Work Order
49							
	65019	1025	Upgrade to new Windows	Wed, Dec 01 2021	$137.50	$0.00	$137.50
	65020	1030	Troubleshoot noisy fan	Thu, Dec 02 2021	$75.00	$72.50	$147.50
	65021	1035	Customer has blue screen upon boot	Thu, Dec 02 2021	$178.75	$0.00	$178.75
	65022	1040	Customer reports screen is fuzzy	Fri, Dec 03 2021	$82.50	$400.00	$482.50
	65023	1045	Upgrade RAM	Sat, Dec 04 2021	$62.50	$100.00	$162.50
Work Order Count:	5					Weekly Total:	$1,108.75

11. With the report displayed in Design view, select the Sum function control object in the *Report Footer* section and move up the control until it is positioned at the same horizontal position as the Count function below the *WO* column.

12. Click to select the *Work Order Count:* label control object, press and hold down the Shift key, click to select the *Weekly Total:* label control object, release the Shift key, click the Home tab, and then click the Copy button. Click the *Report Footer* section bar and then click the Paste button. Move and align the copied labels as shown below. Edit the *Weekly Total:* label control object to *Grand Total:* as shown.

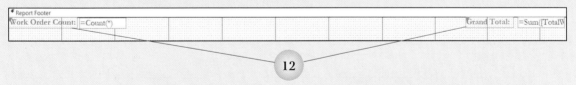

13. Display the report in Report view. Scroll to the bottom of the page to view the labels next to the grand totals. If necessary, return to Design view to further adjust the size and alignment and then save and close the report.

Check Your Work

Adding a Calculated Field to a Report

Several options are available for adding a calculated field to a report. Either create a query that includes a calculated column and then create a new report based on the query or create a calculated control object in an existing report using Design view. A calculated value can be placed in any section of the report. To create a calculated control object in a report, follow the same steps learned in Chapter 4 for adding a calculated control object to a form. Remember that field names in a formula are enclosed in square brackets.

A calculated control object can be used as a field in another formula. To do this, reference the calculated control object in the formula by its Name property (found on the Other tab of the Property Sheet task pane). If necessary, change the name of the calculated control object to a more descriptive or succinct name.

Activity 3c Adding a Calculated Field to a Report

Part 3 of 3

1. With **5-RSRCompServ** open, display the WorkOrdersbyWeek report in Design view. In the following steps you will create a calculated control object to project the value of January's work orders. Assume that the value for January will be a 10% increase over the value for December noted in the grand total.
2. Change the name of the text box control object that contains the summed value of December's work orders to a shortened name by completing the following steps:
   a. Click the control object that contains the summed monthly value (displays part of the formula =*Sum([TotalWorkOrder])*) in the *Report Footer* section.
   b. Click the Property Sheet button in the Tools group on the Report Design Tools Design tab.
   c. Click the Other tab in the Property Sheet task pane.
   d. Select and delete the existing text (displays *AccessTotalsTotal Work Order*) in the *Name* property box, type MonthlyTotal, and then close the Property Sheet task pane.
3. Add a calculated text box control object to the *Report Footer* section by completing the following steps:
   a. Click the Text Box button in the Controls group.
   b. Position the crosshairs with the text box icon attached at the 6-inch mark on the horizontal ruler in the *Report Footer* section. Drag to create a control object that is approximately 2 inches wide by 0.25 inch tall at the same vertical position as the other control objects in the *Report Footer* section and then release the mouse.
   c. Click in the text box control object (displays *Unbound*), type =[MonthlyTotal]*1.1, and then press the Enter key.

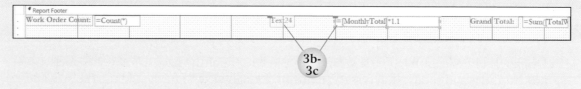

   d. Format the calculated control object using the Currency format.

e. Select the entry in the label control object (displays *Textxx,* where *xx* is the text box control object number) and then type January's Projected Total:. The two control objects will be resized and moved in steps below.

f. Apply bold formatting and the red font color to the text in both control objects.

g. Resize the text box control that contains the new formula until the left edge of the control object is aligned at approximately the 7-inch position on the horizontal ruler.

h. Resize the label control object that contains the text *January's Projected Total:* until the right edge of the control object meets the left edge of the text box control object, as shown below. Right-align the text in the label control object.

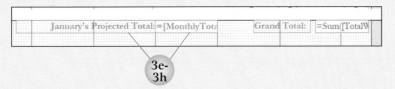

4. Display the report in Report view. Scroll to the bottom of the page to view the calculated field. If necessary, return to Design view to further adjust the size and alignment including decreasing the height of the *Report Footer* section and then save and close the report. Further changes to the report will be made in Activity 4.

5. Redisplay the Navigation pane.

 Check Your Work

---

Activity 4 Modify Section and Group Properties   1 Part

You will change a report's page setup and then modify section and group properties to control print options.

 Tutorial

Modifying Section Properties

Modify Section Properties

1. Open report in Design view.
2. Double-click white section bar.
3. Change properties.
4. Close Property Sheet task pane.

Modifying Section Properties

A report has a Property Sheet, each control object within a report has a Property Sheet, and each section within a report has a Property Sheet. Section properties control whether the section is visible when printed, along with the section's height, background color, special effects, and so on.

Figure 5.11 displays the Format tab in the Property Sheet task pane for the *Report Header* section. Some of the options can be changed without opening the Property Sheet task pane. For example, increase or decrease the height of a section by dragging the top or bottom of a white section bar in Design view. The background color can also be set using the Fill/Back Color button in the Font group on the Report Design Tools Format tab.

Figure 5.11 Report Header Section Property Sheet Task Pane with the Format Tab Selected

Use the Keep Together property to ensure that a section does not split over two pages because of a page break. If necessary, Access prints the section at the top of the next page. However, if the section is longer than can fit on one page, Access continues printing the section on the following page. In that case, decrease the margins and/or apply a smaller font size to fit all the text for the section on one page.

Use the Force New Page property to insert a page break before a section begins *(Before Section)*, after a section ends *(After Section)*, or before and after a section *(Before & After)*.

Tutorial

Keeping a Group Together on the Same Page

Quick Steps

Keep Group Together on One Page
1. Open report in Design view or Layout view.
2. Click Group & Sort button.
3. Click More Options button.
4. Click arrow next to *do not keep group together on one page.*
5. Click print option.
6. Close Group, Sort, and Total pane.

Keeping a Group Together on the Same Page

Open the Group, Sort, and Total pane and click the More Options button for a group to specify whether to keep a group together on the same page. By default, Access does not keep a group together. Click the option box arrow next to *do not keep group together on one page* and then click a print option, as shown in Figure 5.12.

Figure 5.12 Group, Sort, and Total Pane with Keep Group Together Print Options

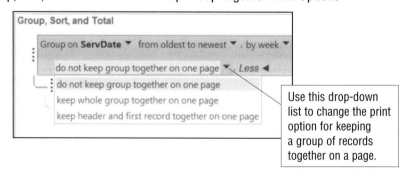

Use this drop-down list to change the print option for keeping a group of records together on a page.

1. With **5-RSRCompServ** open, display the WorkOrdersbyWeek report in Print Preview and then click the Zoom button (not the button arrow) in the Zoom group to change the zoom to view an entire page within the window.
2. Click the Next Page button in the Page Navigation bar to view page 2 of the report, which shows the grand total. Notice that week 52 is split over two pages.
3. Switch to Design view and then minimize the Navigation pane.
4. Change the section properties for the *ServDate Footer* and *Report Footer* sections displaying the Count and Sum functions and control objects by completing the following steps:
   a. Double-click the *ServDate Footer* section bar to open the section's Property Sheet task pane.
   b. Click in the *Back Color* property box on the Format tab and then click the Build button to open the color palette.
   c. Click *Blue-Gray, Accent 3, Lighter 80%* (seventh column, second row in the *Theme Colors* section).
   d. Close the Property Sheet task pane.
   e. Select the Count function control object and Sum function control object in the *ServDate Footer* section and then change the font color to red and apply bold formatting.

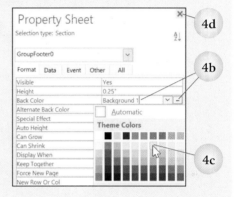

   f. With the Count and Sum function control objects still selected, right-click one of the selected control objects, point to *Fill/Back Color* at the shortcut menu, and then click *Transparent*. The background color applied in Step 4c will now display behind the calculations.
   g. Double-click the *Report Footer* section bar, change the Back Color property to the same color as the *ServDate Footer* section (see Steps 4b and 4c), and then close the Property Sheet task pane.
   h. Apply the formatting in Steps 4e and 4f to the Count and Sum function control objects in the *Report Footer* section.
   i. Right-click the text box control object containing the formula *=[MonthlyTotal]*1.1* and then click *Properties* at the shortcut menu. Change the Back Style property to *Transparent* and the Border Style property to *Transparent*. Close the Property Sheet task pane.
5. Print the work orders on a separate page by completing the following steps:
   a. Click the Group & Sort button on the Report Design Tools Design tab.
   b. Click the More Options button in the Group, Sort, and Total pane.
   c. Click the option box arrow next to *do not keep group together on one page* and then click *keep whole group together on one page* at the drop-down list.
   d. Close the Group, Sort, and Total pane.
6. Create a label control object at the top right of the *Report Header* section with your first and last names. Apply bold formatting to the label control object and click the *Blue-Gray, Accent 3, Darker 50%* font color (seventh column, last row in the *Theme Colors* section).
7. Display the report in Print Preview and then change the zoom to view an entire page within the window. Compare the report with the one shown in Figure 5.13. Scroll to page 2 to view all of work orders for week 52 on the same page.
8. Save, print, and then close the report.
9. Redisplay the Navigation pane.

Check Your Work

Figure 5.13 Page 1 of the Completed Report in Activity 4

Work Orders by Week							Student Name
Serv Date by Week	Work Order	CustID	Descr	Service Date	Total Labor	Parts	Total Work Order
49							
	65019	1025	Upgrade to new Windows	Wed, Dec 01 2021	$137.50	$0.00	$137.50
	65020	1030	Troubleshoot noisy fan	Thu, Dec 02 2021	$75.00	$72.50	$147.50
	65021	1035	Customer has blue screen upon boot	Thu, Dec 02 2021	$178.75	$0.00	$178.75
	65022	1040	Customer reports screen is fuzzy	Fri, Dec 03 2021	$82.50	$400.00	$482.50
	65023	1045	Upgrade RAM	Sat, Dec 04 2021	$62.50	$100.00	$162.50
Work Order Count:	5					Weekly Total:	$1,108.75
50							
	65024	1005	Install malware protection	Mon, Dec 06 2021	$45.00	$125.00	$170.00
	65025	1010	Troubleshoot hard drive noise	Mon, Dec 06 2021	$75.00	$0.00	$75.00
	65026	1025	Upgrade RAM	Tue, Dec 07 2021	$37.50	$100.00	$137.50
	65027	1010	Replace hard drive with SSD	Tue, Dec 07 2021	$75.00	$110.00	$185.00
	65028	1030	Reinstall operating system	Wed, Dec 08 2021	$112.50	$0.00	$112.50
	65029	1035	Set up automatic backup	Fri, Dec 10 2021	$22.50	$0.00	$22.50
	65031	1045	Customer reports noisy hard drive	Sat, Dec 11 2021	$87.50	$0.00	$87.50
Work Order Count:	7					Weekly Total:	$790.00
51							
	65030	1000	Clean malware from system	Mon, Dec 13 2021	$40.00	$0.00	$40.00
	65032	1008	Install second storage drive	Mon, Dec 13 2021	$100.00	$75.00	$175.00
	65033	1000	Install Windows 10	Tue, Dec 14 2021	$162.50	$335.75	$498.25
	65034	1035	File management training	Tue, Dec 14 2021	$75.00	$0.00	$75.00
	65035	1020	Office 365 training	Thu, Dec 16 2021	$125.00	$0.00	$125.00
	65036	1008	Set up home network	Fri, Dec 17 2021	$135.00	$85.22	$220.22
	65037	1015	Biannual computer maintenance	Fri, Dec 17 2021	$62.50	$8.75	$71.25
	65039	1030	Set up automatic backup	Sat, Dec 18 2021	$30.00	$0.00	$30.00
Work Order Count:	8					Weekly Total:	$1,234.72

Tuesday, January 19, 2021   Page 1 of 2

Activity 5 Create and Format a Chart   1 Part

You will create a chart in a customer report to show the total parts and labor on work orders by customer for a month.

Inserting, Editing, and Formatting a Chart in a Report

A chart can be added to a report to graphically display the numerical data from another table or query. The chart is linked to a field in the report that is common to both objects. Access summarizes and graphs the data from the charted table or query based on the fields selected for each record in the report.

Tutorial

Inserting a Chart

Insert Chart

Inserting a Chart

With a report open in Design view, increase the height or width of the *Detail* section to make room for the chart, click the Insert Chart button in the Controls group on the Report Design Tools Design tab, and then drag the crosshairs with the chart icon attached to create the approximate height and width of the chart. Release the mouse and Access launches the Chart Wizard, which has six dialog boxes that guide the user through the steps of creating a chart. The first Chart Wizard dialog box is shown in Figure 5.14.

**Insert Chart
in Report**

1. Open report in Design view.
2. Click Insert Chart button.
3. Drag to create control object of required height and width.
4. Select table or query for chart data.
5. Click Next.
6. Add fields to use in chart.
7. Click Next.
8. Click chart type.
9. Click Next.
10. Add fields as needed to chart layout.
11. Click Preview Chart.
12. Close Sample Preview window.
13. Click Next.
14. Select field to link report with chart.
15. Click Next.
16. Type chart name.
17. Click Finish.

Figure 5.14 First Chart Wizard Dialog Box

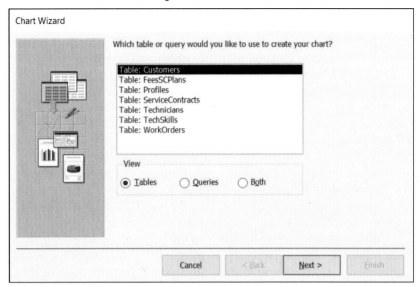

Activity 5 uses the Chart Wizard to insert a chart in a customer report that depicts the total value of work orders for each customer by month. The data for the chart will be drawn from a related query. A chart can also be inserted into a form and formatted by completing steps similar to those in Activity 5.

Activity 5 Creating a Report and Inserting a Chart

Part 1 of 1

1. With **5-RSRCompServ** open, create a new report using the Report Wizard by completing the following steps:
   a. Click *Customers* in the Tables group in the Navigation pane, click the Create tab, and then click the Report Wizard button.
   b. At the first Report Wizard dialog box with *Table: Customers* selected in the *Tables/Queries* option box, move the *CustID, FName, LName,* and *ServCont* fields from the *Available Fields* list box to the *Selected Fields* list box and then click the Next button.
   c. At the second Report Wizard dialog box, with no group field selected, click the Next button.
   d. Click the Next button at the third Report Wizard dialog to choose not to sort the report.
   e. Click *Columnar* at the fourth Report Wizard dialog box and then click the Next button.
   f. Click at the end of the current text in the *What title do you want for your report?* text box, type WOChart so that the report title is *CustomersWOChart*, and then click the Finish button.
2. Minimize the Navigation pane and display the report in Design view.
3. To minimize the number of pages printed, change the page layout of the report to two columns by completing the following steps:
   a. Delete the date and page number control objects in the *Page Footer* section.
   b. Drag the right side of the grid to the left to meet the right edge of the *LName* text box control object. The grid will be approximately 3.8 inches wide.
   c. Click the Columns button in the Page Layout group on the Report Design Tools Page Setup tab, change *1* to *2* in the *Number of Columns* text box, and then click OK.
   d. Switch to Print Preview and review the report.

4. There is not enough room to place a chart showing the value of the work orders for each customer. Switch the report back to one column by completing the following steps:
   a. Close Print Preview and press the keys Ctrl + Z three times or until the two text box control objects are deleted and the change in the grid size is undone.
   b. Click the Columns button in the Page Layout group on the Report Design Tools Page Setup tab, change *2* to *1* in the *Number of Columns* text box, and then click OK.
5. Select the text in the report title and then type December 2021 Work Orders by Customer.
6. Drag down the top of the *Page Footer* section bar until the bottom of the *Detail* section is positioned at the 2-inch mark on the vertical ruler.
7. Insert a chart at the right side of the report to show the value of the work orders for each customer by month by completing the following steps:
   a. Click the More button in the Controls group on the Report Design Tools Design tab and then click the Chart button in the last row. (Depending on the size of your screen, the Chart button may be in the first row).
   b. Position the crosshairs with the chart icon attached in the *Detail* section at the 5-inch mark on the horizontal ruler aligned near the top of the *CustID* control object and then drag down and to the right to create a chart control object of the approximate height and width shown below.

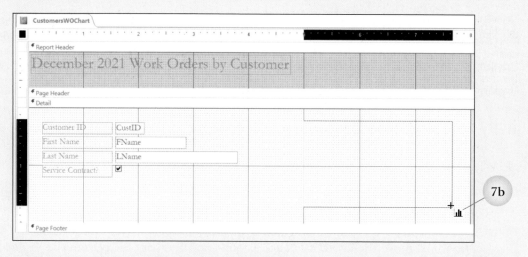

   c. At the first Chart Wizard dialog box, click *Queries* in the *View* section, click *Query: TotalWorkOrders* in the list box, and then click the Next button.

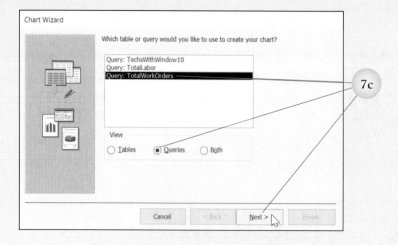

d. At the second Chart Wizard dialog box, double-click *ServDate* and *TotalWorkOrder* in the *Available Fields* list box to move the fields to the *Fields for Chart* list box and then click the Next button.

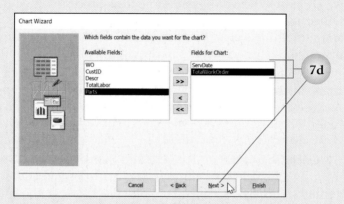

e. At the third Chart Wizard dialog box, click the second chart type in the first row (*3-D Column Chart*) and then click the Next button.

f. At the fourth Chart Wizard dialog box, look at the fields that Access has already placed to lay out the chart. Since only two fields were added, Access automatically used the numeric field with a Sum function as the data series for the chart and the date field as the *x*-axis category. Click the Next button.

g. At the fifth Chart Wizard dialog box, notice that Access has correctly detected *CustID* as the field to link records to in the Customers report with the chart (based on the TotalWorkOrders query). Click the Next button.

h. At the last Chart Wizard dialog box, click the *No, don't display a legend option* and then click the Finish button. Access inserts a chart within the height and width of the chart control. The chart displayed in the control object in Design view is not the actual chart based on the query data; it is only a sample to show the chart elements.

8. Display the report in Print Preview and scroll through the four pages. Customers for which an empty chart displays have no work order data to be graphed.

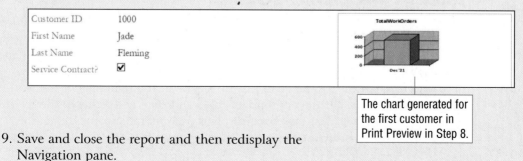

The chart generated for the first customer in Print Preview in Step 8.

9. Save and close the report and then redisplay the Navigation pane.

Check Your Work

Editing and Formatting a Chart

The Chart feature within Access is not the same Chart tool available in Word, Excel, and PowerPoint. Access uses the Microsoft Graph application for charts, which means the editing and formatting processes are different than in other Office applications. Depending on your version of Access, use care when editing and formatting a chart, as changes can result in the chart dimensions becoming distorted. If this occurs, delete the control object and start again.

Open a report in Design view and then double-click a chart object to edit the chart. In chart-editing mode, a Menu bar and a toolbar display at the top of the Access window, as well as a datasheet for the chart in the work area. Use these tools to change the chart type; add, remove, or change chart options; and format chart elements.

Click Chart on the Menu bar and then click *Chart Options* at the drop-down menu to add, delete, or edit text in chart titles and add or remove chart axes, gridlines, the legend, data labels, or a data table at the Chart Options dialog box. Click *Chart Type* at the Chart drop-down menu to open the Chart Type dialog box and choose a different type of chart, such as a bar chart or pie chart.

Right-click an object within a chart—such as the chart title, legend, chart area, or data series—and a format option displays in the shortcut menu for the selected chart element. Click the format option to open a Format dialog box for the selected element. Make the required changes and then click OK.

After editing the chart, click outside the chart object to exit chart-editing mode. Sometimes, Access displays a sample chart within the control object in chart-editing mode instead of the actual chart, which can make editing specific chart elements difficult if the chart does not match the sample. If this occurs, exit chart-editing mode, close and reopen the report in Design view, or switch views to cause Access to update the chart displayed in the control object.

Quick Steps

Change Chart Options
1. Open report in Design view.
2. Double-click chart.
3. Click Chart on Menu bar.
4. Click *Chart Options*.
5. Click tab.
6. Change options.
7. Click OK.

Change Chart Type
1. Open report in Design view.
2. Double-click chart.
3. Click Chart on Menu bar.
4. Click *Chart Type*.
5. Click chart type in list box.
6. Click chart subtype.
7. Click OK.

Format Chart Element
1. Open report in Design view.
2. Double-click chart.
3. Right-click chart element.
4. Click *Format*.
5. Change format options.
6. Click OK.

Activity 6 Create a Blank Report with Hyperlinks, and a List Box   1 Part

You will use the Blank Report tool to create a new report for viewing technician certifications. In the report, you will reorder the tab fields, create a list box inside a tab control object, change the shape of the tab control object, and add hyperlinks.

Creating a Report Using the Blank Report Tool

A blank report created with the Blank Report tool begins with no control objects or formatting and displays as a blank white page in Layout view, similar to a form created with the Blank Form tool, as learned in Chapter 4. Click the Create tab and then click the Blank Report button in the Reports group to begin a new report. Access opens the Field List task pane at the right side of the work area. Expand the list for the appropriate table and then add fields to the report as needed. If the Field List task pane displays with no table names, click the Show all tables hyperlink at the top of the task pane.

Quick Steps

Create Blank Report
1. Click Create tab.
2. Click Blank Report button.
3. Expand field list for table.
4. Drag fields to report.
5. Save report.

Adding a Tab Control Object to a Report

 Tab Control

Recall from Chapter 4 that a tab control object can be added to a form to display fields from different tables on separate pages. In Form view, a page is displayed by clicking the page tab. Similarly, a tab control object can be used in a report to display fields from the same table or a different table in pages. To create a tab control object in a report, follow the steps from Chapter 4 for adding a tab control object to a form.

Adding a List Box or Combo Box to a Report

Tutorial

Adding a List Box to a Report

 List Box

Like a list box in a form, a list box in a report displays a list of values for a field within the control object. In Report view, the entire list for the field can easily be seen. If a list is too long for the size of the list box control object, scroll bars display that can be used to scroll up or down the list when viewing the report. Although the data cannot be edited in a report, a list box can be used to view all the field values and see which values have been selected for the current record.

 Combo Box

A combo box added to a report does not display as a list. However, the field value that was entered into the field from the associated table, query, or form is shown in the combo box. Since the data cannot be edited in a report, the combo box field is not shown as a drop-down list. A combo box can be changed to display as a list box within the report. In this case, the list box displays all the field values and the value stored in the current record shown is selected within the list. A list box or combo box can be added to a report by following the steps from Chapter 4 for adding a list box or combo box to a form.

Adding Hyperlinks to a Report

Quick Steps

Add Hyperlink to Report
1. Open report in Layout view or Design view.
2. Click Hyperlink button in Controls group.
3. Click within report.
4. Type text in *Text to display* text box.
5. Type URL in *Address* text box.
6. Click OK.

Change Shape of Control Object
1. Open report in Layout view.
2. Click to select control object.
3. Click Report Layout Tools Format tab.
4. Click Change Shape button.
5. Click shape.

Use the Hyperlink button in the Controls group on the Report Layout Tools Design tab to create a link in a report to a web page, graphic, email address, or program. Click the Hyperlink button and then click within the report to open the Insert Hyperlink dialog box, in which the user provides the text to display in the control object and the address to which the object should be linked. Use the Places bar left of the Insert Hyperlink dialog box to choose to link to an existing file or web page, another object within the database, or an email address.

Changing the Shape of a Control Object

The Change Shape button in the Control Formatting group on the Report Layout Tools Format tab contains a drop-down list with eight shape options. Use the shape options to modify the appearance of a command button, toggle button, navigation button, or tab control object. Select the control object to be modified, click the Change Shape button, and then click a shape at the drop-down list.

Changing the Tab Order of Fields

 Hyperlink

 Change Shape

 Tab Order

Recall from Chapter 4 how to open the Tab Order dialog box and change the order in which the Tab key moves from field to field. In a report, the order in which the Tab key moves from field to field in Report view can also be changed. Although data is not added, deleted, or edited in Report view, use the Tab key to move within a report. Display the report in Design view, click the Tab Order button in the Tools group on the Report Design Tools Design tab, and then drag the fields up or down the *Custom Order* list.

1. With **5-RSRCompServ** open, click the Create tab and then click the Blank Report button in the Reports group. Save the report with the name *TechCertifications*.
2. If the Field List task pane at the right side of the work area does not display the table names, click the <u>Show all tables</u> hyperlink. If the table names display, proceed to Step 3.
3. Add fields from the Technicians table and a tab control object to the report by completing the following steps:
   a. Click the plus symbol next to the *Technicians* table name to expand the field list.
   b. Click *TechID* in the Field List task pane and then drag the field to the top left of the report.
   c. Right-click in the *Technician ID* column, point to *Layout* at the shortcut menu, and then click *Stacked*. A stacked layout is better suited to this report because it shows each technician's certifications next to his or her name in a tab control object.
   d. Drag the *FName* field from the Field List task pane below the first *Technician ID* text box control object. Release the mouse when the pink bar is below the *01* text box control object next to *Technician ID*.

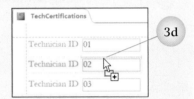

3d

   e. Drag the *LName* field from the Field List task pane below the first *First Name* text box control object. Release the mouse when the pink bar displays below *Pat*.
   f. Select the *HPhone* and *CPhone* fields in the Field List task pane and then drag the two fields below the *Last Name* text box control object. Release the mouse when the pink bar displays below *Hynes*.

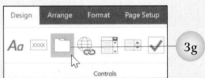

3g

   g. Click the Tab Control button in the Controls group on the Report Layout Tools Design tab.
   h. Position the mouse pointer with the tab control icon attached right of the first *Technician ID* text box control object in the report. Click the mouse when the pink bar displays right of the *01* text box control object.

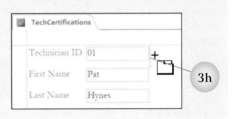

3h

   i. Right-click the selected tab control object, point to *Layout*, and then click *Remove Layout* at the shortcut menu.
   j. With the first *Pagexx* tab selected (where *xx* is the page number) point to the bottom orange border of the selected tab control object until the pointer displays as an up-and-down-pointing arrow and then drag down the bottom of the object until it aligns with the bottom of the *Cell Phone* control object.

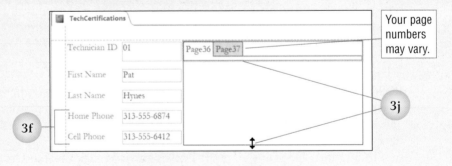

3f

3j

Your page numbers may vary.

4. Add a field from the TechSkills table to the tab control object and change the control to a list box by completing the following steps:

   a. Click the plus symbol next to the *TechSkills* table name in the *Fields available in related tables* section of the Field List task pane to expand the list.

   b. Click to select the *Certifications* field name and then drag the field to the first page in the tab control object, next to *Technician ID 01*.

   c. Access inserts the field in the page with both the label control object and text box control object selected. To change the field to display as a list box, click to select only the text box control object (displays *CCNA Cloud, CCNA Wireless,* and *Microsoft MCT* in the first record).

   d. Right-click the selected text box control object, point to *Change To*, and then click *List Box* at the shortcut menu.

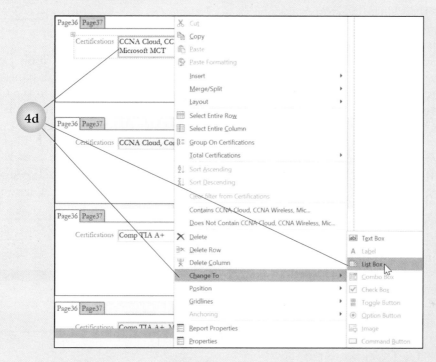

5. Add a hyperlink to the bottom of the tab control object by completing the following steps:

   a. Click the Hyperlink button in the Controls group on the Report Layout Tools Design tab.

   b. Position the mouse pointer with the hyperlink icon attached below *Certifications* in the tab control object. Click the mouse when the pink bar displays.

   c. At the Insert Hyperlink dialog box, click in the *Text to display* text box and then type Cisco Certifications.

   d. Click in the *Address* text box, type https://www.cisco.com/c/en/us/training-events/training-certifications/overview.html, and then click OK.

   e. Drag the right orange border of the hyperlink control object to the right until all the text displays within the control object.

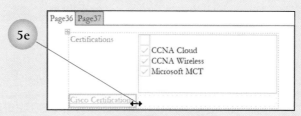

6. Click the second page in the tab control to select it, right-click, and then click *Delete Page* at the shortcut menu. Do not be concerned if Access displays the first page with an empty list box. The screen will refresh at the next step.

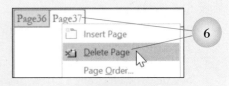

7. Double-click over *Pagexx* (where *xx* is the page number) in the tab control object to select the entire tab control object. Click the Report Layout Tools Format tab, click the Change Shape button in the Control Formatting group, and then click *Rectangle: Single Corner Snipped* at the drop-down list (third column, second row).

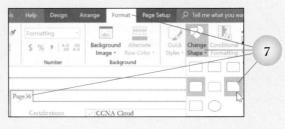

8. Click to select *Pagexx* (where *xx* is the page number), click the Report Layout Tools Design tab, and then click the Property Sheet button to open the Property Sheet task pane. With *Selection type: Page* displayed at the top of the Property Sheet task pane, click in the *Caption* property box, type Technician's Certifications, and then close the Property Sheet task pane.

9. Right-click the *Technician ID* text box control object (displays *01*) and then click *Select Entire Column* at the shortcut menu. Click the Report Layout Tools Format tab, click the Shape Outline button, and then click *Transparent* at the drop-down list.

10. Switch to Report view. Click in the first *Technician ID* text box and press the Tab key four times to see how active field moves through the report in order starting at the left side of the report.

11. Switch to Design view. You want the Tab key to move to the technician's certifications after their cell phone. The *TabCtrlxx* field (where *xx* is the control number) will be moved to the last field in the order. Change the tab order of the fields by completing the following steps:

    a. Click the Tab Order button in the Tools group on the Report Design Tools Design tab.

    b. At the Tab Order dialog box, click in the bar next to *TabCtrlxx* field (where *xx* is the control number) in the *Custom Order* section to select the *TabCtrlxx* field and then drag the field to the bottom of the list until the black line displays below *CPhone*.

    c. Click OK.

12. Save the revised report and then switch to Report view. Press the Tab key. Notice that the first field selected is the *Technician ID* field. Press the Tab key a second time. Notice that the selected field moves to *First Name*. Press the Tab key six more times to watch the selected fields move through the fields.

13. Click the <u>Cisco Certifications</u> hyperlink to open a Microsoft Edge window and display the Cisco Training & Certifications web page.

14. Close the browser window.
15. Display the report in Print Preview, print only the first page, and then close Print Preview.
16. Save and close the TechCertifications report and then close **5-RSRCompServ**.

Chapter Summary

- Click the Create tab and then click the Report Design button to create a custom report using Design view.
- A report can contain up to five sections: *Report Header, Page Header, Detail, Page Footer,* and *Report Footer*.
- *Group Header* and *Group Footer* sections can be added to group records that contain repeating values in a field, such as a department or city.
- Connect a table or query to a report using the *Record Source* property box in the Data tab of the report's Property Sheet task pane.
- Display the Field List task pane and then drag individual fields or a group of fields from the table or query to the *Detail* section of the report, which represents the body of the report.
- Place label control objects to be used as column headings in a tabular report within the *Page Header* section.
- Apply a theme to a report to maintain a consistent look to all reports.
- A related table or query can be inserted as a subreport within a main report. A subreport is stored as a separate object outside the main report.
- Click the Page Numbers button in the Header / Footer group to open the Page Numbers dialog box, where the format, position, and alignment of page numbers in a report are specified.
- The current date and/or time can be added as a control object within the *Report Header* section using the Date and Time button in the Header / Footer group.
- Add online graphics and draw lines in a report using the same techniques for adding graphics and drawing lines in a form.
- A report can be grouped by a field at the Report Wizard or by opening the Group, Sort, and Total pane.
- Functions such as Sum, Average, and Count can be added to each group within a report and grand totals can be added to the end of a report by expanding the group options in the Group, Sort, and Total pane.
- Create a calculated control object in a report using the Text Box button in the Controls group.
- Each section within a report has a set of properties that can be viewed or changed by opening the section's Property Sheet task pane.
- Use the Keep Together property to prevent a section from splitting across two pages.
- Use the Force New Page property to automatically insert a page break before a section begins, after a section ends, or before and after a section.
- At the Group, Sort, and Total pane, specify whether to keep an entire group together on the same page.

- A chart can be added to a report to graphically display numerical data from another table or query.

- A report can be formatted into multiple columns using the Columns button in the Report Design Tools Page Setup tab.

- To create a chart in a report, open the report in Design view, click the Chart button in the Controls group, and then use the Chart Wizard to generate the chart.

- Open a report in Design view and then double-click a chart control object to edit the chart using Microsoft Graph by changing the chart type; adding, removing, or changing chart options; and formatting chart elements.

- Use the Blank Report tool in the Reports group on the Create tab to create a new report with no control objects or formatting applied. The report opens as a blank white page in Layout view; the Field List task pane opens right of the work area.

- A tab control object, list box, and combo box can be added to a blank report using the same techniques for adding these control objects to a form.

- Use the Hyperlink button to create a link in a report to a web page, graphic, email address, or program.

- Modify the shape of a command button, toggle button, navigation button, or tab control object using the Change Shape button in the Control Formatting group on the Report Layout Tools Format tab.

- Change the order in which the Tab key moves from field to field within a report at the Tab Order dialog box. Display a report in Design view and click the Tab Order button in the Tools group on the Report Design Tools Design tab.

- As you become more comfortable with reports, explore other tools available in Layout View and Design view using the Design, Arrange, Format, and Page Setup tabs. More features are available to assist you with creating professional-quality reports.

Commands Review

FEATURE	RIBBON TAB, GROUP	BUTTON	KEYBOARD SHORTCUT
add existing fields	Report Design Tools Design, Tools		
blank report	Create, Reports		
change shape of selected control	Report Layout Tools Format, Control Formatting		
date and time	Report Design Tools Design, Header / Footer		
Design view	Report Design Tools Design, Views		
group and sort	Report Design Tools Design, Grouping & Totals		
insert chart	Report Design Tools Design, Controls		
insert hyperlink	Report Layout Tools Design, Controls		
insert image	Report Design Tools Design, Controls		
page numbers	Report Design Tools Design, Header / Footer		
Property Sheet task pane	Report Design Tools Design, Tools		F4
report design	Create, Reports		
Report view	Report Design Tools Design, Views		
Report Wizard	Create, Reports		
subreport	Report Design Tools Design, Controls		
tab control object	Report Layout Tools Design, Controls		
tab order	Report Design Tools Design, Tools		
theme	Report Design Tools Design, Themes		
title control object	Report Design Tools Design, Header / Footer		

Microsoft®

Access®

Using Access Tools and Managing Objects

Performance Objectives

Upon successful completion of Chapter 6, you will be able to:

1 Create a new database using a template

2 Save a database as a template

3 Add prebuilt objects to a database using an Application Parts template

4 Create a new form using an Application Parts Blank Form

5 Create a form to be used as a template in a database

6 Create a table by copying the structure of another table

7 Evaluate a table using the Table Analyzer Wizard

8 Evaluate a database using the Performance Analyzer

9 Split a database

10 Print documentation about a database using the Database Documenter

11 Rename and delete objects within a database file

12 Use SQL to modify a query

Access provides tools to assist with creating and managing databases. Use a template to create a new database or a new table and/or a related group of objects. A blank form template provides a predefined layout and may include a form title and command buttons. If none of the predefined templates is suitable, create a custom template. Access provides wizards to assist with analyzing tables and databases to improve performance. A database can be split into two files to store the tables separately from the queries, forms, and reports. Use the Database Documenter to print a report that provides details about objects and their properties. In this chapter, you will learn how to use these Access tools and how to rename and delete objects in the Navigation pane.

 Data Files

Before beginning chapter work, copy the AL2C6 folder to your storage medium and then make AL2C6 the active folder.

The online course includes additional training and assessment resources.

Activity 1 Create a New Contact Database Using a Template   3 Parts

You will create a new database using one of the database templates supplied with Access and create your own template.

Tutorial

Creating a New Database Using a Template

Quick Steps

Create Database from Template
1. Start Access.
2. Click template.
3. Click Browse button.
4. Navigate to drive and/or folder.
5. Edit file name as required.
6. Click OK.
7. Click Create button.

Creating a New Database Using a Template

At the Recent or New backstage area, create a new database using one of the professionally designed templates provided by Microsoft. The database templates provide a complete series of objects, including predefined tables, forms, reports, queries, and relationships. Use a template as provided and immediately start entering data or base a new database on a template and modify the objects as needed. If a template exists for a database application that is needed, save time by basing the database on one of the template designs.

To create a new database using a template, start Access and then click one of the available templates like the ones shown in Figure 6.1 in the Recent backstage area. If Access is already open, click one of the available templates at the New backstage area. The templates are constantly being updated, so the options you see may vary. When the template is clicked, a preview appears, similar to what is shown in Figure 6.2. The name of the template provider, a description of the template, and the download size are displayed in the preview. Use the directional arrows on either side of the preview to move through the other available templates. Once a template has been chosen, click the Browse button to navigate to the drive and/or folder where the database is to be stored and then type a file name at the *File Name* text box. Click the Create button to create the database.

If none of the sample templates is suitable for the type of database to be created, close the preview and search for templates online by clicking one of the hyperlinked suggested searches (such as <u>Database</u>, <u>Business</u>, or <u>Logs</u>). Another option is to type search words in the search text box. Templates are downloaded from Office.com.

Figure 6.1 Available Templates

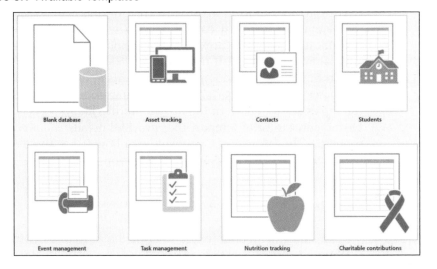

Figure 6.2 Available Templates in the New Backstage Area

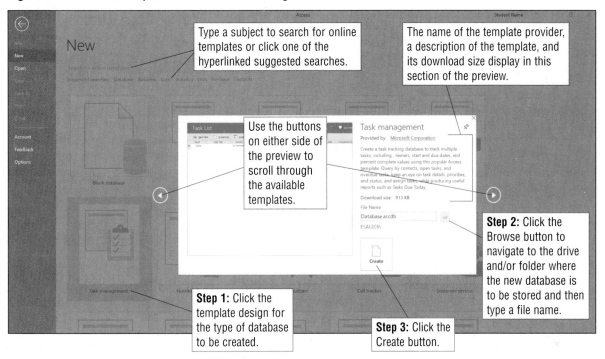

Type a subject to search for online templates or click one of the hyperlinked suggested searches.

The name of the template provider, a description of the template, and its download size display in this section of the preview.

Use the buttons on either side of the preview to scroll through the available templates.

Step 2: Click the Browse button to navigate to the drive and/or folder where the new database is to be stored and then type a file name.

Step 1: Click the template design for the type of database to be created.

Step 3: Click the Create button.

Activity 1a Creating a New Contacts Database Using a Template

Part 1 of 3

1. Start Microsoft Access 365.
2. At the New or Recent backstage area, click the *Contacts* template. *Note: The templates are constantly being updated, so the objects and steps may vary.*
3. Click the Browse button (which displays as a file folder icon) next to the *File Name* text box in the template preview.
4. At the File New Database dialog box, navigate to the AL2C6 folder on your storage medium, select the current entry in the *File name* text box, and then type 6-Contacts.
5. Click OK.

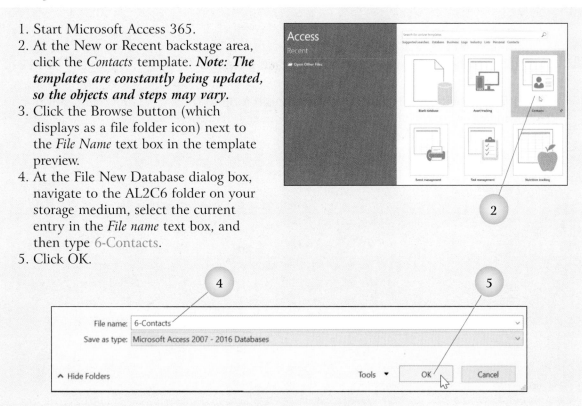

6. Click the Create button.
7. The database is created with all the objects from the template loaded into the current window and the Contact List form open.
8. Click the Enable Content button in the message bar.
9. A Welcome to the Contacts Database form appears that provides information on using this database. Click the *Show Welcome when this database is opened* check box to remove the check mark. If the box is left checked, this form will open every time the database is opened.
10. Close the Welcome form. Notice that the Title bar contains the text *Contact Management Database* rather than the name of the database, *6-Contacts*. (How to change the Title bar to display a more descriptive database title will be discussed in Chapter 7.)
11. Review the list of objects that Access has created in the Navigation pane.
12. Double-click to open the Contacts table.
13. Scroll right to view all the fields in the Contacts table and then close the table.
14. Review the Contact List form that is open in the work area and then close it.

Activity 1b Entering and Viewing Data in the Contacts Database

Part 2 of 3

1. With **6-Contacts** open, open the Contact Details form in Form view, and then add the following record to modify the data source file using the form:

First Name	Ariel
Last Name	Grayson
Company	Grayson Accounting Services
Job Title	Accountant
E-mail	ariel@ppi-edu.net **Note: Press the Tab key four times after this field to move to the Street field.**
Street	17399 Windsor Avenue
City	Detroit
State/Province	MI
Zip/Postal Code	48214-3274
Country/Region	USA
Business Phone	800-555-4988
Home Phone	313-555-9684
Mobile Phone	313-555-6811
Fax Number	800-555-3472
Notes	Ariel recommended to RSR by Pat Hynes.

2. Add a picture of the contact using the *Attachments* field by completing the following steps:
   a. Click the Edit Picture button at the top left part of the form.

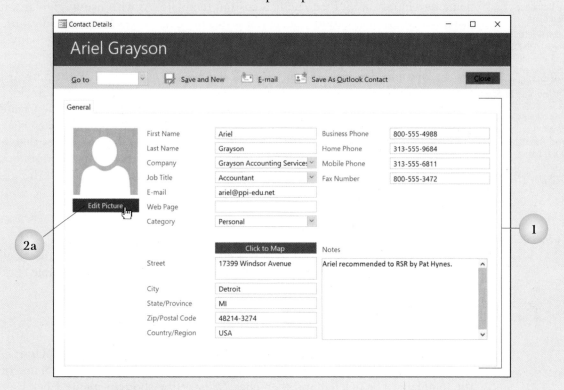

b. At the Attachments dialog box, click the Add button.
c. At the Choose File dialog box, navigate to the AL2C6 folder on your storage medium.
d. Double-click **ArielGrayson.jpg** to add the file to the Attachments dialog box and then click OK.

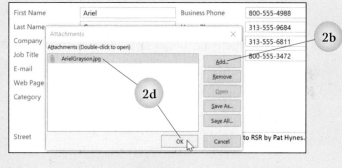

e. Notice that the picture now displays in the form. If a Word document had been selected in the Attachments dialog box, a Word icon would display.

3. Click the Close button in the upper right corner of the form to close the form.
   Note: Do not use the purple Close button right of **Save as Outlook Contact,** *as macro errors may occur.*
4. Double-click *Directory* in the Reports group in the Navigation pane. Review the report.
5. Display the report in Print Preview and then print the report.
6. Close Print Preview and then close the Directory report.

Check Your Work

Saving a Database as a Template

Quick Steps

Save Database as Template

1. Open database.
2. Make changes.
3. Click File tab.
4. Click *Save As*.
5. Click Browse button.
6. Change *Save as type* to *Template (*.accdt)*.
7. Type information in Create New Template from This Database dialog box.
8. Click OK.
9. Click OK.

If none of the existing templates is suitable and the same database structure is being developed often, then create a custom template. As with using templates in other Microsoft applications, using templates in Access can save time and effort. To create a database template, either create an entirely new database or modify an existing database based on a template and then save it as a new template.

To save a database as a template, click the File tab and then click the *Save As* option. With *Save Database As* selected in the *File Types* section, click *Template* in the *Database File Types* section and then click the Save As button. Fill in the Create New Template from This Database dialog box, shown in Figure 6.3, and then click OK. Click OK at the Contact Management Database dialog box that states that the template has been saved as [c:]\Users*username*\AppData\Roaming\Microsoft\ Templates\Access\TemplateName.accdt. Note the default location in which the templates are stored.

Using a Custom Template

To start a user-created template, click the File tab and then click *New*. At the New backstage area, click Personal. This opens the Personal template area. Double-click the name of a template to open it.

Deleting a Custom Template

To delete a custom template, use the Open dialog box to navigate to [c:]\Users\ *username*\AppData\Roaming\Microsoft\Templates\Access\. Right-click the name of the template to be deleted and then click *Delete* at the shortcut menu. Click the Cancel button to close the Open dialog box.

Figure 6.3 Create New Template from This Database Dialog Box

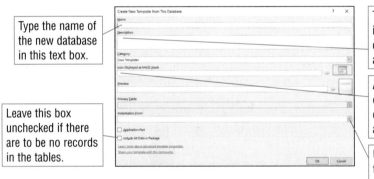

Type the name of the new database in this text box.

Leave this box unchecked if there are to be no records in the tables.

Type a description of the database in this text box, including who created it, when it was created, and other relevant information.

Add an image (png, jpg, jpeg, gif, or bmp) that will represent the new database in the Personal template area of the New backstage area.

Use this option box to specify the form that will display when a database created from this template is opened.

1. With **6-Contacts** open, open the Contact Details form in Design view, click the Close button that contains a close command macro, and then press the Delete key. (Macros will be discussed in Chapter 7.) Save and close the form.

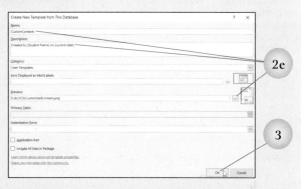

2. Save the revised database as a template by completing the following steps:
   a. Click the File tab.
   b. Click *Save As*.
   c. Click *Template* in the *Database File Types* section.
   d. Click the Save As button.
   e. At the Create New Template from This Database dialog box, enter data as follows:

Name	CustomContacts
Description	Created by [Student Name] on [current date]. (Substitute your name for *[Student Name]* and today's date for *[current date]*.)
Preview	Click the Browse button, navigate to the AL2C6 data folder, and then double-click *CustomizedContacts*.

3. Click OK.
4. Click OK.
5. Close **6-Contacts**.
6. To view the new template, click the File tab and then click Personal. The CustomContacts template created is shown with **CustomizedContacts** as the image. Click the Back button.

Activity 2 Create Objects Using a Template   3 Parts

You will use Application Parts templates to create a series of objects in an existing database. You will also define a form as a template for all the new forms in a database.

Creating Objects Using an Application Parts Template

Application Parts

Access 365 provides templates for prebuilt objects that can be inserted into an existing database using the Application Parts button in the Templates group on the Create tab. The *Quick Start* section of the Application Parts button drop-down list includes the options *Comments*, which creates a table; *Contacts*, which creates a table, a query, forms, and reports; and *Issues*, *Tasks*, and *Users*, each of which creates a table and two forms.

Creating Objects Using Quick Start

If any of these kinds of tables need to be created in an existing database, consider creating it using the Application Parts template because related objects, such as forms and reports, will be generated automatically. Once the application part is added, any part of an object's design can be modified as needed. To create a group of objects based on a template, click the Create tab and then click the Application Parts button in the Templates group. Click a template in the *Quick Start* section of the drop-down list, as shown in Figure 6.4.

Access opens the Create Relationship Wizard to guide the user through creating the relationship for the new table. Decide in advance of creating the new table what relationship, if any, will exist between the new table and an existing table in the database. At the first Create Relationship dialog box, shown in Figure 6.5, click the first option if the new table will be the "many" table in a one-to-many relationship. Use the drop-down list to choose the "one" table and then click the Next button. If the new table will be the "one" table in a one-to-many relationship, click the second option, use the drop-down list to choose the "many" table, and then click the Next button.

At the second Create Relationship dialog box, enter the settings for the lookup column between the two tables. Choose the field to use to join the tables, choose a sort order if needed, assign the name of the lookup column, and then click the Create button. If the new table will not be related to any of the existing tables in the database, choose the *There is no relationship* option at the first Create Relationship dialog box and then click the Create button.

Quick Steps

Create Objects Using Application Parts Template
1. Open database.
2. Click Create tab.
3. Click Application Parts button.
4. Click template.
5. Choose relationship options.
6. Add data or modify objects as required.
7. Click Create button.

Hint Create your own Application Parts template by copying an object that is reused in other databases to a new database and then saving the database as a template at the Save As backstage area.

Figure 6.4 Application Parts Button Drop-Down List

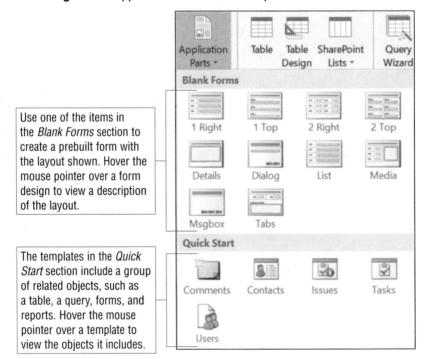

Use one of the items in the *Blank Forms* section to create a prebuilt form with the layout shown. Hover the mouse pointer over a form design to view a description of the layout.

The templates in the *Quick Start* section include a group of related objects, such as a table, a query, forms, and reports. Hover the mouse pointer over a template to view the objects it includes.

Figure 6.5 First Dialog Box in the Create Relationship Wizard

Choose this option if the new table will be the "many" table in a one-to-many relationship.

Choose this option if the new table will be the "one" table in a one-to-many relationship.

Choose this option if the new table will not be related to any existing tables.

Create Relationship	? ×

Create a simple relationship

You can specify a relationship that would be created when Microsoft Access imports the new template.

One 'Customers' to many 'Contacts'.
⦿ Customers

One 'Contacts' to many 'Customers'.
○ Customers

○ There is no relationship.

< Back   Next >   Create   Cancel

Activity 2a Creating a Table, a Query, Forms, and Reports Using Application Parts Quick Start Part 1 of 3

1. Open **6-RSRCompServ** and enable the content.
2. Use a template to create a new table, a query, forms, and reports related to contacts by completing the following steps:
   a. Click the Create tab.
   b. Click the Application Parts button in the Templates group and then click *Contacts* in the *Quick Start* section of the drop-down list.

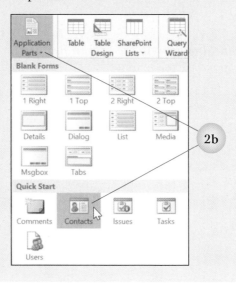

Application Parts | Table | Table Design | SharePoint Lists | Query Wizard

Blank Forms
1 Right | 1 Top | 2 Right | 2 Top
Details | Dialog | List | Media
Msgbox | Tabs

Quick Start
Comments | Contacts | Issues | Tasks
Users

2b

c. The Create Relationship Wizard starts. At the first Create Relationship dialog box, click the *There is no relationship* option and then click the Create button. Access imports a Contacts table; a ContactsExtended query; ContactDetails, ContactDS, and ContactList forms; and ContactAddressBook, ContactList, and ContactPhoneBook reports into the database.

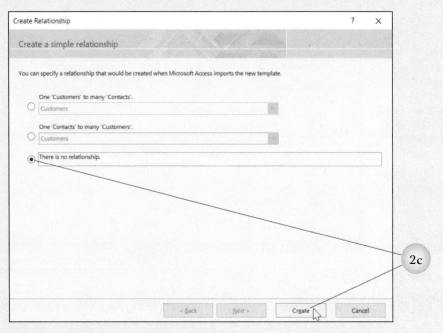

3. Double-click *Contacts* in the Tables group in the Navigation pane. Scroll right to view all the fields in the new table and then close the table.
4. Double-click *ContactDetails* in the Forms group in the Navigation pane.
5. Enter the following record using the ContactDetails form:

First Name	Terry
Last Name	Silver
Job Title	Sales Manager
Company	Cityscape Electronics **Note: Press the Tab key two times after this field to move to the** E-mail *field.*
E-mail	terry_s@ppi-edu.net
Web Page	(leave blank)
Business Phone	800-555-4968
Fax	800-555-6941
Home Phone	(leave blank)
Mobile Phone	313-555-3442
Address	3700 Woodward Avenue
City	Detroit
State/Province	MI
ZIP/Postal Code	48201-2006
Country/Region	(leave blank)
Notes	(leave blank)

6. Click the Save & Close button at the top right of the form.

7. Open the ContactList report.
8. Display the report in Print Preview and then print it.
9. Close Print Preview and then close the ContactList report.

 Check Your Work

Tutorial

Creating a New Form Using an Application Parts Blank Form

Creating a New Form Using an Application Parts Blank Form

Quick Steps

Create Form Using Blank Application Parts Form
1. Click Create tab.
2. Click Application Parts button.
3. Click blank form layout.
4. Add fields to form.
5. Customize form as needed.
6. Save form.

The Application Parts button drop-down list also includes various blank form layouts. Using one of them makes the task of creating a new form easier because the layout has already been defined. The *Blank Forms* section of the Application Parts button drop-down list contains 10 prebuilt blank forms. Most of the forms contain command buttons that perform actions such as saving changes or saving and closing the form. Hover the mouse pointer over a blank form option at the Application Parts button drop-down list to display a description of the form's layout in a ScreenTip. When a blank form option is clicked, Access creates the form object using a predefined form name. For example, if the *1 Right* is clicked, Access creates a form named *SingleOneColumnRightLabels*. Locate the form name in the Navigation pane and then open the form in Layout view or Design view to customize it as needed.

Each Application Parts form has a control layout applied, which means all the form's control objects move and resize together. Remove the control layout to make individual size adjustments. To do this in Design view, select all the control objects and then click the Remove Layout button in the Table group on the Form Design Tools Arrange tab.

Activity 2b Creating a New Form Using an Application Parts Blank Form

1. With **6-RSRCompServ** open, use an
   Application Parts blank form to create a new
   form for maintaining records in the Parts table by
   completing the following steps:
   a. If necessary, click the Create tab.
   b. Click the Applications Parts button in the
      Templates group and then click *1 Right* in the
      Blank Forms section of the drop-down list. Access
      creates a form named *SingleOneColumnRightLabels*.
   c. If necessary, position the mouse pointer on the
      right border of the Navigation pane until the
      pointer changes to a left-and-right-pointing arrow
      and then drag right to widen the Navigation pane
      until all the object names are visible.
   d. Double-click *SingleOneColumnRightLabels* in the Forms
      group in the Navigation pane.
2. Switch to Layout view.
3. Click to select the *Field1* label control object. Press and
   hold down the Shift key; click to select the *Field2*, *Field3*,
   and *Field4* label control objects, release the Shift key, and
   then press the Delete key.
4. Associate the Parts table with the form and add fields
   from the Parts table by completing the following steps:
   a. If the Field List task pane is not open, click the Add
      Existing Fields button in the Tools group on the Form
      Layout Tools Design tab.
   b. Click the Show all tables hyperlink at the top of the task pane. Skip this step if the
      Field List task pane already displays all the table names in the database.
   c. Click the plus symbol
      next to the *Parts* option
      in the Field List task
      pane to expand the list
      and show all the fields in
      the Parts table.
   d. Drag the *PartNo* field to
      the second column in
      the row, as shown at the
      right.
   e. Drag the remaining fields
      (*PartName*, *Supplier*,
      and *Cost*) below
      PartNo, as shown at
      the right.
   f. Close the Field List
      task pane.
5. With the *Cost* field selected, drag up the bottom orange border of the control to decrease
   the height of the control object so the bottom border is directly below the label text.

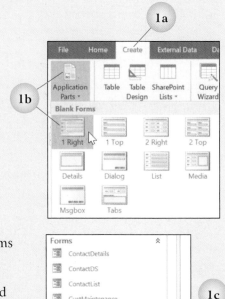

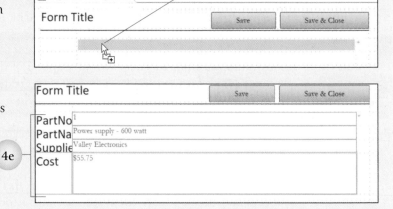

6. Select the four label control objects and drag the right orange border of the selected control objects right to widen the labels until all the label text is visible.

7. With the four label control objects still selected, click the Form Layout Tools Arrange tab, click the Control Padding button in the Position group, and then click *Wide* at the drop-down list.

8. Double-click the form title to place the insertion point inside the title text, delete *Form Title*, type Repair Parts, and then press the Enter key.

9. Make formatting changes to the form as follows:
   a. Click the Save button at the top right of the form. Press and hold down the Ctrl key, click the Save & Close button, and then release the Ctrl key. Click the Form Layout Tools Format tab, click the Quick Styles button, and then click *Intense Effect - Blue-Gray, Accent 3* (fourth column, last row).
   b. Select the *Repair Parts* title control object and apply the Blue-Gray, Accent 3 font color (seventh column, first row in the *Theme Colors* section).
   c. Apply the Blue-Gray, Accent 3 font color to the four label control objects.
   d. Apply the Blue-Gray, Accent 3 Lighter 80% background color (seventh column, second row in the *Theme Colors* section) to the four text box control objects adjacent to the labels.
   e. Click the label control object that displays the text *Supplier*. Use the bottom sizing handle to increase the height of the object so all the text displays.

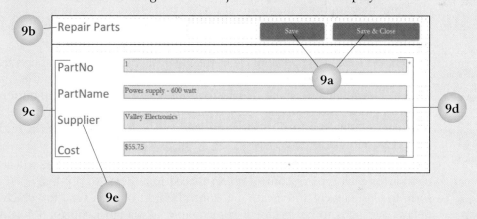

   f. Click in a blank area of the form to deselect the label control object.

10. Click the File tab, click the *Save As* option, click *Save Object As*, and then click the Save As button. Type Parts in the *Save 'SingleOneColumnRightLabels' to* text box at the Save As dialog box and then click OK.

11. Click the Home tab and then switch to Form view.

12. Scroll through a few records in the Parts form and then click the Save & Close button at the top right. Your record may look different than the one shown below.

13. The SingleOneColumnRightLabels form is no longer needed. Click *SingleOneColumnRightLabels* in the Forms group in the Navigation pane and then press the Delete key. Answer Yes to the question *Do you want to permanently delete the form 'SingleOneColumnRightLabels'?*

 Check Your Work

 Tutorial

Creating a User-Defined Form Template

Setting Form Control Defaults and Creating a User-Defined Form Template

Using a standard design for all the forms in a database is an effective way of ensuring consistency and portraying a professional image—for example, formatting all the label control objects as 14-point blue text on a gray background. However, changing these options manually in each form is time consuming and may lead to inconsistencies. To save time and ensure consistency, customize the default settings instead.

The *Set Control Defaults* option at the Controls drop-down list allows the user to change the default properties for all the new labels in a form. To do this, open the form in Design view, format one label control object, click the More button in the Controls group, and then click *Set Control Defaults* at the drop-down list. All the new label control objects added to the form will now be formatted with the options that have been defined.

To further customize the database, create a form template that will set the default for each type of control object placed in a form. To do this, open a new form in Design view and create one control object of each type for which a default setting is to be specified. For example, add a label control object, text box control object, command button, and list box, making sure to format each with colors and backgrounds. When each control object is finished, use the *Set Control Defaults* option to change the default settings. Another option is to select all the control objects after the form is finished and then perform one *Set Control Defaults* command. Save the form using the name *Normal*. The Normal form becomes the template for all the new forms in the database. Existing forms retain their initial formatting, unless they are changed manually.

Quick Steps

Create User-Defined Form Template
1. Create new form in Design view.
2. Add control object to form.
3. Format control object.
4. Click More button in Control group.
5. Click *Set Control Defaults*.
6. Repeat Steps 2–5 for each type of control object to be used.
7. Save form, naming it *Normal*.

Hint Delete the Normal form if you want to go back to using the standard default options for control objects in forms.

1. With **6-RSRCompServ** open, create a form template for the database by completing the following steps:
   a. Click the Create tab and then click the Form Design button in the Forms group.
   b. Click the Themes button in the Themes group. Notice that the database uses the Organic theme for any new forms. Click in a blank area of the form to close the Themes list.
   c. Click the Label button in the Controls group on the Form Design Tools Design tab, draw a label in the *Detail* section, type Sample Label Text, and then press the Enter key. Do not be concerned with the position and size of the label control object at this time, since this control object will be used only to set new default formatting options.
   d. Click the Form Design Tools Format tab. Apply bold formatting and change the font color to Blue-Gray, Accent 3 (seventh column, first row in the *Theme Colors* section) and the background color to Blue-Gray, Accent 3, Lighter 80% (seventh column, second row in the *Theme Colors* section).

   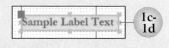
   1c-1d

   e. Click the Form Design Tools Design tab, click the Text Box button in the Controls group, and then draw a text box control object in the *Detail* section. Format the text box control object and its associated label control object as follows:

      1) To the label control object, apply bold formatting and the same font and background colors that were applied to the label control object in Step 1d.

      2) Select the text box control object (displays *Unbound*) and apply the same background color applied to the label control object in Step 1d.

      Your text box and combo box numbers may vary.

      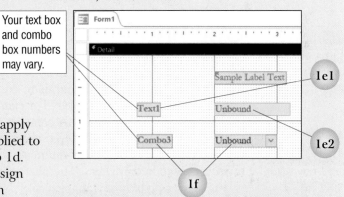
      1e1
      1e2
      1f

   f. Click the Form Design Tools Design tab, click the Combo Box button in the Controls group, and then draw a combo box in the *Detail* section. If the Combo Box Wizard begins, click Cancel. Format the combo box control objects using the same options applied to the text box control object in Steps 1e1 and 1e2.

   g. Press Ctrl + A to select all the control objects in the form.

   h. Click the Form Design Tools Design tab, click the More button in the Controls group, and then click *Set Control Defaults*.

   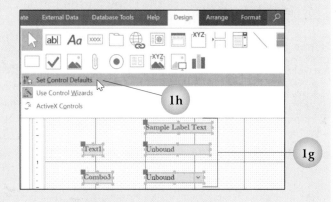
   1h
   1g

2. Save the form with the name *Normal*.
3. Close the Normal form. Normal becomes the form template for **6-RSRCompServ**. Any new form created will have labels, text boxes, and combo boxes formatted as specified in Step 1.

   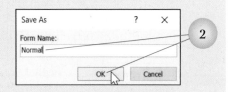
   2

4. Click *WorkOrders* in the Tables group in the Navigation pane, click the Create tab, and then click the Form button in the Forms group. The new WorkOrders form uses the formatting applied to the labels, text boxes, and combo boxes in the Normal form template.

5. With the first record displayed in the WorkOrders form in Form view, click the File tab, click *Print*, and then click *Print Preview*. Click the Columns button in the Page Layout group. Select the current value in the *Width* text box in the *Column Size* section, type 7.5, and then click OK. Close Print Preview.

6. Close the WorkOrders form. Click Yes when prompted to save changes to the design of the form and then click OK to accept the default form name *WorkOrders*.

 Check Your Work

---

Activity 3 **Copy Table Structure**   **1 Part**

You will create a new table to store contact information for manufacturer sales representatives by copying an existing table's field names and field properties.

 Tutorial

Copying a Table
Structure to a New
Table

Quick Steps

Copy Table Structure
1. Select table.
2. Click Copy button.
3. Click Paste button.
4. Type new table name.
5. Click *Structure Only*.
6. Click OK.

Hint If you are creating a new table that will be similar to an existing table, save time by copying the existing table's structure and then adding, deleting, and modifying fields in Design view.

Copying a Table Structure to a New Table

Use the copy and paste commands to create a new table that uses the same or similar fields as an existing table. For example, in Activity 3, the structure of the Contacts table will be copied to create a new table for manufacturer contacts that need to be maintained separately from other contact records. Since the fields needed for the manufacturer contact records are the same as those for the existing contact records, base the new table on the existing table.

To copy a table structure, click the existing table name in the Navigation pane, click the Copy button in the Clipboard group on the Home tab, and then click the Paste button in the Clipboard group. When a table has been copied to the Clipboard, clicking the Paste button causes the Paste Table As dialog box to appear, as shown in Figure 6.6. Type the name for the new table in the *Table Name* text box, click *Structure Only* in the *Paste Options* section, and then click OK. Once the table has been created, fields can be added, deleted, and modified as needed.

Figure 6.6 Paste Table As Dialog Box

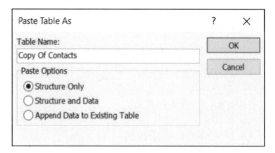

1. With **6-RSRCompServ** open, copy the structure of the Contacts table by completing the following steps:
   a. Click *Contacts* in the Tables group in the Navigation pane.
   b. Click the Home tab and then click the Copy button.
   c. Click the Paste button. (Do not click the Paste button arrow.)
   d. At the Paste Table As dialog box, type MfrContacts in the *Table Name* text box and then click *Structure Only* in the *Paste Options* section.
   e. Click OK.

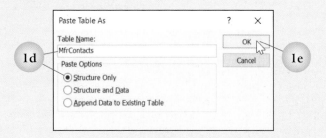

2. Add a description to the table noting that it is based on the Contacts table by completing the following steps:
   a. Right-click MfrContacts in the Navigation Pane.
   b. Click *Table Properties* at the shortcut menu.
   c. Type Based on the Contacts table. in the *Description* text box and then click OK
3. Open the MfrContacts table. Note that the table structure contains the same fields as the Contacts table.
4. Enter the following data in a new record using Datasheet view. Press the Tab key to move past the remaining fields after *ZIP/Postal Code*.

ID	(AutoNumber)
Company	Dell Inc.
Last Name	Haldstadt
First Name	Cari
Email Address	haldstadt@ppi-edu.net
Job Title	Northeast Sales Manager
Business Phone	800-555-9522
Home Phone	(leave blank)
Mobile Phone	800-555-4662
Fax Number	800-555-7781
Address	One Dell Way
City	Round Rock
State/Province	TX
ZIP/Postal Code	78682

5. Close the table.

6. With *MfrContacts* selected in the Tables group in the Navigation pane, click the Create tab and then click the Form button in the Forms group. The form should appear similar to the one shown below.

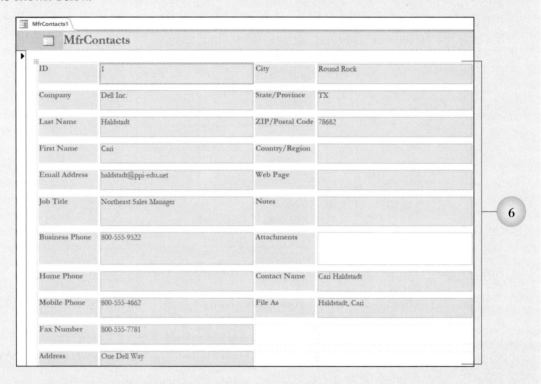

7. With the form in Layout view, select the label control object and text box control object for the *Attachments* field. Press the Delete key.
8. To move the *Contact Name* field and *File As* field up one spot, select the label and text box control objects for both of these fields and then click the Move Up button in the Move group on the Form Layout Tools Arrange tab.
9. Click the File tab, click the *Print* option, and then click *Print Preview* to display the form in Print Preview. Change to landscape orientation, change the left and right margins to 1 inch, and then close Print Preview.
10. Save the form using the default name *MfrContacts*.
11. Print the selected record and then close the form.

Check Your Work

Tutorial

Evaluating a Table
Using the Table
Analyzer Wizard

Quick Steps

**Evaluate Table with
Table Analyzer Wizard**
 1. Click Database
    Tools tab.
 2. Click Analyze Table
    button.
 3. Click Next.
 4. Click Next.
 5. Click table name.
 6. Click Next.
 7. If necessary, click
    *Yes, let wizard
    decide.*
 8. Click Next.
 9. Confirm grouping
    of fields in
    proposed tables.
10. Rename each
    table.
11. Click Next.
12. Confirm and/or set
    primary key field in
    each table.
13. Click Next.
14. If necessary, click
    *Yes, create the
    query.*
15. Click Finish.
16. Close Help
    window.
17. Close query.

[icon] Analyze
Table

Hint Use the Table
Analyzer Wizard to help
normalize a table.

Evaluating a Table Using the Table Analyzer Wizard

Repeating information within a table can result in inconsistencies and wasted storage space. Use the Table Analyzer Wizard to examine a table and determine if duplicate information can be split into smaller related tables to improve the table design. The wizard suggests fields that can be separated into a new table and remain related to the original table with a lookup field. The user can either accept the proposed solutions or modify the suggestions.

Activity 4a uses the Table Analyzer Wizard in a new Parts table. The table was created to store information about the parts commonly used by technicians at RSR Computer Services. Access will examine the table and propose moving the *Supplier* field to a separate table. Making this change will improve the design because several parts records can be associated with the same supplier. In the current table design, the supplier name is typed into a field in each record. Since several parts are associated with each supplier name, the field contains many duplicate entries that take up more storage space than necessary. Furthermore, entering the same data more than once increases the potential for introducing errors in records, which could result in a query that produces incorrect results.

To begin the Table Analyzer Wizard, click the Database Tools tab and then click the Analyze Table button in the Analyze group. This opens the first Table Analyzer Wizard dialog box, shown in Figure 6.7.

Figure 6.7 First Table Analyzer Wizard Dialog Box

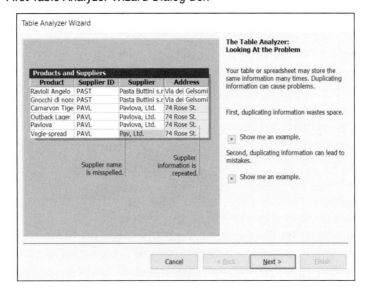

At the first two dialog boxes, review the information about what the Table Analyzer can do to improve the table design. At the third dialog box in the wizard, select the table to be analyzed. At the fourth dialog box, either choose to let the wizard decide which fields to group in the smaller tables or split the tables manually by dragging and dropping fields.

The wizard looks for fields with repetitive data and then suggests a solution. Confirm the grouping of fields and primary key fields in the new tables. At the final step in the wizard, elect to have Access create a query so the fields in the split tables are presented together in a datasheet that resembles the original table.

Activity 4a   Splitting a Table Using the Table Analyzer Wizard   Part 1 of 5

1. With **6-RSRCompServ** open, open the Parts table in Datasheet view and review the table structure and data. Notice that the table includes four fields: *PartNo*, *PartName*, *Supplier*, and *Cost*. Also notice that the supplier names are repeated in the *Supplier* field.
2. Close the Parts table.
3. Use the Table Analyzer Wizard to evaluate the Parts table design by completing the following steps:
   a. Click the Database Tools tab.
   b. Click the Analyze Table button in the Analyze group.
   c. Read the information at the first Table Analyzer Wizard dialog box and then click the Next button.
   d. Read the information at the second Table Analyzer Wizard dialog box and then click the Next button.
   e. At the third Table Analyzer Wizard dialog box with *Parts* selected in the *Tables* list box, click the Next button.

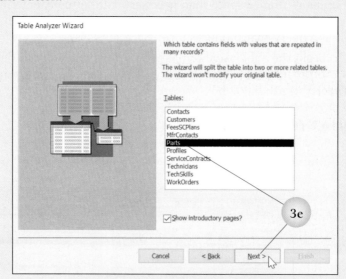

   f. At the fourth Table Analyzer Wizard dialog box with *Yes, let the wizard decide* selected for *Do you want the wizard to decide what fields go in what tables?*, click the Next button.
   g. At the fifth Table Analyzer Wizard dialog box, look at the two tables the wizard is proposing. Notice that the *Supplier* field has been moved to a new table and that a one-to-many relationship has been created between the tables. Access names the new tables *Table1* and *Table2* and asks two questions: *Is the wizard grouping information correctly?* and *What name do you want for each table?* **Note: If necessary, resize the table field list boxes to see all the proposed fields.**

h. The proposed tables have the fields grouped correctly. Double-click the *Table1* Title bar to give the table a new name.

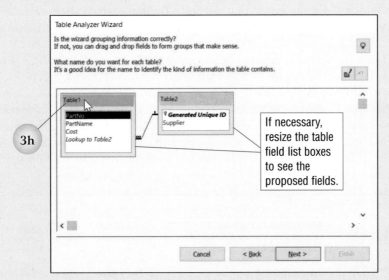

i. Type PartsAndCosts in the *Table Name* text box and then click OK.

j. Click the *Table2* Title bar and then click the Rename Table button near the top right of the dialog box, above the table field list boxes. Type PartsSuppliers in the *Table Name* text box and then press the Enter key.

k. Click the Next button.

l. At the sixth Table Analyzer Wizard dialog box, the unique identifier, or primary key field, for each table is set and/or confirmed. The primary key field is displayed in bold in each table field list box. Notice that the PartsAndCosts table does not have a primary key field defined. Click the *PartNo* field in the PartsAndCosts table field list box and then click the Primary Key button near the top right of the dialog box. Access sets the *PartNo* field as the primary key field, displays a key icon, and applies bold formatting to the field name.

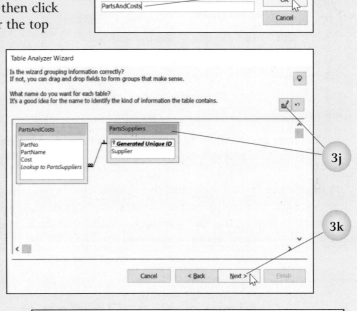

m. Click the Next button.

n. At the last Table Analyzer Wizard dialog box, you can choose to have Access create a query with the original table name that includes the fields from the new tables. Creating the query means that existing forms or reports that were based on the original table will still operate. With *Yes, create the query* selected, click the Finish button. Access renames the original table *Parts_OLD*, creates the query with the name *Parts*, and opens the Parts query results datasheet with the Access Help Task pane.

o. Close the Help task pane.

4. Examine the Parts query datasheet and the object names added in the Navigation pane, including the new tables, PartsAndCosts and PartsSuppliers, along with the original table, Parts_OLD. The Parts query looks just like the original table opened in Step 1, with the exception of the additional field named *Lookup to PartsSuppliers*. The lookup field displays the supplier name, which is also displayed in the original *Supplier* field.

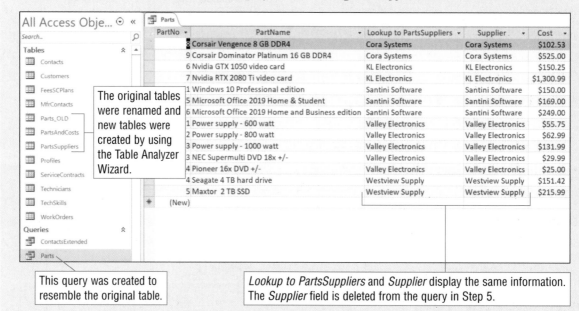

The original tables were renamed and new tables were created by using the Table Analyzer Wizard.

This query was created to resemble the original table.

Lookup to PartsSuppliers and *Supplier* display the same information. The *Supplier* field is deleted from the query in Step 5.

5. Switch to Design view and then delete the *Supplier* field.

6. Save the revised query. Switch to Datasheet view, adjust all the column widths to best fit, and then print the query results datasheet in landscape orientation.

7. Close the query, saving changes to the layout.

 Check Your Work

 Tutorial

Using the Performance Analyzer

Using the Performance Analyzer

The Performance Analyzer can evaluate an individual object, a group of objects, or an entire database for ways that objects can be modified to optimize the use of system resources (such as memory) and improve the speed of data access. If a database seems to run slowly, consider running the tables, queries, forms, reports, or entire database through the Performance Analyzer.

 Analyze Performance

To use the Performance Analyzer, click the Database Tools tab and then click the Analyze Performance button in the Analyze group to open the Performance Analyzer dialog box, shown in Figure 6.8. Select an object type tab, click the check box next to an object to have it analyzed, and then click OK. Select multiple objects or click the Select All button to select all the objects on the

Optimize Database Performance

1. Click Database Tools tab.
2. Click Analyze Performance button.
3. Click All Object Types tab.
4. Click Select All button.
5. Click OK.
6. Review *Analysis Results* items.
7. Optimize relevant recommendations or suggestions.
8. Click Close button.

Hint Before running the Performance Analyzer, make sure all the objects to be analyzed are closed. If they are open, the Performance Analyzer will skip them.

Figure 6.8 Performance Analyzer Dialog Box

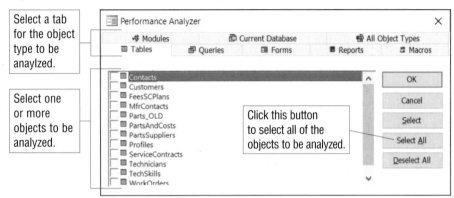

current tab for analysis. To evaluate the entire database, click the All Object Types tab and then click the Select All button. Click OK to begin the analysis.

Three types of results are presented to optimize the selected objects: recommendation, suggestion, and idea. Click an item in the *Analysis Results* list box to read a description of the proposed optimization method in the *Analysis Notes* section. Click a recommendation or suggestion in the *Analysis Results* list box and then click the Optimize button to instruct Access to carry out the recommendation or suggestion. Access will modify the object and mark the item as fixed when completed. The Performance Analyzer may provide ideas for improving the design, such as assigning a different data type for a field based on the type of data that has been entered into records or creating relationships between tables that are not already related.

Activity 4b Analyzing a Database to Improve Performance
Part 2 of 5

1. With **6-RSRCompServ** open, use the Performance Analyzer to evaluate the database by completing the following steps:
   a. If necessary, click the Database Tools tab.
   b. Click the Analyze Performance button in the Analyze group.
   c. At the Performance Analyzer dialog box, click the All Object Types tab.
   d. Click the Select All button.
   e. Click OK. The Performance Analyzer displays the name of each object as it is evaluated and then presents a full list of results when the analysis is finished.

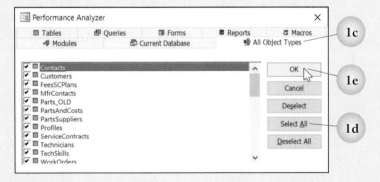

2. Review the items in the *Analysis Results* list box and then optimize a relationship by completing the following steps:

   a. Click the first entry in the *Analysis Results* list box (contains the text *Application: Save your application as an MDE file*) and then read the description of the idea in the *Analysis Notes* section. (Saving an application as an MDE file will be discussed in Chapter 7.)

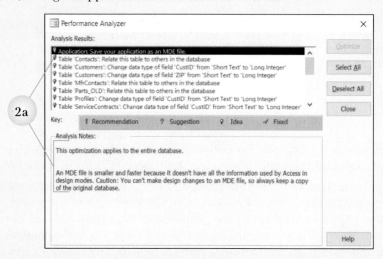

   b. Click the fifth entry in the *Analysis Results* list box (contains the text *Table 'MfrContacts': Relate this table to others in the database*) and then read the description of the idea in the *Analysis Notes* section. The contact information stored in this table is for manufacturer sales representatives and cannot be related to any other tables.

   c. Scroll down the *Analysis Results* list box and then click the item with the green question mark, which identifies a suggestion (contains the text *Table 'WorkOrders': Relate to table 'WorkOrders'*). Read the description of the suggestion in the *Analysis Notes* section. Note that the optimization will benefit the TotalWorkOrders query. This optimization refers to a query that contains a subquery with two levels of calculations. The suggestion is to create a relationship to speed up the query calculations.

   d. Click the Optimize button.

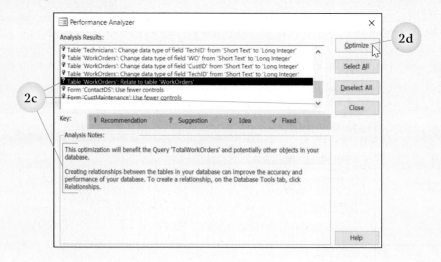

e. Access creates the relationship and changes the question mark next to the item in the *Analysis Results* list box to a check mark (✓). The check mark indicates that the item has been fixed.

f. Click the last item in the *Analysis Results* list box (contains the text *Form 'CustMaintenance': Use fewer controls*) and then read the description of the idea. The idea is to break the form into multiple forms, retaining information that is used often in the existing form. Information viewed less often would be split out into individual forms. To implement this optimization idea would require redesigning the form.

3. Click the Close button to close the Performance Analyzer dialog box.

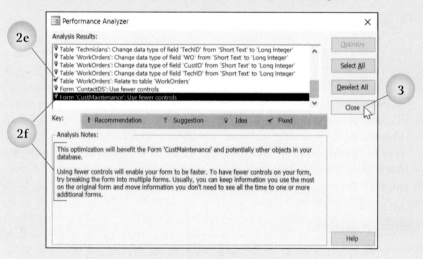

Tutorial

Splitting a Database

Splitting a Database

If a database is placed on a network where multiple users can access it simultaneously, the speed with which the data is accessed may decrease. One way to improve performance is to split the database into two files.

Quick Steps

Split Database
1. Click Database Tools tab.
2. Click Access Database button.
3. Click Split Database button.
4. If necessary, navigate to drive and/or folder.
5. If necessary, edit file name.
6. Click Split button.
7. Click OK.

Access Database

The file containing the tables (called the *back-end database*) is stored in the network share folder, while the file containing the queries, forms, and reports (called the *front-end database*) is stored on individual users' computers. The users can create and/or customize their own queries, forms, and reports to serve their individual purposes. The front-end database contains tables linked to the back-end data, so all the users update a single data source file.

To split an existing database into back-end and front-end databases, Access provides the Database Splitter Wizard. Click the Database Tools tab and then click the Access Database button in the Move Data group to open the first Database Splitter Wizard dialog box, shown in Figure 6.9.

Click the Split Database button to open the Create Back-end Database dialog box and navigate to the drive and/or folder where the database file containing the original tables is to be stored. By default, Access uses the original database file name with *_be* added to the end (before the file extension). Change the file name if required and then click the Split button. Access moves the table objects to the back-end file, creates links to the back-end tables in the front-end file, and then displays a message stating that the database was successfully split.

Figure 6.9 First Database Splitter Wizard Dialog Box

💡 **Hint** Consider making a backup copy of the database before splitting the file, in case the database needs to be restored to its original state.

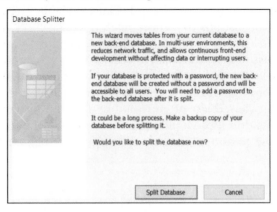

Another reason to split a database is to overcome the file size restriction in Access 365. Database specifications for Access 365 set the maximum file size at 2 gigabytes (GB). This size includes the space needed by Access to open system objects while working with the database, which means the actual maximum file size is even less. However, the size restriction does not include links to external data source files. Splitting a database extends the size beyond the 2-gigabyte limitation.

Activity 4c Splitting a Database

Part 3 of 5

1. With **6-RSRCompServ** open, split the database to create back-end and front-end databases by completing the following steps:
   a. If necessary, click the Database Tools tab.
   b. Click the Access Database button in the Move Data group.
   c. At the first Database Splitter Wizard dialog box, click the Split Database button.

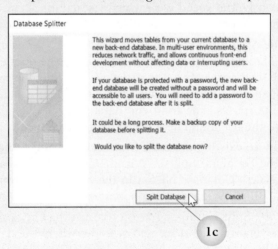

1c

d. At the Create Back-end Database dialog box, navigate to the same folder as the original database (AL2C6). To accept the default file name **6-RSRCompServ_be** in the *File name* text box, click the Split button.

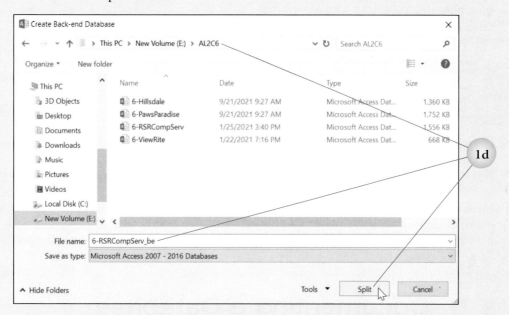

e. Click OK at the Database Splitter message box stating that the database was successfully split.

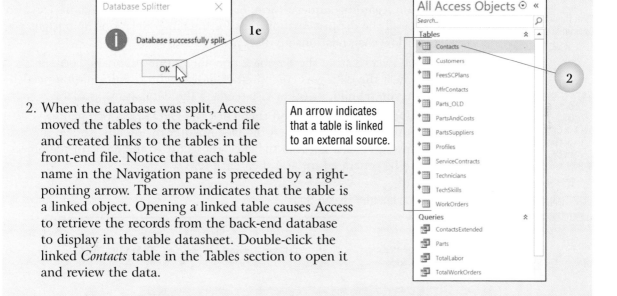

2. When the database was split, Access moved the tables to the back-end file and created links to the tables in the front-end file. Notice that each table name in the Navigation pane is preceded by a right-pointing arrow. The arrow indicates that the table is a linked object. Opening a linked table causes Access to retrieve the records from the back-end database to display in the table datasheet. Double-click the linked *Contacts* table in the Tables section to open it and review the data.

An arrow indicates that a table is linked to an external source.

3. Switch to Design view. Access displays a message stating that the table is a linked table whose design can't be modified and that if you want to add or remove fields or change properties that you will have to do it in the source database the Click the No button and then close the table.

4. Close **6-RSRCompServ**.
5. Open **6-RSRCompServ_be** and enable the content.
6. Notice that the back-end database file contains only the tables. Open the Customers table in Datasheet view and review the data.
7. Switch to Design view. Note that in the back-end database, switching to Design view to make changes does not prompt display of the message box that displayed in Step 3. The message does not display because this database contains the original source table.
8. Close the table.

 Tutorial

Documentating a Database

Quick Steps

Print Object Documentation
1. Click Database Tools tab.
2. Click Database Documenter button.
3. Click Options button.
4. Choose report options.
5. Click OK.
6. Click object name.
7. Click OK.
8. Print report.
9. Close report.

Database Documenter

Documenting a Database

Another Access feature, the Database Documenter feature, prepares and prints a report with details about the definition of a database object. This report can be used as hard-copy documentation of the table structure, including field properties and information regarding a query, form, or report. Relationship diagrams can also be included for all the defined relationships for the table. Relationship options are documented below each relationship diagram.

It is a good idea to store the database documentation report in a secure place. That way, if the data gets corrupted or destroyed, the information needed to manually repair, rebuild, or otherwise recreate the database is available.

Click the Database Tools tab and then click the Database Documenter button in the Analyze group to open the Documenter dialog box, shown in Figure 6.10. Insert a check mark in the check box next to the name of the object the report is to be generated for and then click OK.

Figure 6.10 Documenter Dialog Box

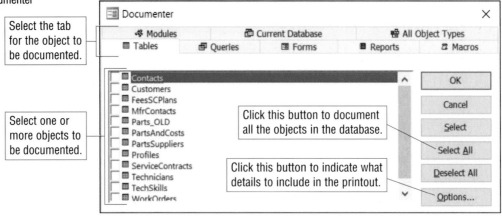

1. With **6-RSRCompServ_be** open, generate a report providing details of the table structure, field properties, and relationships for an individual table by completing the following steps:
   a. Click the Database Tools tab.
   b. In the Analyze group, click the Database Documenter button.
   c. At the Documenter dialog box, click the Options button.
   d. At the Print Table Definition dialog box, click the *Permissions by User and Group* check box in the *Include for Table* section to remove the check mark.
   e. Make sure *Names, Data Types, Sizes, and Properties* is selected in the *Include for Fields* section.
   f. Click *Nothing* in the *Include for Indexes* section.
   g. Click OK.
   h. With Tables the active tab in the Documenter dialog box, click the *PartsSuppliers* check box to insert a check mark.
   i. Click OK.

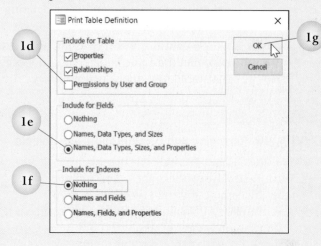

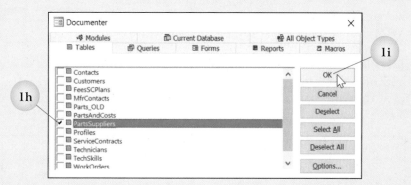

2. Access generates the table definition report and displays it in Print Preview. Print the report.
3. Notice that the Save option is dimmed. A report generated by the Database Documenter cannot be saved, but it can be exported as a PDF, XPS, or one of the other file options found in the Data group on the Print Preview tab. Click the Close Print Preview button in the Close Preview group on the Print Preview tab.

Check Your Work

Tutorial

Renaming and
Deleting Objects

Renaming and Deleting Objects

As part of managing a database, objects may need to be renamed or deleted within the database file. To do this, right-click the object name in the Navigation pane and then click *Rename* or *Delete* at the shortcut menu. An object may also be deleted from a database by selecting the object in the Navigation pane and then pressing the Delete key. Click Yes at the Microsoft Access message box asking you to confirm deletion of the object. Consider making a backup copy of a database before renaming or deleting objects in case the database needs to be restored to its previous state.

Be cautious when deleting objects that have dependencies to other objects. For example, if a table is deleted and a query exists that is dependent on fields within that table, the query will no longer run.

Quick Steps

Rename Object
1. Right-click object in Navigation pane.
2. Click *Rename*.
3. Type new name.
4. Press Enter key.

Delete Object
1. Right-click object in Navigation pane.
2. Click *Delete*.
3. Click Yes.

Activity 4e Renaming and Deleting Database Objects

Part 5 of 5

1. With **6-RSRCompServ_be** open, rename the MfrContacts table by completing the following steps:
   a. Right-click *MfrContacts* in the Tables group in the Navigation pane.
   b. Click *Rename* at the shortcut menu.
   c. Type ManufacturerContacts and then press the Enter key.
2. Delete the original table that was split using the Table Analyzer Wizard in Activity 4a by completing the following steps:
   a. Right-click *Parts_OLD* in the Tables group in the Navigation pane.
   b. Click *Delete* at the shortcut menu.
   c. Click Yes at the Microsoft Access message box asking if you want to delete the table *Parts_OLD*.

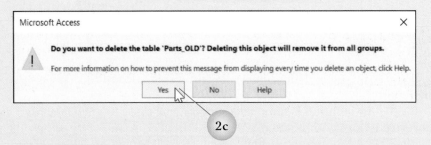

3. Close **6-RSRCompServ_be**.
4. Open **6-RSRCompServ** (the front-end database) and enable the content.

5. Double-click *MfrContacts* in the Tables group in the Navigation pane. Because the table was renamed in Step 1, Access can no longer find the source data. At the Microsoft Access message box stating that the database engine cannot find the input table, click OK. The link will have to be recreated to establish a new connection to the renamed table. (Creating a link to an external table will be discussed in Chapter 8.)

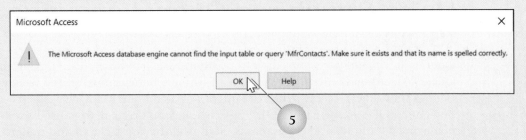

6. Double-click *Parts_OLD* in the Tables group in the Navigation pane. Because this table was deleted in Step 2, the same message appears. Click OK to close the message box.
7. Right-click *Parts_OLD* in the Tables group in the Navigation pane and then click *Delete* at the shortcut menu. Click Yes at the Microsoft Access message box asking if you want to remove the link.

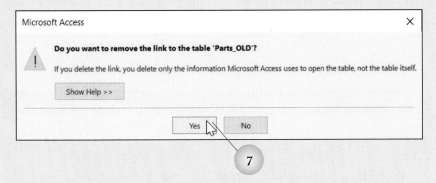

8. Delete the ContactPhoneBook report by completing the following steps:
   a. Click *ContactPhoneBook* in the Reports group in the Navigation pane.
   b. Press the Delete key.
   c. Click Yes at the Microsoft Access message box asking if you want to permanently delete the report.

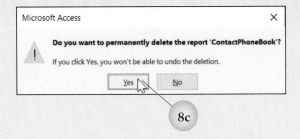

Activity 5 Use Structured Query Language   **1 Part**

You will modify a query using Structured Query Language (SQL).

Using Structured Query Language (SQL)

Hint A typical query may look like this:

SELECT field_1
FROM table_1
WHERE criterion_1;

Note that Access ignores line breaks and uses a semicolon to indicate the end of a statement.

Queries are created to extract data from a database. They are created using the Query Wizard and Query Design buttons in the Queries group on the Create tab. Behind the scenes, Access writes the queries in a computer language called Structured Query Language (SQL). SQL is the computer language used for managing data in relational databases. SQL is fairly easy to learn and understand, but the correct syntax, or order of instructions, is very important.

Each SQL statement begins with a command, such as SELECT, INSERT, and so on. The most common command is SELECT. A SELECT statement is used to retrieve specific data from the database. Each SELECT statement includes several clauses, which give specific instructions about where to find the data, what criteria to use, and how to sort the data. Some common clauses can be found in Table 6.1.

SQL clauses are made up of SQL terms that are comparable to the parts of speech. See Table 6.2 for some common SQL terms.

Table 6.1 Common SQL Clauses

SQL Clause	Description	Required
SELECT	Specifies the fields from which to select the data	Yes
FROM	Lists the tables that contain the fields required	Yes
WHERE	Specifies additional criteria that must be met	No
ORDER BY	Indicates how to sort the data	No

Table 6.2 Common SQL Terms

SQL Term	Comparable Part of Speech	Description	Example
identifier	noun	Used to identify a database object, or content of object such as a field	*Technician.[FName]*
operator	verb or adverb	Represents an action, such as addition (+), or describes how to perform an action (AS)	*SELECT [Technicians] AS [Supervisors]*
constant	noun	A non-changing or null value	*25*
expression	adjective	A combination of identifiers, operators, constants, and functions that produces a result	*>=WorkOrders.[Rate]*

Figures 6.11, 6.12, and 6.13 show a TotalLabor query in SQL, Design, and Datasheet view. Figure 6.11 shows a SELECT statement, beginning with a SELECT clause that selects each of the fields and creates a new field. The FROM clause identifies the table where the fields are located, and the WHERE clause gives additional criteria that must be met in order for the record to be returned.

Figure 6.11 TotalLabor Query SQL View

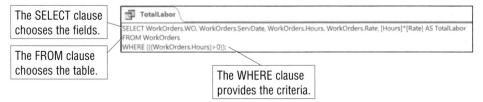

The SELECT clause chooses the fields.

The FROM clause chooses the table.

The WHERE clause provides the criteria.

TotalLabor

SELECT WorkOrders.WO, WorkOrders.ServDate, WorkOrders.Hours, WorkOrders.Rate, [Hours]*[Rate] AS TotalLabor
FROM WorkOrders
WHERE (((WorkOrders.Hours)>0));

Figure 6.12 TotalLabor Query Design View

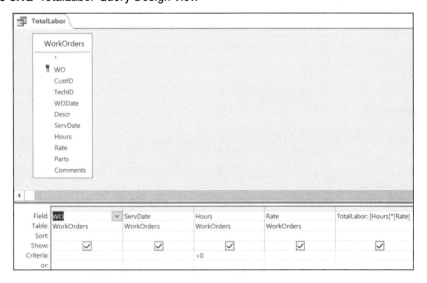

Figure 6.13 TotalLabor Query Datasheet View

Work Order	Service Date	Hours	Rate	Total Labor
65019	Wed, Dec 01 2021	2.50	55.00	$137.50
65020	Thu, Dec 02 2021	1.50	50.00	$75.00
65021	Thu, Dec 02 2021	3.25	55.00	$178.75
65022	Fri, Dec 03 2021	1.50	55.00	$82.50
65023	Sat, Dec 04 2021	1.25	50.00	$62.50
65024	Mon, Dec 06 2021	1.00	45.00	$45.00
65025	Mon, Dec 06 2021	1.50	50.00	$75.00
65026	Tue, Dec 07 2021	0.75	50.00	$37.50

1. With **6-RSRCompServ** open, modify the TotalLabor query to add the customer's ID using SQL view by completing the following steps:
   a. Double-click the *TotalLabor* query in the Navigation pane.
   b. Click the View option arrow in the Views group on the Home tab and then click *SQL View*.
   c. Click directly after the first comma in the SELECT statement.
   d. Press the space bar and then type WorkOrders. CustID,. The first line of code should now read *SELECT WorkOrders.WO, WorkOrders.CustID, WorkOrders.ServDate, WorkOrders.Hours, WorkOrders.Rate, [Hours]*[Rate] AS TotalLabor*

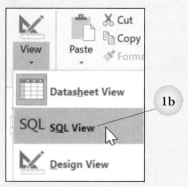
1b

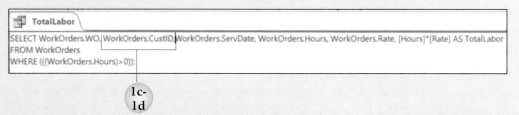

1c-1d

2. Click the Run button. The customer's ID appears between the work order number and the service date.
3. Save the query and then close it.
4. Close **6-RSRCompServ**.

Chapter Summary

- Access provides professionally designed database templates that include predefined tables, queries, forms, and reports for use in creating a new database.
- Choose a database template from the sample templates stored on your computer or download a database template from Office.com.
- Predefined table and related object templates for Comments; Contacts; and Issues, Tasks, and Users are available in the *Quick Start* section of the Application Parts button drop-down list in the Templates group on the Create tab.
- The *Blank Forms* section of the Application Parts button drop-down list contains 10 prebuilt blank forms. Most of these forms include command buttons that perform actions such as saving changes and saving and closing the form.
- Define a custom form template by creating a form named *Normal* that includes a sample of each control object with the formatting options required for future forms. Select all the controls and use the *Set Control Defaults* option to save the new settings.
- When a table has been copied to the Clipboard from the Navigation pane, clicking the Paste button causes the Paste Table As dialog box to open. At this dialog box, choose one of the three options: *Structure Only*, *Structure and Data*, or *Append Data to Existing Table*.

- Use the Table Analyzer Wizard to evaluate a table for repeated data and determine if the table can be split into smaller related tables.
- Use the Performance Analyzer to evaluate an individual object, a group of objects, or an entire database for ways to optimize the use of system resources and improve the speed of data access.
- The Performance Analyzer provides three types of results in the *Analysis Results* list box: recommendation, suggestion, and idea.
- Click a recommendation or suggestion in the *Analysis Results* list box and then click the Optimize button to instruct Access to carry out the recommendation or suggestion.
- A database can be split into two individual files—a back-end database and a front-end database—to improve performance for a multiuser database or overcome the maximum database file size restriction.
- To split a database using the Database Splitter Wizard, begin by clicking the Access Database button in the Move Data group on the Database Tools tab.
- Use the Database Documenter feature to print a hard-copy report with details about the definition of a database object, query, form, or report or the structure of a table.
- Rename an object by right-clicking the object name in the Navigation pane, clicking *Rename* at the shortcut menu, typing the new name, and then pressing the Enter key.
- Delete an object by right-clicking the object name in the Navigation pane, clicking *Delete* at the shortcut menu, and then clicking Yes at the message box asking you to confirm deletion of the object.
- Use SQL view to modify queries using common SQL clauses such as SELECT, FROM, and WHERE.

Commands Review

FEATURE	RIBBON TAB, GROUP	BUTTON	KEYBOARD SHORTCUT
Application Parts template	Create, Templates		
Database Documenter	Database Tools, Analyze		
Paste Table As	Home, Clipboard		Ctrl + V
Performance Analyzer	Database Tools, Analyze		
split database	Database Tools, Move Data		
Table Analyzer Wizard	Database Tools, Analyze		

Microsoft®
Access®

Automating, Customizing, and Securing Access

Performance Objectives

Upon successful completion of Chapter 7, you will be able to:

1 Create, run, edit, and delete a macro

2 Create a command button to run a macro

3 View an embedded macro in the Property Sheet task pane

4 Convert a macro to Visual Basic for Applications

5 Create and edit a Navigation form

6 Customize database startup options

7 Limit access to options in ribbons and menus

8 Customize the Navigation pane

9 Define error-checking options

10 Import and export customized settings

11 Customize the ribbon and the Quick Access Toolbar

Macros are used to automate repetitive tasks and to store these actions so that they can be executed by clicking a button in a form. A Navigation form is used as a menu to provide an interface between the user and objects within the database file. In this chapter, you will learn how to automate a database using macros and a Navigation form. You will also learn how to customize the Access environment.

 Data Files

Before beginning chapter work, copy the AL2C7 folder to your storage medium and then make AL2C7 the active folder.

The online course includes additional training and assessment resources.

Activity 1 Create Macros and Assign Macros to
Command Buttons

8 Parts

You will create macros to automate routine tasks and add macros to command
buttons in forms that run the macros.

 Tutorial

Creating and
Running a Macro

💡 *Hint* To create a
complex macro, begin
by working through
the steps to be saved.
Write down all the
parameters before
attempting to record
the macro.

Creating and Running a Macro

A macro is a series of instructions stored in sequence that can be recalled and
carried out as needed. A macro is generally created for a task that never varies and
that is frequently repeated. For example, a macro could be created to open a query,
form, or report. The macro object stores a series of instructions (called *actions*) in
the sequence in which they are to be performed. Macros appear as objects within
the Navigation pane. A macro that opens a query, form, or report can be run by
double-clicking the macro name in the Navigation pane.

Not all macros can be run by this method, however. Some need to be assigned
to buttons that are clicked in order to run them. For example, a user can create a
macro in a form that automates the process of searching through the records for
a specific client's last name. The macro will contain two instructions: to move to
the field in which the last name is stored and to open the Find and Replace dialog
box. In general, a macro that does not include an instruction containing the query,
form, or report name must be assigned to a button. Otherwise, an error message will
appear stating that the object is not currently selected or is not in the active view.

To create a macro, click the Create tab and then click the Macro button in
the Macros & Code group. This opens the Macro Builder window, shown in
Figure 7.1. Click the arrow at the right side of the *Add New Action* option box and
then click an instruction at the drop-down list. As an alternative, add an action
using the Action Catalog task pane, as described in Figure 7.1.

Figure 7.1 Macro Builder Window

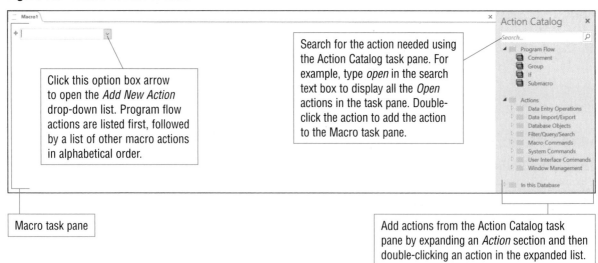

Click this option box arrow
to open the *Add New Action*
drop-down list. Program flow
actions are listed first, followed
by a list of other macro actions
in alphabetical order.

Search for the action needed using
the Action Catalog task pane. For
example, type *open* in the search
text box to display all the *Open*
actions in the task pane. Double-
click the action to add the action
to the Macro task pane.

Macro task pane

Add actions from the Action Catalog task
pane by expanding an *Action* section and then
double-clicking an action in the expanded list.

Create Macro
1. Click Create tab.
2. Click Macro button.
3. Click *Add New Action* option box arrow.
4. Click action.
5. Enter arguments as required in Macro task pane.
6. Click Save button.
7. Type name for macro.
8. Press Enter key.
9. Repeat Steps 3–6 as needed.

Run Macro
Double-click macro name in Navigation pane.
OR
1. Right-click macro name.
2. Click *Run*.
OR
With macro open in Design view, click Run button.

 Hint Hover the mouse pointer over the entry box for an argument to reveal a description of the argument and the available parameters in a ScreenTip.

! Run

Macro

Each new action entered into the Macro task pane is associated with a set of arguments that displays once the action has been added. Similar to the field properties displayed in Table Design view, the arguments displayed in the Macro task pane vary depending on the active action that has been expanded. Figure 7.2 displays the action arguments for the OpenForm action.

The OpenForm action is used to open a form. Within the Macro task pane, specify the name of the form to open and the view in which the form is to be presented. Choose to open the form in Form view, Datasheet view, Layout view, Design view, or Print Preview. Use the Filter Name or Where Condition argument to restrict the records displayed in the report.

The Data Mode argument is used to place editing restrictions on records while the form is open. Choose to open the form in Add mode to allow only adding new records (users cannot view existing records); Edit mode to allow adding, editing, and deleting records; or Read Only mode to allow viewing records. The Window Mode argument is used to instruct Access to open the form in Normal mode (the way forms are normally viewed in the work area), Hidden mode (form is hidden), Icon mode (the form opens minimized), or Dialog mode (the form opens in a separate window similar to a dialog box).

To create a macro that performs multiple actions, use the *Add New Action* option box to add the second action below the first action. Access executes the actions in the order they appear in the Macro task pane. Activity 1a demonstrates how to create a macro with multiple actions that will instruct Access to open a form, make active a control within the form, and then open the Find and Replace dialog box to search for a record.

The GoToControl action is used to make active a control within a form or report, and the RunMenuCommand action is used to execute an Access command. For each action, a single argument specifies the name of the control to move to and the name of the command to run.

As actions are added to the Macro task pane, expand and collapse them as needed. When several actions have been added to the Macro task pane, collapsing them allows the user to focus on the current action being edited.

Figure 7.2 Macro Task Pane with Action Arguments for the OpenForm Action

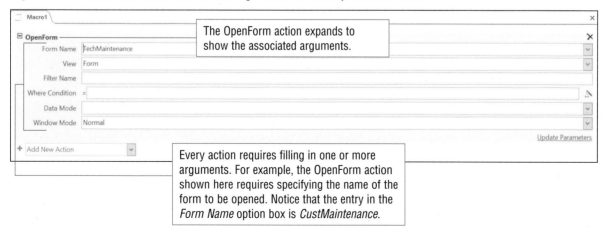

1. Open **7-RSRCompServ** and enable the content. Create a macro to open the
   TechMaintenance form by completing the following steps:
   a. Click the Create tab.
   b. Click the Macro button in the Macros & Code group.
   c. At the Macro task pane, click the *Add New Action*
      option box arrow, scroll down the list, and then click
      OpenForm. Access adds the action and the arguments
      associated with the action. Most actions require at
      least one argument.
   d. Click the *Form Name* option box arrow in
      the *OpenForm* action section and then click
      TechMaintenance at the drop-down list.

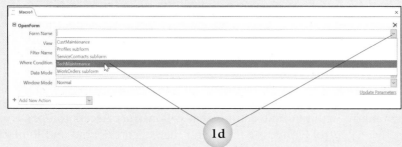

2. Add additional instructions to move to the field for the technician's last name and then
   open the Find and Replace dialog box by completing the following steps:
   a. Click the *Add New Action* option box arrow, scroll down the list, and then click
      GoToControl. Notice that only one action argument is required for the GoToControl
      action.
   b. With the insertion point in the *Control Name* text box in the GoToControl action,
      type LName. (Entering the name of the field that is to be made active in the form is
      required.)

   c. Click the *Add New Action* option box arrow, scroll down the list, and then click
      RunMenuCommand.

d. Click the *Command* option box arrow, scroll down the list, and then click *Find*.

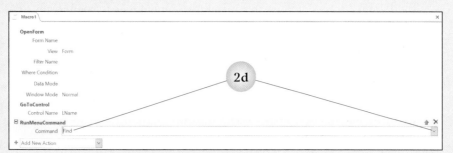

3. Click the Save button on the Quick Access Toolbar, type FormFindTech in the *Macro Name* text box at the Save As dialog box, and then click OK.

4. Click the Run button (which displays as a red exclamation point) in the Tools group on the Macro Tools Design tab. The Run button instructs Access to carry out the instructions in the macro. The TechMaintenance form opens, the *Last Name* field is active, and the Find and Replace dialog box appears with the last name of the first technician entered in the *Find What* text box. Type Eastman and then click the Find Next button. ***Note: If the Find and Replace dialog box overlaps the*** Last Name ***field in the form, drag the dialog box to the bottom or right edge of the work area.***

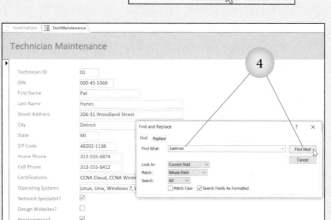

5. Access moves to record 10. Close the Find and Replace dialog box. Notice that the last name text is selected in the form. Read the data displayed in the form for the technician named Kelsey Eastman.

6. Close the form.

7. At the Macro Builder window, click the Close button. This closes the FormFindTech macro. Notice that a Macros group has been added to the Navigation pane and that FormFindTech appears as an object below this new group name. You may need to scroll down the Navigation pane to view the Macros group.

A macro can also be created by dragging and dropping an object name from the Navigation pane into the *Add New Action* option box in a Macro task pane. By default, Access creates an OpenTable, OpenQuery, OpenForm, or OpenReport action depending on the object that was dragged to the window. The object name is also automatically entered in the Object Name argument list box.

Activity 1b Creating a Macro by Dragging and Dropping an Object

Part 2 of 8

1. With **7-RSRCompServ** open, use the drag and drop method to create a macro to open the CustMaintenance form by completing the following steps:
   a. Click the Create tab.
   b. Click the Macro button in the Macros & Code group.
   c. Position the mouse pointer on the CustMaintenance form name in the Navigation pane, click and hold down the left mouse button, drag the object name into the *Add New Action* option box in the Macro task pane, and then release the mouse button. Access inserts an OpenForm action with *CustMaintenance* entered in the *Form Name* option box.
2. Click the Save button, type FormCustMaint, and then click OK.
3. Click the Run button in the Tools group on the Macro Tools Design tab.
4. Close the form.
5. Close the macro by clicking the Close button.

Activity 1c Creating a Macro Using the Action Catalog Task Pane

Part 3 of 8

1. With **7-RSRCompServ** open, use the Action Catalog task pane to create a macro to find a record in a form by making active the home telephone field and then opening the Find and Replace dialog box by completing the following steps. ***Note: This macro will be assigned to a button in Activities 1e and 1f. An error will occur if you try to run the macro by double-clicking it in the Navigation pane.***
   a. Click the Create tab.
   b. Click the Macro button.
   c. Click the expand button (which displays as a white triangle) next to *Database Objects* in the *Actions* list in the Action Catalog task pane to expand the category and display the actions available for changing controls or objects in the database. ***Note: If the Action Catalog task pane was accidentally closed and does not redisplay in the new Macro Builder window, restore the task pane by clicking the Action Catalog button in the Show/Hide group on the Macro Tools Design tab.***

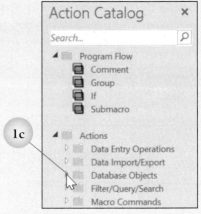

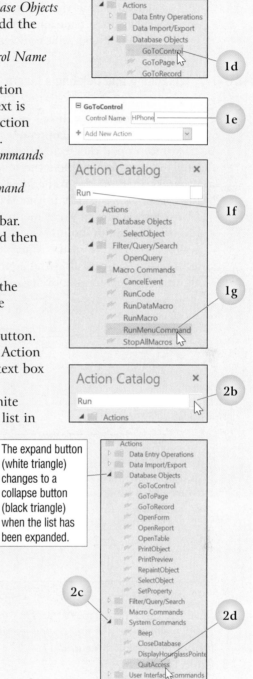

d. Double-click *GoToControl* at the expanded *Database Objects* actions list in the Action Catalog task pane to add the action to the Macro task pane.

e. With the insertion point positioned in the *Control Name* text box in the Macro task pane, type HPhone.

f. Click in the search text box at the top of the Action Catalog task pane and then type Run. As this text is typed, Access displays available actions in the Action Catalog task pane that begin with the same text.

g. Double-click *RunMenuCommand* in the *Macro Commands* list in the Action Catalog task pane.

h. With the insertion point positioned in the *Command* text box in the Macro task pane, type Find.

i. Click the Save button on the Quick Access Toolbar.

j. At the Save As dialog box, type HPhoneFind and then click OK.

k. Close the HPhoneFind macro.

2. Use the Action Catalog to create a macro to close the current database and exit Access by completing the following steps:

a. Click the Create tab and then click the Macro button.

b. Click the box right of the search text box in the Action Catalog task pane to clear *Run* from the search text box and redisplay all the Action Catalog categories.

c. Click the expand button (which displays as a white triangle) next to *System Commands* in the *Actions* list in the Action Catalog task pane.

d. Double-click *QuitAccess* in the expanded *System Commands* actions list.

The expand button (white triangle) changes to a collapse button (black triangle) when the list has been expanded.

e. With *Save All* the default argument in the *Options* option box, click the Save button on the Quick Access Toolbar.

f. Type ExitRSRdb at the Save As dialog box and then click OK.

3. Close the ExitRSRdb macro.

4. Double-click *ExitRSRdb* in the Navigation pane.

 Tutorial

Editing a Macro

Editing and Deleting a Macro

To edit a macro, right-click the macro name in the Navigation pane and then click *Design View* at the shortcut menu or right-click the macro name and press Ctrl + Enter. The macro opens in the Macro Builder window. Edit an action and/or its arguments, insert new actions, and/or delete actions as required. Save the revised macro and close the Macro Builder window when finished.

To delete a macro, right-click the macro name in the Navigation pane and then click *Delete* at the shortcut menu. A macro can also be deleted by selecting it in the Navigation pane and clicking the Delete key. At the Microsoft Access dialog box asking if you want to delete the macro, click Yes.

Activity 1d Editing and Deleting a Macro

1. Open **7-RSRCompServ**. It has been decided that the macro to find a technician record will begin with the TechMaintenance form already opened. This means that the first macro instruction to open the TechMaintenance form in the FormFindTech macro needs to be deleted. To do this, complete the following steps:
   a. If necessary, scroll down the Navigation pane to view the macro object names.
   b. Right-click *FormFindTech* in the Macros group in the Navigation pane and then click *Design View* at the shortcut menu. The macro opens in the Macro Builder window.
   c. Position the mouse pointer over the *OpenForm* action in the Macro task pane. As an action is pointed to in the Macro task pane, Access displays a collapse indicator at the left of the action to allow the arguments to be collapsed; a green arrow at the right to allow the action to be moved; and a black *X* to delete the action.

   d. Click the black *X* at the right side of the Macro task pane across from *OpenForm*. The action is removed from the Macro task pane. *Note: If the buttons at the right side of the Macro task pane disappear as you move the mouse to the right, move the pointer up so it is on the same line as **OpenForm** to redisplay the buttons.*

2. Save the revised macro. *Note: This macro can no longer be run by double-clicking it in the Navigation pane. As mentioned earlier, an error will occur. The macro will be assigned to a button in Activities 1e and 1f.*
3. Close the macro.
4. The revised macro contains two instructions that activate the *LName* control and then open the Find and Replace dialog box. This macro can be used in any form that contains a field named *LName*; therefore, the macro should be renamed. To do this, right-click *FormFindTech* in the Navigation pane, click *Rename* at the shortcut menu, type LNameFind, and then press the Enter key.

5. Delete the FormCustMaint macro by completing the following steps:

   a. Right-click *FormCustMaint* in the Navigation pane and then click *Delete* at the shortcut menu.

   b. At the Microsoft Access dialog box asking if you want to delete the macro, click Yes.

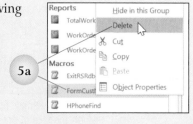

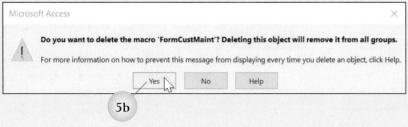

Creating a Command Button to Run a Macro

A macro can be assigned to a button in a form so it can be executed with a single mouse click. This method of running macros makes them more accessible and efficient.

To add a button to a form for running a macro, open the form in Design view. Click the Button button in the Controls group on the Form Design Tools Design tab and then drag to create a button in any form section. When the mouse is released, the Command Button Wizard launches (if the Use Control Wizards feature is active). Recall from Chapter 4 that the Use Control Wizards button displays with a gray background when the feature is active.

At the first Command Button Wizard dialog box, shown in Figure 7.3, begin by choosing the type of command to assign to the button. Click *Miscellaneous* in the *Categories* list box, click *Run Macro* in the *Actions* list box, and then click the Next button. At the second Command Button Wizard dialog box, choose the name of the macro to assign to the button.

Quick Steps

Create Command Button in Form

1. Open form in Design view.
2. Click Button button.
3. Drag to create button.
4. Click *Miscellaneous*.
5. Click *Run Macro*.
6. Click Next.
7. Click macro name.
8. Click Next.
9. Click *Text*.
10. Select current text in *Text* text box.
11. Type text to appear on button.
12. Click Next.
13. Type name for command button.
14. Click Finish.

Hint An action can be assigned to a command button without using a macro. Explore the categories and actions for each category at the first Command Button Wizard dialog box.

Figure 7.3 First Command Button Wizard Dialog Box

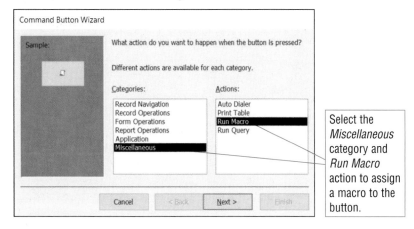

Figure 7.4 Third Command Button Wizard Dialog Box

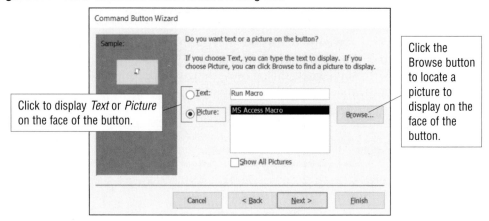

Click to display *Text* or *Picture* on the face of the button.

Click the Browse button to locate a picture to display on the face of the button.

At the third dialog box, shown in Figure 7.4, specify text or a picture to display on the face of the button. When text is entered or a picture file is selected, the button in the *Sample* section of the dialog box updates to show how the button will appear. At the last Command Button Wizard dialog box, assign a name to associate with the command button and then click the Finish button.

Activity 1e Creating a Button and Assigning a Macro to a Button in a Form   Part 5 of 8

1. With **7-RSRCompServ** open, create a command button to run the macro to locate a technician record by last name in the TechMaintenance form by completing the following steps:

   a. Open the TechMaintenance form in Design view. To make room for the new button in the *Form Header* section, click to select the control object with the title text and then drag the right middle sizing handle to the left until the right edge is at approximately the 3.5-inch mark on the horizontal ruler.

   b. By default, the Use Control Wizards button is toggled on in the Controls group. Click the More button in the Controls group on the Form Design Tools Design tab. If the Use Control Wizards button displays with a gray background, the feature is active. If the button is gray, click in a blank area of the form to close the expanded Controls group. If the feature is not active (displays with a white background), click the Use Control Wizards button to turn on the feature.

   c. Click the Button button in the Controls group.

   d. Position the crosshairs with the button icon attached in the *Form Header* section, click and hold down the left mouse button, drag to create a button of the approximate height and width shown below, and then release the mouse button.

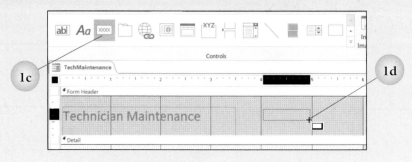

e. At the first Command Button Wizard dialog box, click *Miscellaneous* in the *Categories* list box.

f. Click *Run Macro* in the *Actions* list box and then click the Next button.

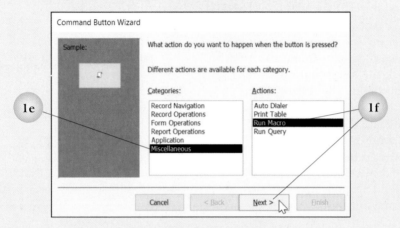

g. At the second Command Button Wizard dialog box, click *LNameFind* in the *What macro would you like the command button to run?* list box and then click the Next button.

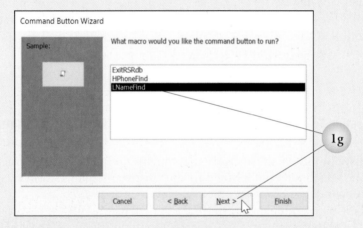

h. At the third Command Button Wizard dialog box, click the *Text* option.

i. Select the current text in the *Text* text box, type Find Technician by Last Name, and then click the Next button.

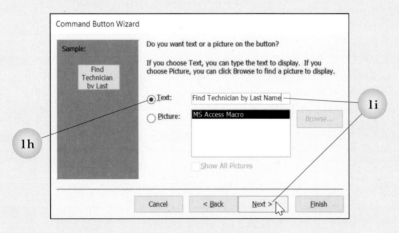

j. With *Command##* (where ## is the number of the command button) already selected in the *What do you want to name the button?* text box, type FindTechRec and then click the Finish button. Access automatically resizes the width of the button to accommodate the text to be displayed on its face.

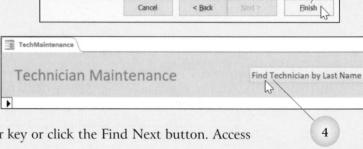

1j

2. Save the revised form.
3. Switch to Form view.
4. Click the Find Technician by Last Name button to run the macro.
5. Type Colacci in the *Find What* text box at the Find and Replace dialog box and then press the Enter key or click the Find Next button. Access makes record 9 active.

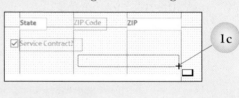

4

6. Close the Find and Replace dialog box.
7. Close the form.

Activity 1f Creating Two Command Buttons and Assigning Macros to the Buttons

Part 6 of 8

1. With **7-RSRCompServ** open, create a command button to run the macro to find a record by the home telephone number field in the CustMaintenance form by completing the following steps:
   a. Open the CustMaintenance form in Design view.
   b. Click the Button button in the Controls group on the Form Design Tools Design tab.
   c. Position the crosshairs with the button icon attached in the *Detail* section below the *Service Contract?* label control object, click and hold down the left mouse button, drag to create a button of the approximate height and width shown at the right, and then release the mouse button.

   1c

   d. Click *Miscellaneous* in the *Categories* list box, click *Run Macro* in the *Actions* list box, and then click the Next button.
   e. Click *HPhoneFind* and then click the Next button.
   f. Click *Text*, select the current text in the *Text* text box, type Find Customer by Home Phone, and then click the Next button.

   1b-g

   g. Type FindByPhone and then click the Finish button.

2. Save the revised form.
3. If necessary, move the tab control group down so the top of the object is at the 2.25-inch mark on the vertical ruler. Create a second button below the button created in Step 1 to run the macro to find a record by the last name by completing steps similar to those in Steps 1b through 1g and using the following additional information:
   - Select *LNameFind* as the macro to assign to the button.
   - Display the text *Find Customer by Last Name* on the button.
   - Name the button *FindByLName*.
4. Select both buttons and apply the Intense Effect - Blue, Accent 1 quick style (second column, last row in the *Theme Styles* gallery of the Quick Styles button). Resize, align, and then position the two buttons as shown below.

5. Save the revised form.
6. Switch to Form view.
7. Click the Find Customer by Home Phone button, type 313-555-7486 at the Find and Replace dialog box, and then press the Enter key. Close the dialog box and review the record for Customer ID 1025.
8. Click the Find Customer by Last Name button, type Antone at the Find and Replace dialog box, and then press the Enter key. Close the dialog box and review the record for Customer ID 1075.
9. Close the CustMaintenance form.

Viewing an Embedded Macro

Quick Steps

View Macro Code for Command Button

1. Open form in Design view.
2. Click to select command button.
3. Display Property Sheet task pane.
4. Click Event tab.
5. Click Build button in *On Click* property box.

The Command Button Wizard used in Activities 1e and 1f created an embedded macro in the Property Sheet task pane for the button. An embedded macro is a macro stored within a form, report, or control that is run when a specific event occurs, such as clicking a button.

Embedded macros are objects that are not seen in the Navigation pane. To view an embedded macro, open the command button's Property Sheet task pane and then click the Event tab. *[Embedded Macro]* will be visible in the *On Click* property box. Click the Build button (which displays three dots) at the right side of the *On Click* property box to open the Macro Builder window with the macro actions displayed. The macro and macro actions were created by the Command Button Wizard.

To delete an embedded macro, open the Property Sheet task pane, select *[Embedded Macro]* in the *On Click* property box, and then press the Delete key. Close the Property Sheet task pane and then save the form.

1. With **7-RSRCompServ** open, view the macro actions embedded in a command button when the Command Button Wizard is used by completing the following steps:

   a. Open the CustMaintenance form in Design view.

   b. Click to select the Find Customer by Home Phone command button.

   c. Click the Property Sheet button in the Tools group on the Form Design Tools Design tab or press the F4 function key.

   d. Click the Event tab in the Command Button Property Sheet task pane.

   e. Notice that the text in the *On Click* property box reads *[Embedded Macro]*.

   f. Click the Build button (which displays three dots) in the *On Click* property box. When the Build button is clicked, Access opens the Macro Builder window for the macro embedded in the command button.

   g. Notice that the name of the macro created by the Command Button Wizard is *CustMaintenance: FindByPhone: On Click*. The macro's name is made up of the form name, followed by the button name with which the macro is associated, and then the event that causes the macro to run (*On Click*).

   h. Review the macro actions in the Macro task pane. Notice that the macro action is *RunMacro* and that the name of the macro selected at the second Command Button Wizard dialog box, *HPhoneFind*, appears in the *Macro Name* option box.

   i. Click the Close button in the Close group on the Macro Tools Design tab.

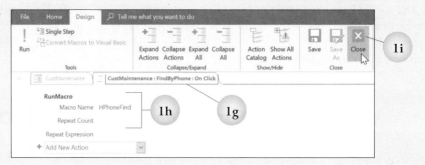

2. With the Property Sheet task pane still open, click to select the Find Customer by Last Name command button and then click the Build button in the *On Click* property box on the Event tab in the Property Sheet task pane.

3. Review the embedded macro name and macro actions in the Macro task pane and then click the Close button in the Close group on the Macro Tools Design tab.

4. Close the form.

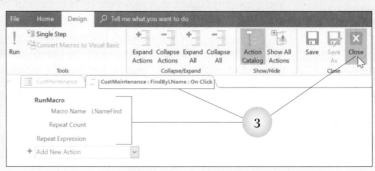

Tutorial

Converting Macros
to Visual Basic for
Applications

Converting Macros to Visual Basic for Applications

Creating and using macros enables the user to add automation and functionality within Access without having to learn how to write programming code. In the Microsoft Office suite, Visual Basic for Applications (VBA) is the programming language used to build custom applications that operate within Word, Excel, PowerPoint, and Access. The macros used so far in this chapter have been simple enough that they do not need VBA programming. However, when automating more complex tasks, a developer may prefer to write a program using VBA.

A quick method for starting a VBA program is to create a macro and then convert it to VBA code. To do this, open the macro in the Macro Builder window and then click the Convert Macros to Visual Basic button in the Tools group on the Macro Tools Design tab. Access opens a Microsoft Visual Basic window with the VBA code for the macro. When converting an embedded macro, Access also changes the property box in the Property Sheet task pane for the form, report, or control to run the VBA procedure instead of the macro.

Quick Steps

Convert Macro to Visual Basic
1. Open macro in Design view.
2. Click Convert Macros to Visual Basic button.
3. Click Convert button.
4. Click OK.

 Convert Macros to Visual Basic

Activity 1h Converting a Macro to Visual Basic for Applications

1. With **7-RSRCompServ** open, convert a macro to Visual Basic for Applications (VBA) by completing the following steps:
   a. Right-click *HPhoneFind* in the Macros group in the Navigation pane and then click *Design View* at the shortcut menu. The Macro Builder window opens with the macro actions and arguments for the HPhoneFind macro.
   b. Click the Convert Macros to Visual Basic button in the Tools group on the Macro Tools Design tab.

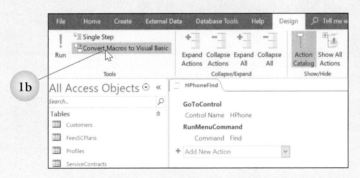

   c. At the Convert macro: HPhoneFind dialog box with the *Add error handling to generated functions* and *Include macro comments* check boxes selected, click the Convert button.
   d. At the *Convert macros to Visual Basic* message box alerting that the conversion has finished, click OK.

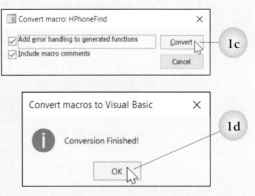

2. Access opens a Microsoft Visual Basic for Applications window when the macro is converted and displays the converted event procedure below the expanded *Modules* folder in the Activity pane. If necessary, drag the right border of the Activity pane to the right to expand the pane so the entire converted macro name is visible and then double-click the macro name to open the event procedure in its class module window.

3. Read the VBA code and then click the Close button.

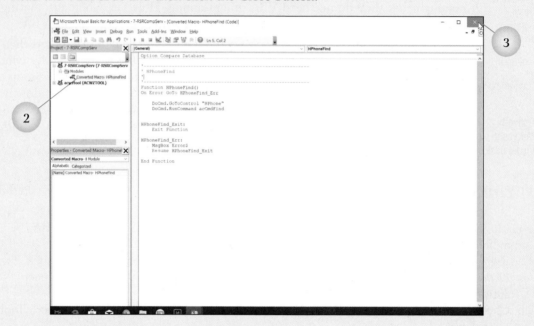

4. Close the HPhoneFind Macro Builder window.

Activity 2 Create a Navigation Form   2 Parts

You will create a Navigation form to be used as a main menu for the RSR Computer Services database.

Tutorial

Creating a
Navigation Form

Creating a Navigation Form

Database files are often accessed by multiple users who use them for specific purposes, such as updating customer records and entering details related to completed work orders. These individuals may not know much about database applications and may want an easy method for completing data entry or maintenance tasks. Users can open the forms and reports needed to update, view, and print data using a Navigation form as a menu. A Navigation form has tabs along the top, left, or right. It can be set to display automatically when the database file is opened so users do not need to choose which objects to open from the Navigation pane.

Quick Steps

Create Navigation Form

1. Click Create tab.
2. Click Navigation button.
3. Click form style.
4. Drag form or report name to *[Add New]* in Navigation Form window.
5. Repeat Step 4 as needed.
6. Click Save.
7. Type form name.
8. Press Enter key.

 Navigation

To create a Navigation form, click the Create tab and then click the Navigation button in the Forms group. At the Navigation button drop-down list, choose the type of form to be created by selecting the option in the drop-down list that positions the tabs where they are to appear. A Navigation Form window opens with a title and a tab bar. Access displays *[Add New]* in the first tab in the form. To add a form or report to a Navigation form, drag it from the Navigation pane and drop it into the to *[Add New]* tab in the Navigation Form window. Continue dragging and dropping form and/or report names from the Navigation pane into the *[Add New]* tab in the Navigation Form window in the order they are to appear. Figure 7.5 illustrates the Navigation form to be created in Activities 2a and 2b.

Figure 7.5 Navigation Form for Activities 2a and 2b

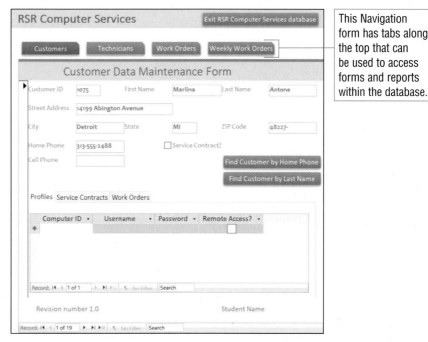

This Navigation form has tabs along the top that can be used to access forms and reports within the database.

Activity 2a Creating a Navigation Form

Part 1 of 2

1. With **7-RSRCompServ** open, create a Navigation form with tabs along the top for accessing forms and reports by completing the following steps:
   a. Click the Create tab.
   b. Click the Navigation button in the Forms group.
   c. Click *Horizontal Tabs* at the drop-down list. Access opens a Navigation Form window with a Field List task pane at the right side of the work area. The horizontal tab at the top of the form is selected and displays *[Add New]*.

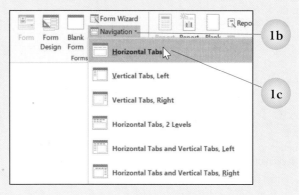

d. Position the mouse pointer on the *CustMaintenance* form name in the Navigation pane, click and hold down the left mouse button, drag the form name to *[Add New]* in the Navigation form, and then release the mouse button. Access adds the CustMaintenance form to the first tab in the form and displays a new tab with *[Add New]* right of the CustMaintenance tab.

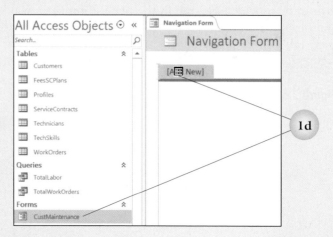

e. Drag the *TechMaintenance* form name from the Navigation pane to the second tab (which displays *[Add New]*).
f. Drag the *WorkOrders* report name from the Navigation pane to the third tab (which displays *[Add New]*).
g. Drag the *WorkOrdersbyWeek* report name from the Navigation pane to the fourth tab (which displays *[Add New]*).

2. Close the Field List task pane at the right side of the work area.
3. Click the Save button on the Quick Access Toolbar. At the Save As dialog box, type MainMenu in the *Form Name* text box and then click OK.
4. Switch to Form view.
5. Click each tab along the top of the Navigation form to view each form or report in the work area. Leave the MainMenu form open for the next activity.

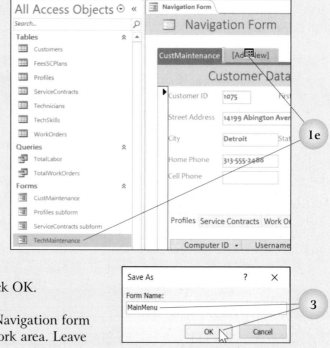

Tutorial

Adding a Command
Button to a
Navigation Form

A Navigation form can be edited in Layout view or Form view using all the tools and techniques discussed in Chapter 4 for working with forms. Consider changing the title, adding a logo and/or a command button, and renaming tabs to customize the Navigation form.

1. With **7-RSRCompServ** open and the MainMenu form displayed in the work area, add a command button to the Navigation form by completing the following steps:
   a. Switch to Design view.
   b. Resize the title control object in the *Form Header* section to align with the right edge of the object at approximately the 2.5-inch mark on the horizontal ruler.
   c. Click the Button button in the Controls group on the Form Design Tools Design tab and then drag to create a button in the Form Header section the approximate height and width shown below.

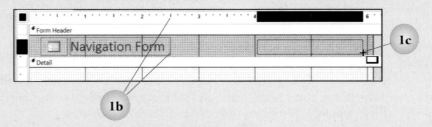

   d. At the first Command Button Wizard dialog box, select *Miscellaneous* in the *Categories* list box, select *Run Macro* in the *Actions* list box, and then click the Next button.
   e. With *ExitRSRdb* already selected in the *Macros* list box at the second Command Button Wizard dialog box, click the Next button.
   f. At the third Command Button Wizard dialog box, click *Text*, select the current entry in the *Text* text box, type Exit RSR Computer Services database, and then click the Next button.
   g. Type Exitdb at the last Command Button Wizard dialog box and then click the Finish button.
   h. With the new command button selected, apply the Intense Effect - Blue, Accent 1 quick style (second column, last row in the *Theme Styles* gallery).
2. Click to select the logo control object left of the title in the *Form Header* section and then press the Delete key.
3. Change the text in the Title control object in the *Form Header* section to *RSR Computer Services* and apply Bold formatting.

4. Relabel the tabs along the top of the Navigation form by completing the following steps:
   a. Click to select the *CustMaintenance* tab and then click the Property Sheet button in the Tools group on the Form Design Tools Design tab.
   b. Click the Format tab in the Property Sheet task pane, select the current text in the *Caption* property box, and then type Customers.
   c. Click the *TechMaintenance* tab, select the current text in the *Caption* property box in the Property Sheet task pane, and then type Technicians.
   d. Click the *WorkOrders* tab, click in the *Caption* property box, and then insert a space between *Work* and *Orders* so the tab name displays as *Work Orders*.
   e. Click the *WorkOrdersbyWeek* tab, select the current text in the *Caption* property box, and then type Weekly Work Orders.
   f. Close the Property Sheet task pane.

5. Select the four named tabs plus the [Add New] tab and then make the following formatting changes:
   a. Change the shape to Rectangle: Single Corner Snipped.
   b. Apply the Intense Effect - Blue, Accent 1 quick style.

Tab names were changed and formatting options were applied in Steps 4 and 5.

6. Switch to Form view. Insert a screenshot of the database window that shows the form in a new Microsoft Word document using either the Screenshot button in the Illustrations group on the Insert tab or the Windows key + Shift + S and the Paste feature. Type your name a few lines below the screen image and add any other identifying information as instructed (for example, the chapter number and activity number).
7. Save the Microsoft Word document with the name 7-MainMenu, close it, and then exit Word.
8. Save and close the revised form.

 Check Your Work

---

Activity 3 Configure Database Options   7 Parts

You will configure database options for the active database and configure error-checking options for all databases. You will also customize the ribbon and Quick Access Toolbar.

 Tutorial

Customizing the Access Environment

Quick Steps

Set Startup Form
1. Click File tab.
2. Click *Options*.
3. Click *Current Database* in left pane.
4. Click arrow next to *Display Form*.
5. Click form.
6. Click OK.

Specify Application Title
1. Click File tab.
2. Click *Options*.
3. Click *Current Database* in left pane.
4. Click in *Application Title* text box.
5. Type title.
6. Click OK.

Customizing the Access Environment

Click the File tab and then click *Options* to open the Access Options dialog box, which can be used to customize the Access environment. Options can be specified for all databases or only the current database. The behaviors of certain keys and the default margins for printing can also be defined. Choose to set a form to display automatically whenever the database file is opened or to hide the Navigation pane for the current database. For example, if a Navigation form has been created that provides access only to a certain group of objects, hide the Navigation pane. A database can be set to open by default in shared use or exclusive use. *Exclusive use* means the file is restricted to one individual user.

Figure 7.6 displays the Access Options dialog box with *Current Database* selected in the left pane. With options for the current database displayed, define a startup form to open automatically when the database is opened. Activity 3a demonstrates how to configure the current database to display the MainMenu form when the database is opened and to specify an application title to display in the Title bar when the database is open. The Navigation pane is customized in Activity 3b.

Figure 7.6 Access Options Dialog Box with *Current Database* Selected

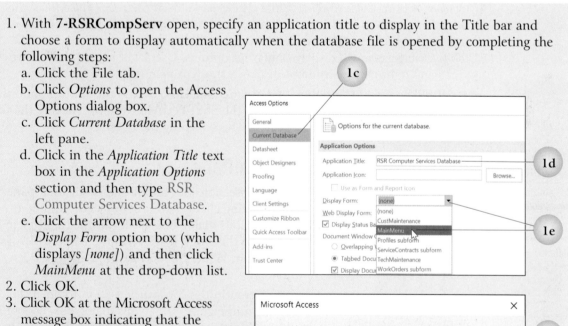

Customize the active database using options in this section.

Click this button to open the Navigation Options dialog box, where the Navigation pane can be customized.

Remove this check mark to hide the Navigation pane in the current database.

Activity 3a Specifying a Startup Form and Application Title

Part 1 of 7

1. With **7-RSRCompServ** open, specify an application title to display in the Title bar and choose a form to display automatically when the database file is opened by completing the following steps:
   a. Click the File tab.
   b. Click *Options* to open the Access Options dialog box.
   c. Click *Current Database* in the left pane.
   d. Click in the *Application Title* text box in the *Application Options* section and then type RSR Computer Services Database.
   e. Click the arrow next to the *Display Form* option box (which displays *[none]*) and then click *MainMenu* at the drop-down list.
2. Click OK.
3. Click OK at the Microsoft Access message box indicating that the database must be closed and reopened for the options to take effect.

4. Close **7-RSRCompServ**.

5. Reopen **7-RSRCompServ**. The MainMenu form displays automatically in the work area and the Title bar displays the application title *RSR Computer Services Database*.

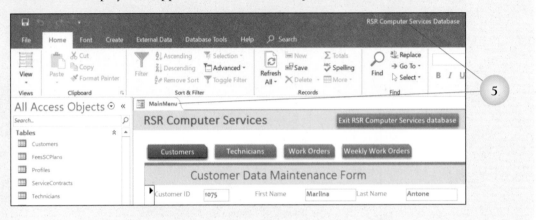

Limiting Ribbon Tabs and Menus in a Database

Limiting Ribbon Tabs and Menus in a Database

In addition to limiting users' access to objects, access to options in the ribbon and menus can be limited. Restricting users from seeing the full ribbon and shortcut menus will prevent accidental changes from being made if someone switches views and edits or deletes objects without knowing the full impact of these changes.

To limit users' access, display the Access Options dialog box and select *Current Database* in the left pane. Scroll down the dialog box to the *Ribbon and Toolbar Options* section. Click to remove the check marks from the *Allow Full Menus* and *Allow Default Shortcut Menus* check boxes and then click OK. When the database is closed and reopened, only the Home tab will display in the ribbon. The File tab backstage area will display only print options. Users will not be able to switch views and right-clicking will not display a shortcut menu.

If full access to the ribbon and menus is needed, bypass the startup options. Do this by pressing and holding down the Shift key while double-clicking the file name in the File Open backstage area when opening the database.

Customizing the Navigation Pane

Customizing the Navigation Pane

Quick Steps

Hide Navigation Pane
1. Click File tab.
2. Click *Options*.
3. Click *Current Database* in left pane.
4. Clear *Display Navigation Pane* check box.
5. Click OK two times.

When using a startup form in a database, consider hiding the Navigation pane to prevent users from accidentally making changes to other objects. To hide the Navigation pane, open the Access Options dialog box, click *Current Database* in the left pane, and then click to remove the check mark from the *Display Navigation Pane* check box in the *Navigation* section. When the Navigation pane is hidden, use the F11 function key to unhide it.

Click the Navigation Options button in the *Navigation* section to open the Navigation Options dialog box, shown in Figure 7.7. At this dialog box, choose to hide individual objects or groups of objects, set display options for the pane, and define whether objects can be opened using a single or double mouse click. For example, to prevent changes from being made to the table design, hide the Tables group.

Quick Steps

Customize Navigation
Pane
1. Click File tab.
2. Click *Options.*
3. Click *Current
   Database* in left
   pane.
4. Click Navigation
   Options button.
5. Select options.
6. Click OK two times.

Figure 7.7 Navigation Options Dialog Box

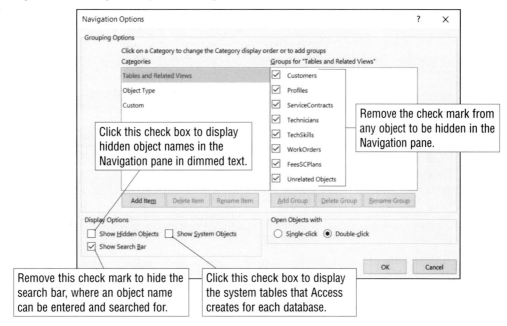

Click this check box to display
hidden object names in the
Navigation pane in dimmed text.

Remove the check mark from
any object to be hidden in the
Navigation pane.

Remove this check mark to hide the
search bar, where an object name
can be entered and searched for.

Click this check box to display
the system tables that Access
creates for each database.

Activity 3b Customizing the Navigation Pane

Part 2 of 7

1. With **7-RSRCompServ** open, customize the Navigation pane to hide all the table, macro,
   and module objects by completing the following steps:
   a. Click the File tab and then click *Options.*
   b. If necessary, click *Current Database* in the left pane. Click the Navigation Options button
      in the *Navigation* section. (You may need to scroll down to view the *Navigation* section.)
   c. Click *Object Type* in the *Categories* list box.
   d. Click the *Tables* check box in the *Groups for "Object Type"* list box to remove the check mark.
   e. Remove the check mark from the *Macros* check box.
   f. Remove the check mark from the *Modules* check box.
   g. Click OK to close the Navigation Options dialog box.

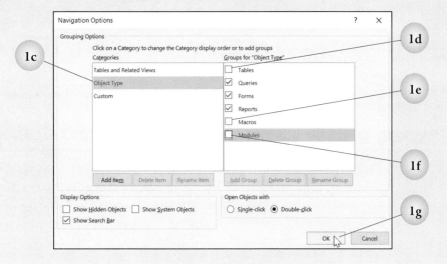

h. Click OK to close the Access Options dialog box.
    i. Click OK at the message box stating that the current database must be closed and reopened for the specified option to take effect.
2. Notice that the Tables, Macros, and Modules groups are hidden in the Navigation pane.

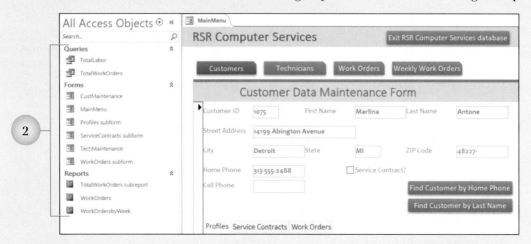

3. After reviewing the customized Navigation pane, you have decided that the database will be more secure if the entire pane is hidden when the database is opened. To do this, complete the following steps:
    a. Click the File tab and then click *Options*.
    b. With *Current Database* already selected in the left pane, click the *Display Navigation Pane* check box in the *Navigation* section to remove the check mark.
    c. Click OK.
    d. Click OK at the message box indicating that the database must be closed and reopened for the option to take effect.

4. Close **7-RSRCompServ**.
5. Reopen **7-RSRCompServ**. The Navigation pane is hidden and the MainMenu form is open in the work area.

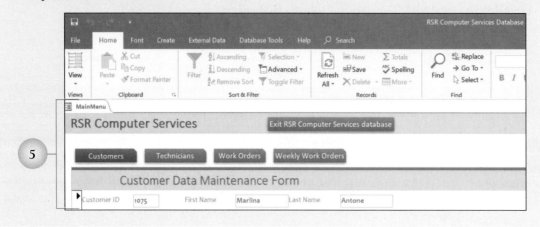

Tutorial

Configuring Error-
Checking Options

Ö̇uick Steps

**Customize Error-
Checking Options**
1. Click File tab.
2. Click *Options*.
3. Click *Object
   Designers* in left
   pane.
4. Scroll down to *Error
   checking in form and
   report design view*
   section.
5. Click to remove
   check marks.
6. Click OK.

Configuring Error-Checking Options

Recall from Chapter 5 that a green diagonal triangle displays in the Report Selector button when the report is wider than the page allows. Clicking the error-checking options button provides access to tools that can fix the report automatically. A green triangle also appears in a new label control object that is added to a report without being associated with another control object. Access flags the label as an error because it is not associated with another control object. Unchecking the *Check for unassociated label and control option* stops Access from flagging the control object.

By default, Access has error checking turned on and all error-checking options active. To configure error-checking options in Access, open the Access Options dialog box and select *Object Designers* in the left pane. Scroll down the right pane to locate the *Error checking in form and report design view* section, shown in Figure 7.8. Remove the check marks in the check boxes for those options to be disabled and then click OK. Table 7.1 provides a description of each option.

Figure 7.8 Error-Checking Options in Access

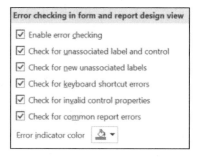

Table 7.1 Error-Checking Options

Error-Checking Option	Description
Enable error checking	Error checking can be turned on or off in forms and reports. An error is indicated by a green triangle in the upper left corner of a control object.
Check for unassociated label and control	A selected label and text box control object are checked to make sure the two control objects are associated. A Trace Error button appears if Access detects an error.
Check for new unassociated labels	Each new label control object is checked for association with a text box control object.
Check for keyboard shortcut errors	Duplicate keyboard shortcuts and invalid shortcuts are flagged.
Check for invalid control properties	Invalid properties, formula expressions, and field names are flagged.
Check for common report errors	Reports are checked for errors, such as invalid sort orders and reports that are wider than the selected paper size.
Error indicator color	A green triangle indicates an error in a control. Click the Color Picker button to change to a different color.

1. With **7-RSRCompServ** open, assume that label control objects that contain explanatory text for users are frequently added to forms and reports. Customize the error-checking options to prevent Access from flagging these independent label control objects as errors. To do this, complete the following steps:
   a. Click the File tab and then click *Options*.
   b. Click *Object Designers* in the left pane.
   c. Scroll down the right pane to the *Error checking in form and report design view* section.
   d. Click the *Check for new unassociated labels* check box to remove the check mark.
   e. Click OK.
   f. Click OK.

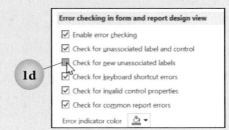

 Tutorial

Customizing the Ribbon

Customizing the Ribbon

Customize the ribbon by creating a new tab. Within the new tab, add groups and then add buttons within the groups. To customize the ribbon, click the File tab and then click *Options*. At the Access Options dialog box, click *Customize Ribbon* in the left pane. Options for customizing the ribbon are shown in Figure 7.9.

💡**Hint** To save mouse clicks, consider creating a custom tab that contains the buttons used on a regular basis.

Figure 7.9 Access Options Dialog Box with *Customize Ribbon* Selected

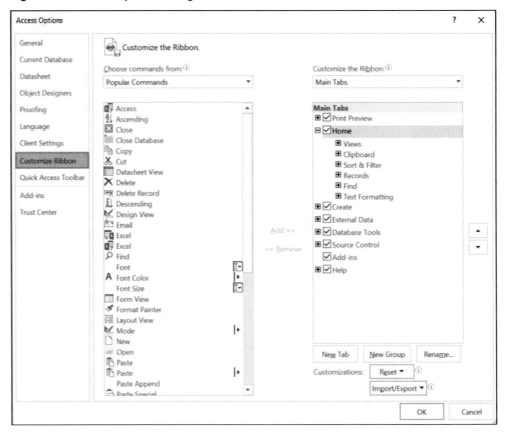

The commands shown in the left list box in Figure 7.9 are dependent on the current option selected in the *Choose commands from* option box above it. Click the arrow right of the current option (*Popular Commands*) to select from a variety of command lists, such as *Commands Not in the Ribbon* and *All Commands*. The tabs shown in the right list box are dependent on the current option selected in the *Customize the Ribbon* option box. Click the arrow right of the current option (*Main Tabs*) to select *All Tabs*, *Main Tabs*, or *Tool Tabs*.

Create a new group in an existing tab or create a new tab along with a new group within the tab. Add buttons that are regularly used to either of the new groups.

Creating a New Tab

Quick Steps

Create New Tab and Group
1. Click File tab.
2. Click *Options*.
3. Click *Customize Ribbon* in left pane.
4. Click tab name to precede new tab.
5. Click New Tab button.

Add a New Group to an Existing Tab
1. Click File tab.
2. Click *Options*.
3. Click *Customize Ribbon* in left pane.
4. Click tab name with which new group is associated.
5. Click New Group button.

To create a new tab, click the tab name in the *Main Tabs* list box that the new tab is to follow and then click the New Tab button below the list box. This inserts a new tab in the list box along with a new group below the new tab, as shown in Figure 7.10. If the wrong tab name was selected before clicking the New Tab button, the new tab can be moved up or down in the list box. To do this, click *New Tab (Custom)* and then click the Move Up or Move Down buttons that display at the right side of the dialog box.

Figure 7.10 New Tab and Group Created in the Customize Ribbon Pane
  at the Access Options Dialog Box

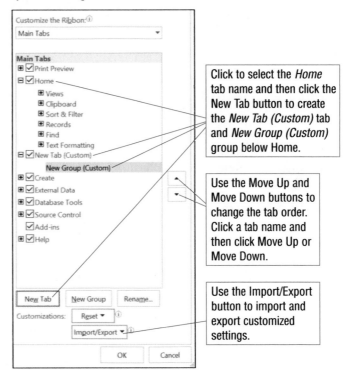

Click to select the *Home* tab name and then click the New Tab button to create the *New Tab (Custom)* tab and *New Group (Custom)* group below Home.

Use the Move Up and Move Down buttons to change the tab order. Click a tab name and then click Move Up or Move Down.

Use the Import/Export button to import and export customized settings.

Exporting Customizations

Your educational institution may already have taken the time to customize the ribbon in Access. To ensure that these customizations can be restored after the ribbon is changed in Activity 3e, the settings will be saved (exported) in Activity 3d. In Activity 3g, the original settings will be reinstalled (imported).

To save the current ribbon settings, click the File tab and then click *Options*. At the Access Options dialog box, click *Customize Ribbon* in the left pane. Click the Import/Export button in the lower right corner of the Access Options dialog box. Click *Export all customizations* to save the file with the custom settings. This file will be used to reinstall the saved settings; it could also be used to install the customized settings on a different computer. To import the saved settings, click the Import/Export button and then click *Import customization file*. Locate the file and reinstall the customized settings.

Quick Steps

**Rename Tab
or Group**

1. Click File tab.
2. Click *Options*.
3. Click *Customize Ribbon* in left pane.
4. Click tab or group to be renamed.
5. Click Rename button.
6. Type new name.
7. Press Enter key.

**Add Buttons
to a Group**

1. Click File tab.
2. Click *Options*.
3. Click *Customize Ribbon* in left pane.
4. Click group name in which to insert new button.
5. Change *Choose commands from* to desired command list.
6. Scroll down and click a command.
7. Click Add button.

Renaming a Tab or Group

Rename a tab by clicking the tab name in the *Main Tabs* list box and then clicking the Rename button below the list box. At the Rename dialog box, type the name for the tab and then press the Enter key or click OK. The Rename dialog box can also be displayed by right-clicking the tab name and then clicking *Rename* at the shortcut menu.

Complete similar steps to rename a group. The Rename dialog box for a group name or command name contains a *Symbol* list box as well as a *Display name* text box. However, the symbols are more useful for identifying new buttons than new groups. Type the new name for the group in the *Display name* text box and then press the Enter key or click OK.

Adding Buttons to a Tab Group

Add commands to a tab by clicking the group name within the tab, clicking the desired command in the list box at the left, and then clicking the Add button between the two list boxes. Remove commands in a similar manner but click the Remove button instead.

Activity 3d Exporting Customizations

Part 4 of 7

1. With **7-RSRCompServ** open, save the current ribbon settings to the desktop by completing the following steps:
   a. Click the File tab and then click *Options*.
   b. Click *Customize Ribbon* in the left pane of the Access Options dialog box.
   c. Click the Import/Export button in the bottom right of the Access Options dialog box.
   d. Click *Export all customizations* at the drop-down list.
   e. Click *Desktop* in the *Favorites* list at the left side of the File Save dialog box.
   f. Change the file name to **7-AccessCustomizations** and then click the Save button.
   g. Click OK two times.

1. With **7-RSRCompServ** open, customize the ribbon by adding a new tab and two new groups on the tab by completing the following steps. *Note: The original ribbon settings will be restored in Activity 3g.*

   a. Click the File tab and then click *Options*.

   b. If necessary click *Customize Ribbon* in the left pane of the Access Options dialog box.

   c. Click *Home* in the *Main Tabs* list box.

   d. Click the New Tab button below the list box. (This inserts a new tab below *Home* and a new group below the new tab.)

   e. With *New Group (Custom)* selected below *New Tab (Custom)*, click the New Group button below the list box. (This inserts another new group below the new tab.)

2. Rename the tab and groups by completing the following steps:

   a. Click to select *New Tab (Custom)* in the *Main Tabs* list box.

   b. Click the Rename button below the list box.

   c. At the Rename dialog box, type your first and last names and then click OK.

   d. Click to select the first *New Group (Custom)* group name below the new tab.

   e. Click the Rename button.

   f. At the Rename dialog box, type Views in the *Display name* text box and then click OK. (Notice that the Rename dialog box for a group or button displays symbols in addition to the *Display name* text box. You will apply a symbol to a button in a later step).

   g. Right-click the *New Group (Custom)* group name below *Views (Custom)* and then click *Rename* at the shortcut menu.

   h. Type Records in the *Display name* text box at the Rename dialog box and then click OK.

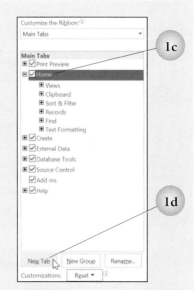

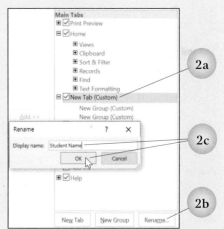

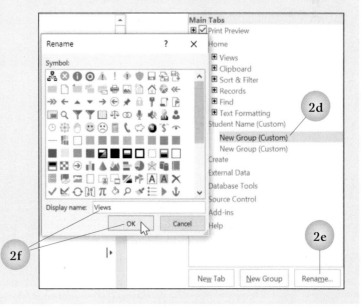

3. Add buttons to the Views (Custom) group by completing the following steps:
   a. Click to select *Views (Custom)* in the *Main Tabs* list box.
   b. With *Popular Commands* selected in the *Choose commands from* option box, click *Form View* in the list box and then click the Add button between the two list boxes. This inserts the command below the *Views (Custom)* group name.

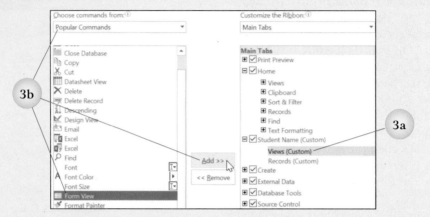

   c. Scroll down the *Popular Commands* list box; click the second *Print Preview* option, which displays the ScreenTip *Popular Commands | Print Preview (FilePrintPreview)*, and then click the Add button. **Note: The commands are organized in alphabetical order.**
   d. Click *Report View* in the *Popular Commands* list box and then click the Add button.
   e. Scroll to the bottom of the *Popular Commands* list box, click *View*, and then click the Add button.
4. Add buttons to the Records (Custom) group by completing the following steps:
   a. Click to select *Records (Custom)* in the *Main Tabs* list box.
   b. Click the *Choose commands from* option box arrow (which displays *Popular Commands*) and then click *All Commands* at the drop-down list.
   c. Scroll down the *All Commands* list box, click *Ascending*, and then click the Add button.
   d. Scroll down the *All Commands* list box (the list displays alphabetically), click *Delete Record*, and then click the Add button.
   e. Scroll down the *All Commands* list box, click *Find*, and then click the Add button.
   f. Scroll down the *All Commands* list box, click the first *New* option (which displays the ScreenTip *Home Tab | Records | New (GoToNewRecord)*), and then click the Add button.
   g. Scroll down the *All Commands* list box, click *Spelling*, and then click the Add button.

5. Change the symbol for the Spelling button by completing the following steps:
   a. Right-click *Spelling* in the *Records (Custom)* group in the *Main Tabs* list box.
   b. Click *Rename* at the shortcut menu.
   c. At the Rename dialog box click the purple book icon in the *Symbol* list box (eleventh column eighth row, position may vary).
   d. Click OK.

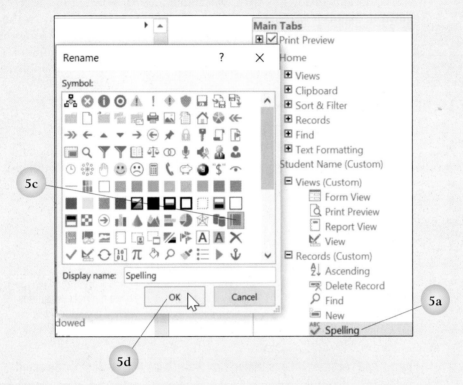

6. Click OK to close the Access Options dialog box. If a message displays stating that the database must be closed and reopened for the option to take effect click OK.
7. Use buttons in the custom tab to change views and spell check a form by completing the following steps:
   a. Click the Work Orders tab at the top of the MainMenu form.
   b. Click the custom tab with your name and then click the Print Preview button in the Views group.
   c. Click the View button in the Views group.
   d. Click the Customers tab in the Main Menu form, click in the *Last Name* field, click the custom tab with your name, and then click the Spelling button.
   e. Click the Cancel button at the Spelling dialog box.
   f. Click the Customers tab at the top of the MainMenu form.
8. Insert a screenshot of the database window that shows the custom tab in a new Microsoft Word document using either the Screenshot button in the Illustrations group on the Insert tab or the Windows key + Shift + S and the Paste feature. Type your name a few lines below the screen image and add any other identifying information as instructed (for example, the chapter number and activity number).
9. Save the Microsoft Word document and name it **7-CustomRibbon**.
10. Print **7-CustomRibbon** and then exit Word.

Customizing the Quick Access Toolbar

Customize Quick Access Toolbar

Click the Customize Quick Access Toolbar button at the right side of the Quick Access Toolbar to open the Customize Quick Access Toolbar drop-down list, as shown in Figure 7.11. Click *More Commands* at the drop-down list to open the Access Options dialog box with *Quick Access Toolbar* selected in the left pane, as shown in Figure 7.12. Change the list of commands that display in the left list box by clicking *Choose commands from* option box arrow and then clicking the appropriate category. Scroll down the list box to locate the command and then double-click the command name to add it to the Quick Access Toolbar.

Quick Steps

Add Button to Quick Access Toolbar

1. Click Customize Quick Access Toolbar button.
2. Click button.
OR
1. Click Customize Quick Access Toolbar button.
2. Click *More Commands*.
3. Click *Choose commands from* option box arrow.
4. Click category.
5. Click command in commands list box.
6. Click Add.
7. Click OK.

Figure 7.11 Customize Quick Access Toolbar Drop-Down List

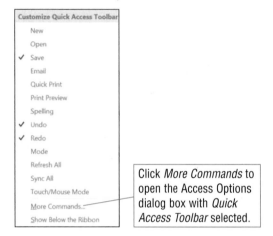

Figure 7.12 Access Options Dialog Box with *Quick Access Toolbar* Selected

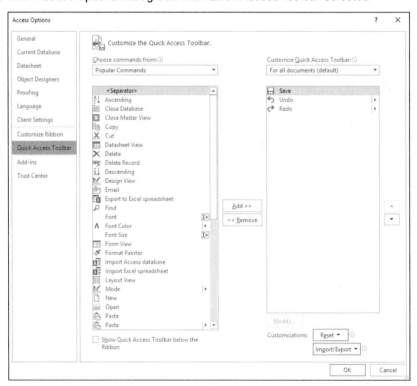

Remove Button from Quick Access Toolbar
1. Click Customize Quick Access Toolbar button.
2. Click button.
OR
1. Click Customize Quick Access Toolbar button.
2. Click *More Commands.*
3. Click command in right list box.
4. Click Remove button.
5. Click OK.

Delete a button from the Quick Access Toolbar by clicking the Customize Quick Access Toolbar button and then clicking the command at the drop-down list. If the command does not appear in the drop-down list, click the *More Commands* option. At the Access Options dialog box, click the command to be removed in the right list box and then click the Remove button.

Activity 3f Adding Commands to the Quick Access Toolbar

1. With **7-RSRCompServ** open, add the Print Preview and Close Window commands to the Quick Access Toolbar by completing the following steps. *Note: The original settings for the Quick Access Toolbar will be restored in Activity 3g.*

   a. Click the Customize Quick Access toolbar button at the right side of the Quick Access Toolbar.

   b. Click *Print Preview* at the drop-down list. The Print Preview button is added to the end of the Quick Access Toobar. *Note: Skip to Step 1d if the Print Preview button already displays on your Quick Access Toolbar.*

   c. Click the Customize Quick Access Toobar button.

   d. Click *More Commands* at the drop-down list.

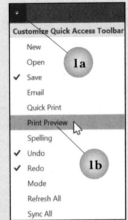

   e. At the Access Options dialog box with *Quick Access Toolbar* selected in the left pane, click the *Choose commands from* option box arrow and then click *Commands Not in the Ribbon.*

   f. Double-click *Close Window* in the *Commands Not in the Ribbon* list box. *Note: The commands are organized in alphabetical order.*

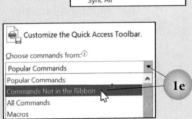

   g. Click OK. If a message displays stating that the database must be closed and reopened for the option to take effect, click OK. The Close Window button is added right of the Save button.

2. Click the WorkOrders tab in the MainMenu Navigation form.

3. Click the PrintPreview button on the Quick Access Toolbar to open the WorkOrders report in Print Preview. Click Close Print Preview.

4. Click the Close Window button on the Quick Access Toolbar.

Quick Steps

Reset Ribbon and Quick Access Toolbar
1. Click the File tab.
2. Click *Options*.
3. Click *Customize Ribbon*.
4. Click Reset button.
5. Click *Reset all customizations*.
6. Click Yes.
7. Click OK.

Resetting the Ribbon and the Quick Access Toolbar

Restore the original ribbon and the Quick Access Toolbar by clicking the Reset button below the *Main Tabs* list box in the Access Options dialog box with *Customize Ribbon* selected in the left pane. Clicking the Reset button displays these two options: *Reset only selected Ribbon tab* and *Reset all customizations*. Click *Reset all customizations* to restore the ribbon and the Quick Access Toolbar to its original settings and then click Yes at the Microsoft Office message box that displays the message *Delete all Ribbon and Quick Access Toolbar customizations for this program?*

To restore the ribbon and the Quick Access Toolbar to your institution's customized settings, import the settings that were exported in Activity 3d. Click the Import/Export button in the lower right corner of the Access Options dialog box and then click *Import customization file* at the drop-down list. Locate the file and reinstall the customized settings.

Activity 3g Restoring the Ribbon and the Quick Access Toolbar

1. Import the customization file saved in Activity 3d to reset the ribbon and the Quick Access Toolbar to your institution's original settings by completing the following steps:
   a. Click the File tab and then click *Options* to open the Access Options dialog box.
   b. If necessary, click *Customize Ribbon* in the left pane.
   c. Click the Import/Export button in the bottom right corner of the dialog box.
   d. Click *Import customization file* at the drop-down list.
   e. Click *Desktop* in the *Favorites* list at the left of the File Open dialog box.
   f. Click **7-AccessCustomizations.exportedUI**.
   g. Click the Open button.
   h. Click Yes at the Microsoft Office message box that asks about replacing all existing ribbon and Quick Access Toolbar customizations.

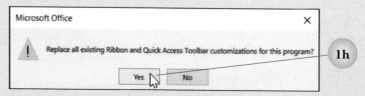

   i. Click OK to close the Access Options dialog box.
   j. Click OK at the message that displays saying that the database must be closed and reopened for the option to take effect.
2. If **7-RSRCompServ** is open, close it.

Chapter Summary

- A macro can be used to automate actions within a database, such as opening a form, query, or report.

- To create a macro, click the Create tab and then click the Macro button in the Macros & Code group, which opens a Macro Builder window. Add a macro action using the *Add New Action* option box or Action Catalog task pane.

- Action arguments are the parameters for the action, such as the object name, mode in which the object opens, and other restrictions placed on the action.

- The arguments displayed vary depending on the active action.

- Most macros can be run by double-clicking the macro name in the Navigation pane, right-clicking the macro name, or clicking the Run button in the Macro Builder window.

- A macro can also be created by dragging and dropping an object name from the Navigation pane into the *Add New Action* option box in a Macro task pane.

- Edit a macro by right-clicking the macro name in the Navigation pane and clicking *Design View* at the shortcut menu.

- A macro can be assigned to a button in a form to allow running the macro with a single mouse click.

- Use the Button button to create a command button in a form.

- Use the Command Button Wizard to create an embedded macro that tells Access which macro to run when the button is clicked. View the macro by opening the command button's Property Sheet task pane, clicking the Event tab, and then clicking the Build button in the *On Click* property box.

- Convert a macro to Visual Basic for Applications (VBA) code by opening it in the Macro Builder window and then clicking the Convert Macros to Visual Basic button in the Tools group on the Macro Tools Design tab.

- A Navigation form has tabs along the top, left, or right and is used as a menu. Users can open the forms and reports in the database by clicking a tab, rather than using the Navigation pane.

- To create a Navigation form, click the Create tab and then click the Navigation button in the Forms group. Choose the form type at the drop-down list and then drag and drop form and/or report names from the Navigation pane into *[Add New]* in the tab bar.

- A form can be set to display automatically when the database is opened. Select the startup form at the *Display Form* option box in the Access Options dialog box with *Current Database* selected in the left pane.

- Change the title that appears in the Title bar for the active database by typing text in the *Application Title* text box at the Access Options dialog box with *Current Database* selected in the left pane.

- Set options for the Navigation pane, such as hiding individual objects and groups of objects, at the Navigation Options dialog box.

- Hide the Navigation pane by removing the check mark from the *Display Navigation Pane* check box at the Access Options dialog box with *Current Database* selected in the left pane.

- Change the default error-checking options at the Access Options dialog box with *Object Designers* selected in the left pane.

- Customize the ribbon by creating a new tab, creating a new group on the new tab, and then adding buttons within the new group.
- To customize the ribbon, open the Access Options dialog box and click *Customize Ribbon* in the left pane.
- To save the current ribbon and Quick Access Toolbar settings, open the Access Options dialog box, click *Customize Ribbon* in the left pane, click the Import/Export button, and then click *Export all customizations.* This file can then be used to reinstall the customizations.
- Create a new tab by clicking the tab name that will precede the new tab in the *Main Tabs* list box and then clicking the New Tab button. A new group is automatically added with the new tab.
- Rename a tab by clicking the tab name in the *Main Tabs* list box, clicking the Rename button, typing a new name, and then pressing the Enter key or clicking OK. Rename a group using a similar process.
- Add buttons to a tab by clicking the group name within the tab, selecting the desired command in the commands list box, and then clicking the Add button between the two list boxes.
- Add buttons to or delete buttons from the Quick Access Toolbar using the Customize Quick Access Toolbar button. To locate a feature to add, click the *More Commands* at the drop-down list to open the Access Options dialog box with *Quick Access Toolbar* selected.
- Restore the ribbon to the default settings by clicking the Reset button below the *Main Tabs* list box in the Access Options dialog box with *Customize Ribbon* selected and then clicking *Reset all customizations* at the drop-down list.

Commands Review

FEATURE	RIBBON TAB, GROUP	BUTTON
convert macros to Visual Basic for Applications	Macro Tools Design, Tools	
create command button	Form Design Tools Design, Controls	
create macro	Create, Macros & Code	
create Navigation form	Create, Forms	
customize Navigation pane	File, Options	
run macro	Macro Tools Design, Tools	

Microsoft®

Access®

Integrating Access Data

Performance Objectives

Upon successful completion of Chapter 8, you will be able to:

1. Create and restore a backup database file
2. Create an ACCDE database file
3. View Trust Center settings
4. Import and merge data from another Access database
5. Link to a table in another Access database
6. Determine when to import or link to external source data
7. Reset or refresh a link using the Linked Table Manager
8. Import data from a text file
9. Save and repeat import specifications
10. Export Access data to a text file
11. Save and repeat export specifications

Integrating Access data with other applications in the Microsoft Office suite is easily accomplished with buttons on the External Data tab. These buttons allow you to export data to and import data from Word and Excel files. Data can be exchanged between the Microsoft programs with the formatting and data structure intact. In some cases, however, you may need to exchange data between Access and a non-Microsoft program. In this chapter, you will learn how to create and restore a backup database file, how to integrate data between individual Access database files, and how to import and export data in a text file format that is recognized by nearly all applications. You will also learn how to prevent changes from being made to the design of objects.

 Data Files

Before beginning chapter work, copy the AL2C8 folder to your storage medium and then make AL2C8 the active folder.

The online course includes additional training and assessment resources.

Review: Backing Up a Database

Quick Steps

Back Up Database
1. Click File.
2. Click *Save As*.
3. Click *Back Up Database*.
4. Click Save As button.
5. Navigate to appropriate folder.
6. Click Save button.

Creating and Restoring a Backup File

Before making any major changes to a database, such as running action queries or modifying table structures, it is a good idea to back it up. Certain changes made to a database cannot be reversed. For example, if a delete query is run to remove the work orders for the month of September, the work orders cannot be reinstated. If field sizes are changed and made too small, the data truncated cannot be restored if the revised table has been saved. The data will be lost unless a backup exists.

Consider backing up your database on a regular basis to minimize any data loss that may occur due to system failures or design mistakes. Once a backup database has been created, use it to restore the entire database or to import a specific object. Activity 1 will demonstrate how to restore the entire database and Activity 2 will demonstrate how to import objects.

To create a backup of a database, click the File tab and then click the *Save As* option. Click *Back Up Database* in the *Advanced* section of the Save As backstage area and then click the Save As button. (Access closes any open objects.) Navigate to or create a folder to store backup databases and then click the Save button. Notice that the current date is added to the end of the file name.

To replace a database with a backup copy, copy the backup database from the backup folder and then paste it in the same folder as the database to be replaced. Delete the database to be replaced and rename the backup database.

Activity 1a **Creating and Restoring a Backup File** **Part 1 of 3**

1. Open **8-RSRCompServ** and enable the content.
2. Create a backup database file by completing the following steps:
   a. Click the File tab and then click the *Save As* option.
   b. Click *Back Up Database* in the *Advanced* section.
   c. Click the Save As button.

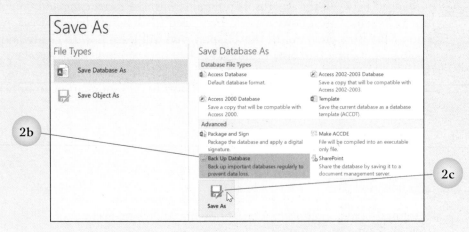

d. The backup file is named **8-RSRCompServ_yyyy-mm-dd**, where *yyyy-mm-dd* represents the year, month, and day the backup was created. (Your date will differ from the one shown below.) Notice that the default location is the AL2C8 folder on your storage medium. Instead of saving the file in this folder, click the New folder button.

e. With *New folder* selected, type DatabaseBackUps and then press the Enter key.

f. Click the Open button to open the folder.

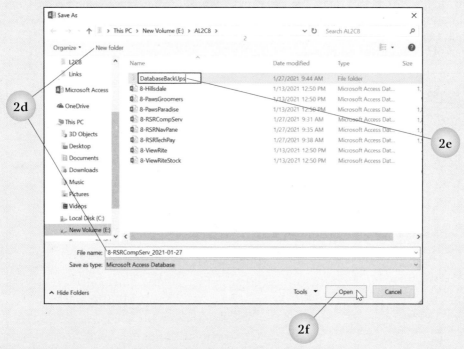

g. Click Save button.

3. With **8-RSRCompServ** open, open the Customers table in Design view, click *City*, and then change the field size from 25 characters to 5 characters. Save the table.

4. Click the Yes button to the Microsoft Access message box stating that some data may be lost.

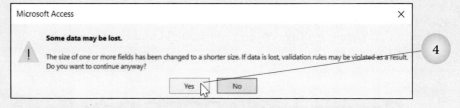

5. Switch to Datasheet view and look at the data in the *City* column. The last two letters were permanently deleted from *Detroit*. Notice that the Undo button on the Quick Access Toolbar is dimmed, indicating that this change cannot be reversed.

6. Close **8-RSRCompServ** and then close Access.

Customer ID	First Name	Last Name	Street Address	City	State
1000	Jade	Fleming	12109 Woodward Avenue	Detro	MI
1005	Cayla	Fahri	12793 Riverdale Avenue	Detro	MI
1008	Leslie	Carmichael	10303 Elmira Street	Detro	MI
1010	Randall	Lemaire	16659 Lawton Street	Detro	MI
1015	Shauna	Friesen	12914 Mitchell Avenue	Detro	MI

The field size for the *City* field has been changed to 5 characters. The last two letters in *Detroit* have been permanently deleted.

7. Restore the backup copy of the database by completing the following steps:
   a. Click the Start button.
   b. At the Start screen, start typing file explorer. When the File Explorer icon displays below the *Apps* heading, press the Enter key. (Depending on your operating system, these steps may vary.)
   c. Navigate to the DatabaseBackUps folder in the AL2C8 folder on your storage medium and then right-click **8-RSRCompServ_yyyy-mm-dd**, where *yyyy-mm-dd* represents the year, month, and day the backup was created.
   d. Click *Copy* at the shortcut menu.
   e. Click *AL2C8* in the Address bar.

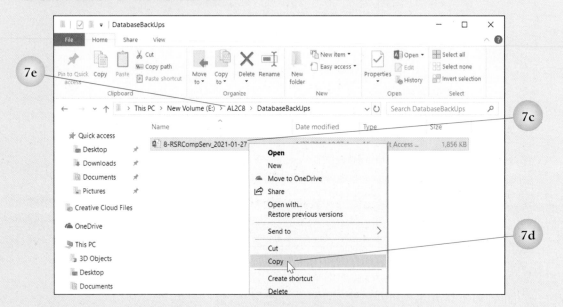

   f. Right-click a blank area of the content pane and then click *Paste* at the shortcut menu.
   g. Right-click **8-RSRCompServ** and then click *Delete* at the shortcut menu.
   h. Right-click **8-RSRCompServ_yyyy-mm-dd** and then click *Rename* at the shortcut menu.
   i. Delete _yyyy-mm-dd_ and then press the Enter key.
8. Open **8-RSRCompServ**, enable the content if necessary, and then open the Customers table. Verify that *Detroit* is displayed in the *City* field and then close the table.
9. Close **8-RSRCompServ**.

Creating an ACCDE
Database File

💡 **Hint** ACCDE stands for "Access Database Execute Only." ACCDE files have the file extension .accde instead of .accdb. To see the different file extensions, turn on the display of file extensions at the File Explorer window.

Creating an ACCDE Database File

In Chapter 6, a database was split into two files to create a front-end database and a back-end database. Using this method improved the performance of the database and protected the table objects from being changed by separating them from the queries, forms, and reports. Another method to protect an Access database is to create an ACCDE file. In an ACCDE file, end users are prevented from making changes to the designs of objects. An Access database stored as an ACCDE file is a locked-down version of the database and therefore does not provide access to Design view or Layout view. In addition, if the database contains any Visual Basic for Applications (VBA) code, the code cannot be modified or changed.

Quick Steps
Create ACCDE File
1. Open database.
2. Click File tab.
3. Click *Save As*.
4. Click *Make ACCDE*.
5. Click Save As button.
6. Navigate to required drive and/or folder.
7. Type name in *File name* text box.
8. Click Save button.

To save an Access database as an ACCDE file, click the File tab and then click the *Save As* option. Click the *Make ACCDE* option in the *Advanced* section of the Save As backstage area and then click the Save As button to open the Save As dialog box. Navigate to the drive and/or folder in which to save the database, type the file name in the *File name* text box, and then click the Save button. Turn on the display of file extensions at the File Explorer window by clicking the View tab and then clicking to add a check mark in the *File extensions* check box in the Show/hide group. Note that the newly saved database file has the extension .accde. Move the original database with the .accdb extension to a secure location and provide end users with the path to the ACCDE file for daily use.

Activity 1b Creating an ACCDE Database File   Part 2 of 3

1. Open **8-RSRNavPane** and enable the content.
2. Create an ACCDE file by completing the following steps:
   a. Click the File tab and then click the *Save As* option.
   b. Click *Make ACCDE* in the *Advanced* section of the Save As backstage area.
   c. Click the Save As button.

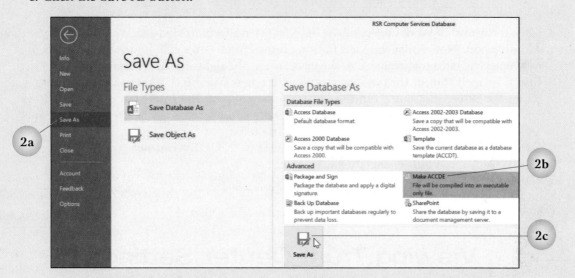

   d. At the Save As dialog box with the default location the AL2C8 folder on your storage medium add the text *NoChanges* to the end of the file name so that it reads **8-RSRNavPaneNoChanges** in the *File name* text box. Click the Save button.

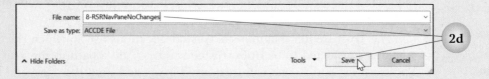

3. Close **8-RSRNavPane**.
4. Open **8-RSRNavPaneNoChanges**.

5. At the Microsoft Access Security Notice dialog box stating that the file might contain unsafe content, click the Open button.
6. With Customers the active tab in the MainMenu form and the Customer Data Maintenance Form open in Form view, click the Home tab if necessary. Notice that the View button in the Views group is dimmed.

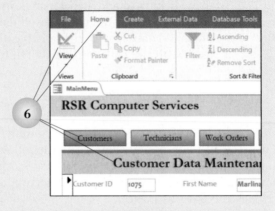

7. Click the Monthly Work Orders tab in the MainMenu form to view the Work Orders by Month report. Notice that the View button on the Home tab is still dimmed. Also notice that only one view button is available in the view area at the right side of the Status bar.
8. Insert a screen shot of the database window in a new Microsoft Word document using either the Screenshot button in the Illustrations group on the Insert tab or the Windows key + Shift + S and the Paste feature. Type your name a few lines below the screen image and add other identifying information as instructed.
9. Save the Microsoft Word document and name it **8-ACCDEWindow**.
10. Print **8-ACCDEWindow** and then exit Word.
11. Close **8-RSRNavPane**.

Viewing Trust Center Settings

Viewing Trust Center Settings

Quick Steps

View Trust Center Options
1. Click File tab.
2. Click *Options*.
3. Click *Trust Center* in left pane.
4. Click Trust Center Settings button.
5. Click desired Trust Center category in left pane.
6. View and/or modify options.
7. Click OK two times.

In Access, the Trust Center is set to block unsafe content when a database is opened. Throughout these chapters, the security warning that appears in the message bar when a database is opened has been closed by clicking the Enable Content button. Access provides the Trust Center to allow viewing and/or modifying the security options that are in place to protect your computer from malicious content.

The Trust Center maintains a Trusted Locations list, in which the content stored is considered to be from trusted sources. Add a path to the Trusted Locations list and Access will treat any file opened from the drive and folder as safe. A database opened from a trusted location does not display the security warning in the message bar and does not have content blocked.

Before the macros are enabled for a database, the Trust Center checks for a valid and current digital signature signed by an entity that is stored in the Trusted Publishers list. The Trusted Publishers list is maintained by the user on the computer being used. A trusted publisher is added to the list when the content is enabled from an authenticated source and the *Trust all content from this publisher* option has been clicked. Depending on the active macro security setting, if the

Trust Center cannot match the digital signature information with an entity in the Trusted Publishers list or if the macro does not contain a digital signature, the security warning displays in the message bar.

Table 8.1 describes the four options available for macro security. The default macro security option is *Disable all macros with notification*. In some cases, it may be decided to change the default macro security setting by opening the Trust Center dialog box. The Trust Center will be explored in Activity 1c.

Table 8.1 Macro Security Settings for a Database Not Opened from a Trusted Location

Macro Setting	Description
Disable all macros without notification	All macros are disabled; security alerts will not appear.
Disable all macros with notification	All macros are disabled; security alert appears with the option to enable the content if the source of the file is trusted. This is the default setting.
Disable all macros except digitally signed macros	A macro that does not contain a digital signature is disabled; security alerts do not appear. If the macro is digitally signed by a publisher in the Trusted Publishers list, the macro is allowed to run. If the macro is digitally signed by a publisher not in the Trusted Publishers list, a security alert appears.
Enable all macros (not recommended; potentially dangerous code can run)	All macros are allowed; security alerts do not appear.

Activity 1c Viewing Trust Center Settings

Part 3 of 3

1. To explore current settings in the Trust Center, complete the following steps:
   a. With Access open, click the File tab and then click *Options or* click the <u>Open Other Files</u> hyperlink at the bottom of the Recent list and then click *Options*.
   b. Click *Trust Center* in the left pane of the Access Options dialog box.
   c. Click the Trust Center Settings button in the *Microsoft Access Trust Center* section.

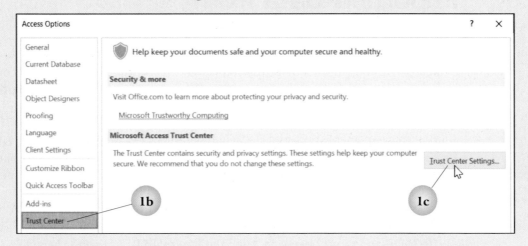

d. At the Trust Center dialog box, click *Macro Settings* in the left pane.

e. Review the options in the *Macro Settings* section. Notice which option is active on your computer. The default option is *Disable all macros with notification*. **Note: The security setting on your computer may be different from the default option. Do not change the security setting without the permission of your instructor.**

f. Click *Trusted Publishers* in the left pane. If any publishers have been added to the list on your computer, their names will appear in the list box. If the list box is empty, no trusted publishers have been added.

g. Click Trusted Locations in the left pane. Review the path and description of any folder in the Trusted Locations list. By default, Access adds the folder created when Microsoft Access is installed that contains the wizard database templates provided by Microsoft. Additional folders that have been added by a system administrator or network administrator may also appear.

h. Click OK to close the Trust Center dialog box.

2. Click OK to close the Access Options dialog box.

Activity 2 Import and Merge Data from External Sources 4 Parts

You will link and import data from a table in another Access database and a comma delimited text file. You will also save import specifications for an import routine that you expect to repeat often.

Tutorial

Importing Data from Another Access Database

Importing Data from Another Access Database

Data stored in another Access database can be merged with data in the active database by importing a copy of the source object. Multiple objects can be copied, including the relationships between tables. When importing, specify to import only the definition or both the definition and the data.

In Chapter 3, an append query was used to merge data from tables that contain the same fields by adding selected records to the end of an existing table. Earlier in this chapter, Activity 1 stressed the importance of backing up a database before making any major changes to it (such as merging the data) because some changes cannot be undone. If problems arise as a result of making changes to a database, either restore the entire database or import only the damaged table from the backup copy that was created.

New Data Source

To begin an import operation, click the External Data tab and then click the New Data Source button in the Import & Link group, hover over *From Database*, and then click *Access* from the drop-down list. The Get External Data - Access Database dialog box opens as shown in Figure 8.1. Specify the source database

Quick Steps

Import Objects from Access Database

1. Open destination database.
2. Click External Data tab.
3. Click New Data Source button.
4. Hover over *From Database*.
5. Click *Access*.
6. Click Browse button.
7. If necessary, navigate to drive and/or folder.
8. Double-click source file name.
9. Click OK.
10. Select import object(s).
11. Click OK.
12. Click Close.

💡 **Hint** When importing a query, form, or report, make sure that the tables associated with the object are also imported.

💡 **Hint** If an object with the same name as an imported table already exists in the destination database, Access does not overwrite the existing object. Instead, it adds *1* to the end of the name of the imported object.

containing the object(s) to be imported by clicking the Browse button to open the File Open dialog box. Navigate to the drive and/or folder containing the source database and then double-click the Access database file name to insert the database file name in the *File name* text box. With the *Import tables, queries, forms, reports, macros, and modules into the current database* option selected by default, click OK. This opens the Import Objects dialog box, shown in Figure 8.2. Select the objects to be imported, change options if necessary, and then click OK.

Click the Options button to display the *Import*, *Import Tables*, and *Import Queries* sections, shown in Figure 8.3. By default, Access imports relationships between tables as well as their structure definitions and data. Access also imports a query as a query, as opposed to importing a query as a table. Select or deselect the options as necessary before clicking OK to begin the import operation.

An object can also be copied by opening two copies of Access: one with the source database opened and the other with the destination database opened. With the source database window active, right-click the source object in the Navigation pane and then click *Copy*. Switch to the window containing the destination database, right-click in the Navigation pane, and then click *Paste*. Close the Access window containing the source database.

Figure 8.2 Import Objects Dialog Box

Click the tab of the object type to be imported, click the object name and then click OK. Use the Shift key (adjacent objects) and the Ctrl key (nonadjacent objects) to select multiple objects.

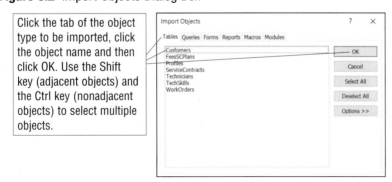

Figure 8.3 Import Objects Dialog Box with Options Displayed

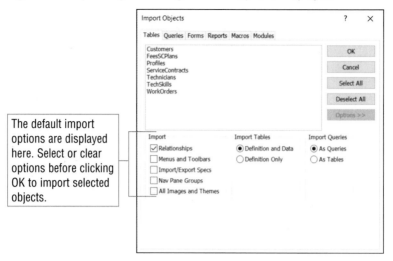

The default import options are displayed here. Select or clear options before clicking OK to import selected objects.

Activity 2a Importing a Form and Table from Another Access Database   Part 1 of 4

1. Open **8-RSRTechPay** and enable the content.
2. Merge the RSRCompServe database with the RSRTechPay database by importing the WorkOrders table and TechMaintenance form from **8-RSRCompServ**. Import the two objects by completing the following steps:
   a. Click the External Data tab.
   b. Click the New Data Source button in the Import & Link group.
   c. Hover over *From Database* and then click *Access* at the drop-down list.
   d. At the Get External Data - Access Database dialog box, click the Browse button.
   e. At the File Open dialog box, double-click **8-RSRCompServ**. *Note: Navigate to the AL2C8 folder on your storage medium, if necessary.*
   f. With the *Import tables, queries, forms, reports, macros, and modules into the current database* option already selected, click OK.

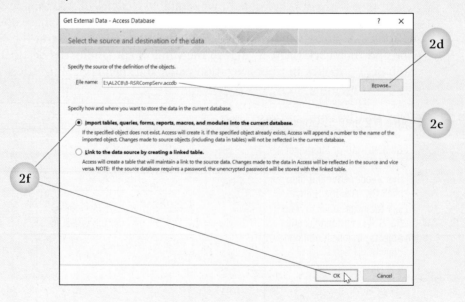

g. At the Import Objects dialog box with the Tables tab selected, click *WorkOrders* in the list box.

h. Click the Forms tab.

i. Click *TechMaintenance* in the list box and then click OK.

j. At the Get External Data - Access Database dialog box with the *Save import steps* check box empty, click the Close button.

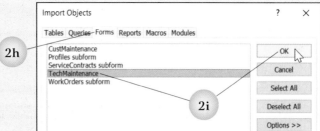

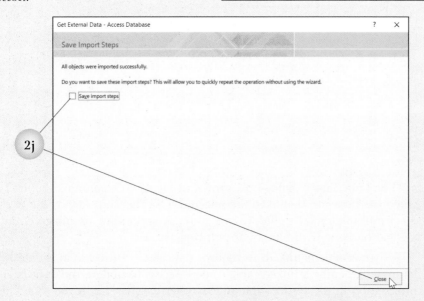

3. Access imports the WorkOrders table and TechMaintenance form and adds the object names to the Navigation pane. The form will not be operational until after Activity 2b because the tables needed to populate data in the form do not yet reside in the database. The dependent tables were not imported in this activity because the tables that contain the records are to be linked in a later activity.

Tutorial

Linking to a Table in Another Access Database

Linking to a Table in Another Access Database

In Activity 2a, a copy of a form was imported from one database to another. If the source form object is modified, the imported copy of the form object will not be altered. To ensure that the table in the destination database will reflect any changes made to the source table, link the data in the object when it is imported.

To create a linked table in the destination database, click the New Source button in the Import & Link group on the External Data tab, hover over *From Database*, and then click *Access* from the drop-down list. Click the Browse button, navigate to the drive and/or folder in which the source database is stored, and then double-click the source database file name. Click the *Link to the data source by creating a linked table* option at the Get External Data - Access Database dialog box and then click OK, as shown in Figure 8.4.

Figure 8.4 Get External Data - Access Database Dialog Box with the Link Option Selected

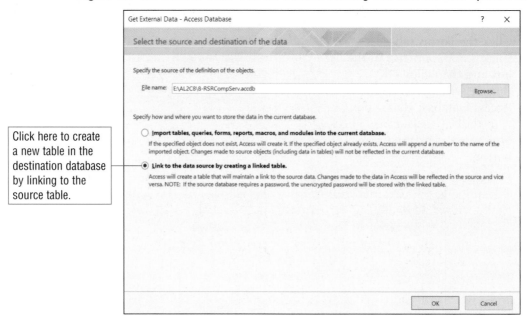

Click here to create a new table in the destination database by linking to the source table.

Quick Steps

Link to Table in Another Database
1. Open destination database.
2. Click External Data tab.
3. Click New Data Source button.
4. Hover over *From Database.*
5. Click *Access.*
6. Click Browse button.
7. If necessary, navigate to drive and/or folder.
8. Double-click source file name.
9. Click *Link to the data source by creating a linked table.*
11. Click OK.
12. Select table(s).
13. Click OK.

The Link Tables dialog box, shown in Figure 8.5, contains the list box, from which the tables to be linked are selected. Use the Shift key or the Ctrl key to select multiple tables to link in one step. Linked tables are indicated in the Navigation pane with blue right-pointing arrows.

When a table is linked, the source data does not reside in the destination database. Opening a linked table causes Access to update the datasheet with the information in the source table in the other database. Edit the source data in either the source table or the linked table in the destination database. Either way, the changes will be reflected in both tables.

Figure 8.5 Link Tables Dialog Box

Choosing Whether to Import or Link to Source Data

If the source data is not likely to change, the table should be imported. Keep in mind, however, that because importing creates copies of the data in two locations, changes or updates to the data must be made in both copies. Making the changes twice increases the risk of making a data entry error or forgetting to make an update in one or the other location.

If the source data is updated frequently, link to it so that changes have to be made only once. Because linked data exists only in the source location, the likelihood of making an error or missing an update is reduced.

Also link to the data source file when several different databases require a common table, such as an Inventory table. To duplicate the table in each database is inefficient and wastes disk space. There is also a potential for error if individual databases are not refreshed with updated data. Given these risks, it makes more sense to link rather than import. In this scenario, a master Inventory table in a separate, shared database will be linked to all the other databases that need to use the data.

Activity 2b Linking to Tables in Another Access Database   Part 2 of 4

1. With **8-RSRTechPay** open, link to two tables in **8-RSRCompServ** by completing the following steps:
   a. With the External Data tab still active, click the New Data Source button.
   b. Hover over *From Database* and then click *Access* at the drop-down list.
   c. Click the Browse button and then double-click **8-RSRCompServ** in the AL2C8 folder.
   d. Click the *Link to the data source by creating a linked table* option and then click OK.

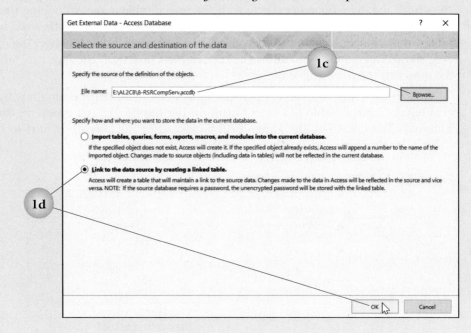

e. At the Link Tables dialog box, click *Technicians* in the list box.
f. Press and hold down the Shift key, click *TechSkills* in the list box, and then release the Shift key.
g. Click OK. Access links the two tables to the source database and adds the table names to the Navigation pane. Each linked table displays with a blue right-pointing arrow next to the table icon.

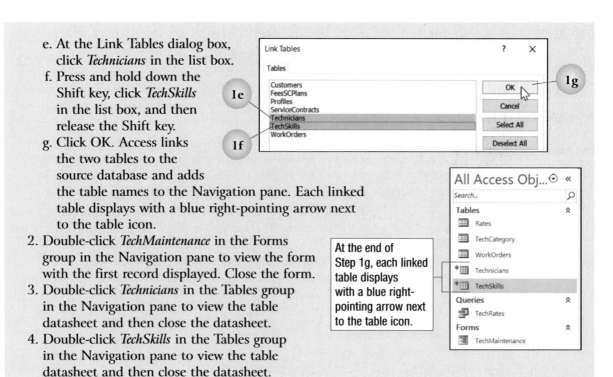

2. Double-click *TechMaintenance* in the Forms group in the Navigation pane to view the form with the first record displayed. Close the form.
3. Double-click *Technicians* in the Tables group in the Navigation pane to view the table datasheet and then close the datasheet.
4. Double-click *TechSkills* in the Tables group in the Navigation pane to view the table datasheet and then close the datasheet.

At the end of Step 1g, each linked table displays with a blue right-pointing arrow next to the table icon.

 Tutorial

Resetting a Link Using the Linked Table Manager

Resetting a Link Using the Linked Table Manager

When a table has been linked to another database, Access stores the full path to the source database file name along with the linked table name. Changing the database file name or folder location for the source database means that the linked table will no longer function. Access provides the Linked Table Manager dialog box, shown in Figure 8.6, to allow the user to reset or refresh the link to a table to reconnect to the data source file.

 Linked Table Manager

To refresh a link to a table, open the Linked Table Manager dialog box by clicking the Linked Table Manager button in the Import & Link group on the External Data tab. Click the Expand All button, click the check box next to the link to be refreshed, and then click the Refresh button.

Quick Steps
Refresh Link
1. Click External Data tab.
2. Click Linked Table Manager button.
3. Click Expand All button.
4. Click check box.
5. Click Refresh.
6. Click Close button.

Figure 8.6 Linked Table Manager Dialog Box

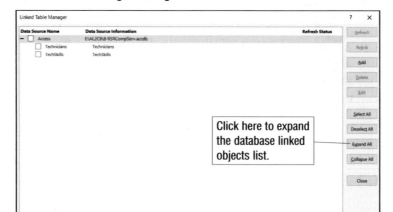

Click here to expand the database linked objects list.

1. With **8-RSRTechPay** open, change the location of **8-RSRCompServ** by completing the following steps:

   a. Click the File tab, click *Open*, click the *Browse* option in the *Recent* section, and then locate the AL2C8 folder on your storage medium.

   b. Right-click **8-RSRCompServ** in the file list box and then click *Cut* at the shortcut menu.

   c. Double-click the *DatabaseBackUps* folder.

   d. Right-click in a blank area of the file list box and then click *Paste* at the shortcut menu.

   e. Close the Open dialog box. Because the location of the source database has been moved, the linked tables are no longer connected to the correct location.

   f. Click the Back button.

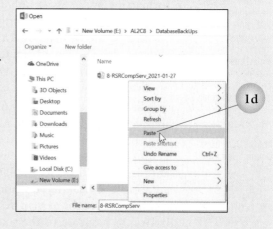

2. Refresh the links to the two tables by completing the following steps:

   a. If necessary, click the External Data tab.

   b. Click the Linked Table Manager button in the Import & Link group.

   c. At the Linked Table Manager dialog box, click the Expand All button

   d. Click the Select All button to select all the linked objects.

   e. Click the Refresh button.

   f. Because the source database has been moved, a message stating that Access could not find the file. Click OK.

   g. Failed now appears the Refresh Status column. Click the Select All button.

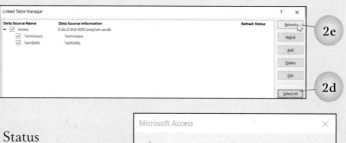

3. Relink the two tables by completing the following steps:

   a. Click the Relink button.

   b. At the Select New Location of Access, double-click **DatabaseBackUps** and then double-click **8-RSRCompServ**.

   c. Click Yes at the message box asking if you want to relink the selected tables in the new data source.

   d. Click OK at the Relink TechSkills dialog box to relink the TechSkills table. ***Note: You may be asked to relink the Technicians table first***.

   e. Click OK at the Relink Technicians dialog box to relink the Technicians table.

   f. With the Refresh Status changed to *Success* click the Close button.

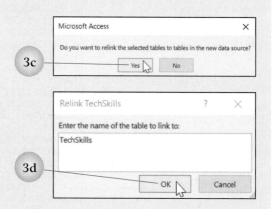

Check Your Work

Importing Data into a Database from a Text File

Text files are often used to exchange data between dissimilar programs because this file format is recognized by nearly all applications. Text files contain no formatting and consist only of letters, numbers, punctuation symbols, and a few control characters. Two commonly used text file formats separate fields with a tab (delimited file format) and a comma (comma separated file format). Figure 8.7 shows a partial view of the text file used in Activity 2d in a Notepad window. If necessary, view and edit a text file in Notepad before importing if the source application inserts characters that will be deleted.

To import a text file into Access, click the External Data tab and then click the New Data Source button in the Import & Link group, hover over *From File*, and then click *Text File* from the drop-down list. Access opens the Get External Data - Text File dialog box, which is similar to the dialog box used to import data from another Access database. When importing a text file, Access provides an append option in addition to the import and link options in the *Specify how and where you want to store the data in the current database* section. Click the Browse button to navigate to the source file and double-click the source file name to launch the Import Text Wizard, which guides the user through the import process using four dialog boxes.

Figure 8.7 Activity 2d Partial View of Text File Content in Notepad

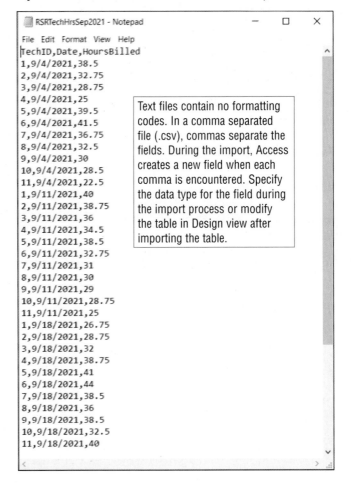

Saving and Repeating Import Specifications

Quick Steps

Save Import Specifications

1. At last Get External Data dialog box, click *Save import steps.*
2. If necessary, edit name in *Save as* text box.
3. Type description in *Description* text box.
4. Click Save Import button.

 Saved Imports

Save import specifications for an import routine that is likely to be repeated. The last step in the Get External Data - Text File dialog box displays a *Save import steps* check box. Click the check box to expand the dialog box to display the *Save as* and *Description* text boxes as shown in Figure 8.8. Type a unique name to assign to the import routine and provide a brief description of the steps. Click the Save Import button to complete the import and store the specifications. Click the *Create Outlook Task* check box to create an Outlook task that can be set up as a recurring item for an import or export operation that is repeated at fixed intervals.

Once an import routine has been saved, repeat the import process by opening the Manage Data Tasks dialog box with the Saved Imports tab selected, as shown in Figure 8.9. To do this, click the External Data tab and then click the Saved Imports button in the Import & Link group. Click the import name and then click the Run button to instruct Access to repeat the import operation.

Figure 8.8 Get External Data - Text File Dialog Box with Save Import Steps

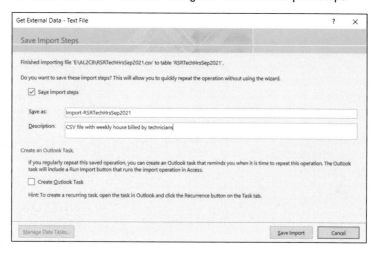

Figure 8.9 Manage Data Tasks Dialog Box with the Saved Imports Tab Selected

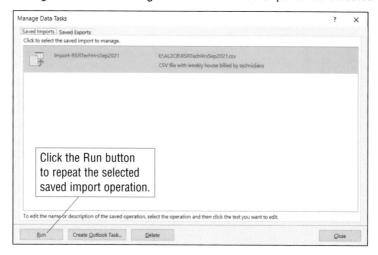

Activity 2d Importing Data from a Comma Separated Text File
and Saving Import Specifications

1. With **8-RSRTechPay** open, select a text file to import that contains the weekly hours billed for each technician for the month of September 2021 by completing the following steps:
   a. If necessary, click the External Data tab.
   b. Click the New Data Source button.
   c. Hover over *From File* and then click *Text File* at the drop-down list.
   d. At the Get External Data - Text File dialog box, click the Browse button.
   e. At the File Open dialog box, navigate to the AL2C8 folder on your storage medium if necessary.
   f. Double-click the file named ***RSRTechHrsSep2021***.
   g. With *Import the source data into a new table in the current database* already selected, click OK. This launches the Import Text Wizard.

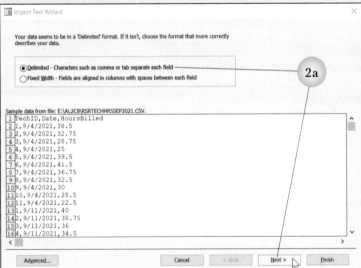

2. Import the comma separated data using the Import Text Wizard by completing the following steps:
   a. At the first Import Text Wizard dialog box with *Delimited* selected as the format, notice that the preview section in the lower half of the dialog box displays a sample of the data in the source text file. Delimited files use commas or tabs as separators while fixed width files use spaces. Click the Next button.

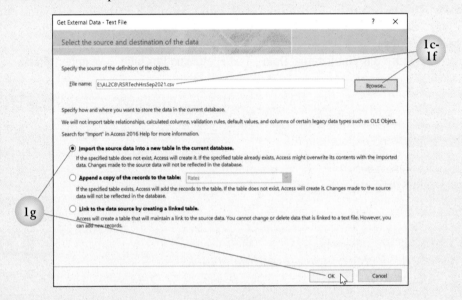

b. At the second Import Text Wizard dialog box with *Comma* already selected as the delimiter, click the *First Row Contains Field Names* check box. Notice that the preview section already shows the data set in columns, similar to a table datasheet. Click the Next button.

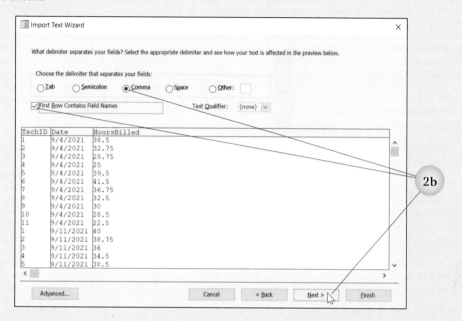

c. At the third Import Text Wizard dialog box with the *TechID* column in the preview section selected, click the *Data Type* option box arrow in the *Field Options* section and then click *Short Text* at the drop-down list.

d. Click the Next button.

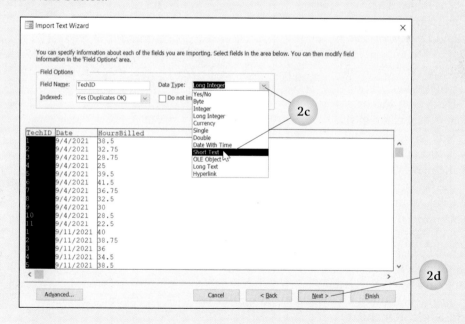

e. At the fourth Import Text Wizard dialog box with *Let Access add primary key* already selected, notice that Access has added a column in the preview section with the field title *ID*. The column added by Access is defined as an AutoNumber data type field, in which each row in the text file is numbered sequentially to make the row unique. The ID for any new record added to

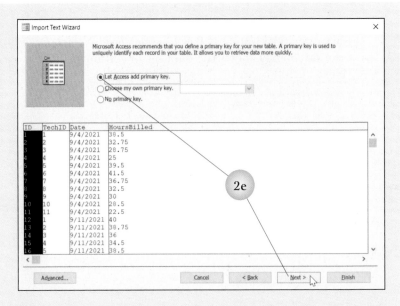

the table will automatically be incremented by one. Click the Next button.

f. At the last Import Text Wizard dialog box, with *RSRTechHrsSep2021* entered in the *Import to Table* text box, click the Finish button.

3. Save the import specifications, in case this import needs to be run again, by completing the following steps:

a. At the Get External Data - Text File dialog box, click the *Save import steps* check box. This causes the *Save as* and *Description* text boxes to appear, as well as the *Create an Outlook Task* section. By default, Access creates a name in the *Save as* text box and *Import* precedes the file name containing the imported data.

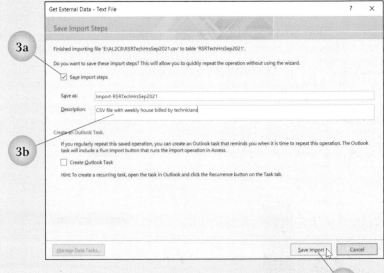

b. Click in the *Description* text box and then type CSV file with weekly hours billed by technicians.

c. Click the Saved Import button.

4. Double-click *RSRTechHoursSep2021* in the Tables group in the Navigation pane to open the table in Datasheet view.

5. Print the datasheet with the bottom margin set to 0.5 inch and then close the datasheet.

6. Close **8-RSRTechPay**.

Check Your Work

You will export a query as a comma delimited text file and another query as a tab delimited text file, including saving the second export steps so you can repeat the export operation.

Tutorial

Exporting Access
Data to a Text File

Exporting Access Data to a Text File

The Export group on the External Data tab contains buttons to export Access data from a table, query, form, or report to other applications, such as Excel and Word. To work with data from other programs, click the More button in the Export group to see if a file format converter exists for the application to be used. For example, the More button contains options to export in Word, SharePoint List, ODBC Database, and HTML Document format.

If a file format converter does not exist for the program to be used, export the data as a text file; most applications recognize and can import a text data file. Access provides the Export Text Wizard, which is launched after an object is selected in the Navigation pane. Click the Export to text file button (which displays the text *Text File*) in the Export group on the External Data tab and then specify the name of the exported text file and where to store it. The steps in the Export Text Wizard are similar to those in the Import Text Wizard, used in Activity 2d.

Export to
text file

Activity 3a Exporting a Query as a Text File
Part 1 of 3

1. Display the Open dialog box and then use the Cut and Paste features to move **8-RSRCompServ** back from the DatabaseBackUps folder to the AL2C8 folder on your storage medium.
2. Open **8-RSRCompServ** and enable the content if necessary.
3. Export the *TotalWorkOrders* query as a text file by completing the following steps:
   a. Select the TotalWorkOrders query in the Navigation pane.
   b. Click the External Data tab.
   c. Click the Export to text file button (which displays the text *Text File*) in the Export group.
   d. At the Export - Text File dialog box, click the Browse button.
   e. At the File Save dialog box, navigate to the AL2C8 folder on your storage medium if necessary.
   f. With the default file name *TotalWorkOrders.txt* in the *File name* text box, click the Save button.
   g. Click OK.

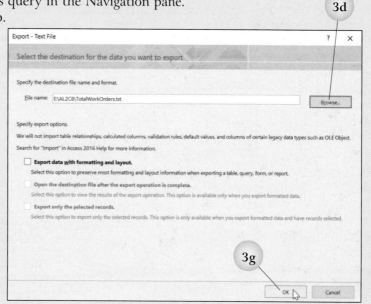

h. At the first Export Text Wizard dialog box with *Delimited* selected as the format, notice in the preview section of the dialog box that commas separate the fields and that data in a field defined with the Text data type is enclosed in quotation marks. Click the Next button.

i. At the second Export Text Wizard dialog box with *Comma* selected as the delimiter character that separates the fields, click the *Include Field Names on First Row* check box to insert a check mark. Note that Access adds a row containing the field names to the top of the data in the preview section. Each field name is enclosed in quotation marks.
j. Click the arrow next to the *Text Qualifier* option box and then click *{none}* at the drop-down list. Access removes all the quotation marks from the text data in the preview section.
k. Click the Next button.

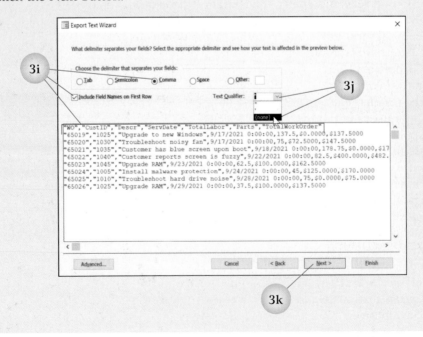

l. At the last Export Text Wizard dialog box, with *[E]:\AL2C8\TotalWorkOrders.txt* (where *[E]* is the drive for your storage medium) entered in the *Export to File* text box, click the Finish button.

m. Click the Close button to close the dialog box without saving the export steps.

4. Open Notepad and view the text file by completing the following steps:

a. Click the Start button, and then start typing notepad. When *Notepad* displays in the *Best match* section, press the Enter key. (Depending on your operating system, these steps may vary).

b. At a blank Notepad window, click the File button and then click *Open*. Navigate to the AL2C8 folder on your storage medium and then double-click *TotalWorkOrders*.

5. Click the File button and then click *Print* to print the exported text file.

6. Exit Notepad.

Saving and
Repeating a Saved
Import or Export
Specifications

Saving and Repeating Export Specifications

Just as import specifications can be saved for reuse, so can export specifications. The last Export - Text File dialog box displays a *Save export steps* check box. Click the check box to expand the dialog box options and display the *Save as* and *Description* text boxes. Type a unique name for the export routine and a brief description of the steps. Click the Save Export button to complete the export operation and store the specifications for later use.

Quick Steps

Save Export Specifications

1. At last Export - Text File dialog box, click *Save export steps* check box.
2. Verify file name.
3. Type description in *Description* text box.
4. Click Save Export button.

Activity 3b Exporting a Query as a Text File and Saving the Export Steps   Part 2 of 3

1. With **8-RSRCompServ** open, double-click the *TotalLabor* query in the Navigation pane. Notice that only the work orders for the month of September are retrieved. Switch to Design view and see that *DatePart("m",[ServDate])* has been added right of the *TotalLabor* field and that *9* displays in the *Criteria* row. With this criteria, the query returns only the results for September, the ninth month. Close the query.

2. Export the TotalLabor query as a text file using tabs as the delimiter characters by completing the following steps:

a. With the TotalLabor query selected in the Navigation pane, click the External Data tab and then click the Text File button in the Export group.

b. With *[E]:\AL2C8\TotalLabor.txt* (where *[E]* is the drive for your storage medium) entered in the *File name* text box, click OK.

c. Complete the steps of the Export Text Wizard as follows:
   1) With *Delimited* selected in the first dialog box, click the Next button.
   2) At the second dialog box, choose *Tab* as the delimiter character, click the *Include Field Names on First Row* check box to insert a check mark, change the *Text Qualifier* option to *{none}*, and then click the Next button.

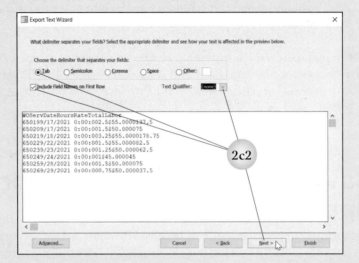

   3) Click the Finish button.
d. At the Export - Text File dialog box, click the *Save export steps* check box to insert a check mark.
e. Click in the *Description* text box and then type TotalLabor query for RSR Computer Service work orders as a text file.
f. Click the Save Export button.

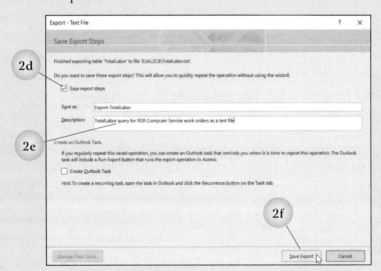

3. Start Notepad.
4. At a blank Notepad window, open **TotalLabor**.
5. Print **TotalLabor** and then exit Notepad.

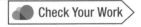
Check Your Work

Saved Exports

Quick Steps

Repeat Export Steps
1. Click External tab.
2. Click Saved Exports button.
3. Click specified saved export.
4. Click Run button.
5. Click OK.

Once an export routine has been saved, repeat the export process by clicking the Saved Exports button in the Export group on the External Data tab. This opens the Manage Data Tasks dialog box with the Saved Exports tab selected, as shown in Figure 8.10. Click the export name in the dialog box and then click the Run button to instruct Access to repeat the export operation. Be careful not to override any previous exported files because the file name and location will be the same.

Figure 8.10 Manage Data Tasks Dialog Box with the Saved Exports Tab Selected

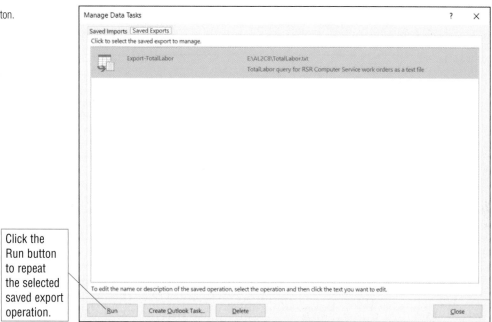

Click the Run button to repeat the selected saved export operation.

Activity 3c Repeating Export Steps

Part 3 of 3

1. With **8-RSRCompServ** open, change the location of **TotalLabor** text file by completing the following steps:
   a. Click the File tab, click *Open*, click the Browse button, and then locate the AL2C8 folder on your storage medium.
   b. Click the *Data Type* option box that currently displays *Microsoft Access* and change the option to *All Files*.
   c. Scroll down, right-click *TotalLabor* in the file list box, and then click *Cut* at the shortcut menu.
   d. Double-click the *DatabaseBackUps* folder.
   e. Right-click in a blank area of the file list box and then click *Paste*.
   f. Close the Open dialog box. The Export procedure can now be run without overwriting the data for the month of September. Click the Back button.

2. Export the results of the TotalLabor query for October by completing the following steps:
   a. Open the TotalLabor query in Design view.
   b. In the *Criteria* row of the *DatePart("m",[ServDate])* field, change the *9* to a *10*. Changing this number will return the data for the month of October.
   c. Save, run, and then close the query.
   d. Click the External Data tab.
   e. Click the Saved Exports button in the Export group.
   f. With *Export-TotalLabor* selected, click the Run button.
   g. Click OK.
   h. Click the Close button.

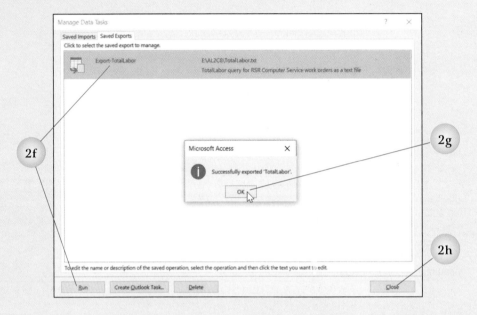

3. Click the Close button in the upper right corner to close Access.

 Check Your Work

Chapter Summary

- Use options at the Save As backstage area to create backup database files on a regular basis. This will help to minimize any data loss due to system failures or design mistakes.

- Create a backup copy of a database before making any major changes to it.

- An ACCDE file is a locked-down version of a database, in which users do not have access to Design view or Layout view and therefore cannot make changes to the design of objects.

- Create an ACCDE database file at the Save As backstage area.

- Open the Access Options dialog box, click *Trust Center* in the left pane, and then click the Trust Center Settings button to view and/or modify Trust Center options.

- An object in another Access database can be imported into the active database using the Import Access database button in the Import & Link group on the External Data tab.

- If the source object is a table, choose to import or link to the source table.

- When a table is linked, the data is not copied into the active database; it resides only in the source database.
- Edit source data in a linked table in the source database or the destination database; the changes will be reflected in both tables.
- Import the table if the source data is not likely to change.
- Link to a data source file that is updated frequently so changes have to be made only once; this reduces the potential for making data entry mistakes or missing updates.
- Link to a source table that is shared among several different databases within an organization.
- When a table has been linked to another database, Access stores the full path to the source database. If the location of the source database is moved or the file name is changed, the links will need to be refreshed.
- A text file is often used to exchange data between dissimilar programs because this file format is recognized by nearly all applications.
- Import a text file into an Access database by clicking the Import text file button in the Import & Link group on the External Data tab.
- When a text file is selected for import, Access launches the Text Import Wizard, which guides the user through the steps to import the text into a table.
- If an import routine is repeated often, consider saving the steps so the routine can be run without having to perform each step every time.
- Open the Manage Data Tasks dialog box to run a saved import by clicking the Saved Imports button in the Import & Link group on the External Data tab.
- Export Access data in a text file format using the Export Text Wizard by clicking the Export to text file button in the Export group on the External Data tab.
- Within the Export Text Wizard, the user is prompted to choose the text format, delimiter character, field names, text qualifier symbols, export path, and file name.
- To repeat an export routine, first save the export steps at the last Export - Text File dialog box.
- Click the Saved Exports button in the Export group on the External Data tab to run a saved export routine.

Commands Review

FEATURE	RIBBON TAB, GROUP	BUTTON
create ACCDE file	File, *Save As*	
export data as text file	External Data, Export	
import data from text file	External Data, Import & Link	
import or link data from Access database	External Data, Import & Link	
Linked Table Manager	External Data, Import & Link	
saved exports	External Data, Export	
saved imports	External Data, Import & Link	

Index

Interior Photo Credits

Page GS-1 (banner image) © lowball-jack/GettyImages; *page GS-1, (in Figure G.1)* all images courtesy of Paradigm Education Solutions; *page GS-2*, © irbis picture/Shutterstock.com; *page GS-3*, © th3fisa/Shutterstock.com; *page GS-4*, © goldyg/Shutterstock.com; *page GS-5*, © Pressmaster/Shutterstock.com.